TELECOMMUNICATIONS MANAGEMENT
BROADCASTING/CABLE AND THE NEW TECHNOLOGIES

McGraw-Hill Series in Mass Communication

CONSULTING EDITOR

Barry L. Sherman

Anderson: Communication Research: Issues and Methods
Carroll and Davis: Electronic Media Programming: Strategies and Decision Making
Dominick: The Dynamics of Mass Communication
Dominick, Sherman, and Copeland: Broadcasting/Cable and Beyond: An Introduction to Modern Electronic Media
Dordick: Understanding Modern Telecommunications
Gross: The International World of Electronic Media
Hickman: Television Directing
Holsinger and Dilts: Media Law
Richardson: Corporate and Organizational Video
Sherman: Telecommunications Management: Broadcasting/Cable and the New Technologies
Walters: Broadcast Writing: Principles and Practice
Whetmore: American Electric: Introduction to Telecommunications and Electronic Media
Wilson: Mass Media/Mass Culture: An Introduction
Wurtzel and Rosenbaum: Television Production

TELECOMMUNICATIONS MANAGEMENT

BROADCASTING/CABLE AND THE NEW TECHNOLOGIES

SECOND EDITION

Barry L. Sherman
College of Journalism and Mass Communication
University of Georgia

McGRAW-HILL, INC.
New York   St. Louis   San Francisco   Auckland   Bogotá
Caracas   Lisbon   London   Madrid   Mexico City   Milan
Montreal   New Delhi   San Juan   Singapore   Sydney   Tokyo   Toronto

This book was set in Optima with Helvetica by Better Graphics, Inc.
The editors were Hilary Jackson, Fran Marino, and David Dunham;
the production supervisor was Paula Keller.
The cover was designed by John Hite.
The photo editor was Anne Manning.
Arcata Graphics/Martinsburg was printer and binder.

Part-Opening Photograph Credits

One: Chuck O'Rear/Woodfin Camp   *Two*: Courtesy of Turner Broadcasting System
Three: Courtesy of TCI   *Four*: Courtesy of Gannett Broadcasting

TELECOMMUNICATIONS MANAGEMENT

Broadcasting/Cable and the New Technologies

Copyright ©1995 by McGraw-Hill, Inc. All rights reserved. Previously published under the title of: *Telecommunications Management: The Broadcast & Cable Industries*. Copyright © 1987 by McGraw-Hill, Inc. All rights reserved. Printed in the United States of America. Except as permitted under the United States Copyright Act of 1976, no part of this publication may be reproduced or distributed in any form or by any means, or stored in a data base or retrieval system, without the prior written permission of the publisher.

This book is printed on recycled, acid-free paper containing 10% postconsumer waste.

2 3 4 5 6 7 8 9 0 AGM AGM 9 0 9 8 7 6 5

ISBN 0-07-056698-4

Library of Congress Cataloging-in-Publication Data

Sherman, Barry L.
  Telecommunications management: broadcasting/cable and the new
technologies / Barry L. Sherman. — 2nd ed.
    p.   cm. — (McGraw-Hill series in mass communication)
  Includes bibliographical references and index.
  ISBN 0-07-056698-4
  1. Broadcasting—Management.   2. Cable television—Management.
3. Telecommunication—Management.   I. Title.   II. Series.
HE8689.6.S55   1995
384'.068—dc20   94-11082

ABOUT THE AUTHOR

Barry L. Sherman is Director of the George Foster Peabody Awards program and Professor of Telecommunications at the Grady College of Journalism and Mass Communication, University of Georgia. Chairman of the Department and Associate Director of the Peabody Awards from 1986 to 1991, he is also Director of the Dowden Center for Telecommunication Studies, founded in 1988 by cable pioneer Thomas C. Dowden.

A graduate of Queens College, the City University of New York (BA 1974; MA 1975), and Penn State (Ph.D. 1979), Dr. Sherman teaches and conducts research in the areas of broadcast and cable management and audience behavior. His research and consulting clients have included CapCities/ABC, the Canadian Broadcasting Company, Cox Enterprises, Connect Research, the National Association of Broadcasters, Paragon, and Burkhart/Douglas, among others. He has been a judge for the Cable Television Administration and Marketing Society's CTAM Awards, the IRIS Awards for the National Association of Television Program Executives, the Best of Gannett, and numerous other media awards programs.

His writings have appeared in a variety of professional and trade publications, including *Journal of Communication, Journal of Broadcasting and Electronic Media, Communication Education, Journalism Quarterly*, and *Channels*.

Dr. Sherman's passions include pick-up basketball and personal computing. He lives with his wife Candice and their two children (Eric and Jessica) in the new-music mecca of Athens, Georgia.

For Candice,
Eric, and Jessica

CONTENTS

PREFACE — xv

PART 1 CORE CONCEPTS — 1

1 The Electronic Media Landscape — 3

ELECTRONIC MEDIA: A DEFINITION — 3
MAPPING THE MEDIA LANDSCAPE — 10
SUMMARY — 19

2 Theories of Telecommunications Management — 21

MANAGEMENT: A DEFINITION — 22
APPROACHES TO THE STUDY OF MANAGEMENT — 22
WHAT DO MANAGERS DO? COMPONENTS OF TELECOMMUNICATIONS MANAGEMENT — 23
HOW DO MEDIA MANAGERS MANAGE? THEORIES X, Y, AND Z — 32
A MODEL OF TELECOMMUNICATIONS MANAGEMENT — 36
EXTERNAL CHANGES — 38
INTERNAL CHANGES — 41
CONTEMPORARY TELECOMMUNICATIONS MANAGEMENT: MYTH AND REALITY — 45
SUMMARY — 45

3 Telecommunications Industry Structure — 51

THE AMERICAN TELECOMMUNICATIONS INDUSTRY — 51
ELEMENTS OF MARKET STRUCTURE — 52
ELEMENTS OF MARKET CONDUCT — 60
ELEMENTS OF MARKET PERFORMANCE — 62
THE MEDIA ARE MATURE: THE PRODUCT LIFE-CYCLE CURVE — 67
A FINAL WORD — 70
SUMMARY — 70

PART 2 CORE BUSINESSES 75

4 The Radio Business 77

RADIO: A MATURE MEDIA BUSINESS 78
NONCOMMERCIAL RADIO 82
RADIO RECEIVERS 83
RADIO LISTENING: DAYPARTS AND DEMOGRAPHIC TRENDS 84
RADIO PROGRAMMING FORMATS 85
THE BROADCAST FINANCIAL PICTURE 89
THE FINANCIAL STATUS OF RADIO 92
THE RADIO EMPLOYMENT PICTURE 98
SUMMARY 101

5 The Television Business 103
THE GROWTH OF COMMERCIAL TELEVISION 104
THE GROWTH OF NONCOMMERCIAL TELEVISION 109
TELEVISION RECEIVERS 111
TELEVISION PROGRAMMING 112
VIEWING TRENDS 114
TV REVENUE TRENDS 117
TV EXPENSE TRENDS 123
PROFITS AND PROFIT MARGINS 125
THE TV EMPLOYMENT PICTURE 126
SUMMARY 129

6 The Cable Business 132
GROWTH OF CABLE 132
CABLE PROGRAMMING 139
CABLE REVENUES 141
CABLE EXPENSES 147
PROFITS AND PROFIT MARGINS 151
THE CABLE EMPLOYMENT PICTURE 153
COMPETITION AND COOPERATION IN THE LAST QUARTER MILE: THE TELEPHONE COMPANIES 155
SUMMARY 158

PART 3 CORE PROCESSES 163

7 Patterns of Telecommunications Ownership 165
RESTRICTIONS ON OWNERSHIP 165
CRITERIA FOR BROADCAST OWNERSHIP 167

	SPECIFIC OWNERSHIP REGULATIONS	170
	FORMS OF MEDIA OWNERSHIP	174
	THE TELECOMMUNICATIONS PYRAMID	176
	INTERNATIONAL MEDIA CONGLOMERATES: THREE PERSPECTIVES	182
	CURRENT ISSUES IN MEDIA OWNERSHIP	188
	SUMMARY	191
8	**Entering the Telecommunications Marketplace**	**194**
	ACQUISITION STRATEGIES	195
	MEDIA VALUATION	208
	RAISING MONEY: CAPITALIZATION AND ACQUISITION FINANCING	214
	FORMAL ACQUISITION: TRANSFER OF CONTROL	219
	MARKET ANALYSIS	220
	SUMMARY	223
9	**Telecommunications Financial Management**	**228**
	FINANCIAL OFFICERS IN TELECOMMUNICATIONS FIRMS	229
	BASIC FINANCIAL ASPECTS OF TELECOMMUNICATIONS MANAGEMENT	229
	METHODS OF FINANCIAL REPORTING	237
	FINANCIAL PLANNING AND PROJECTIONS	243
	THE COMPUTER IN MEDIA MANAGEMENT	244
	SUMMARY	249
10	**Personnel Management and Employee Relations**	**251**
	MEDIA PERSONNEL: AN OVERVIEW	252
	FUNCTIONS OF PERSONNEL ADMINISTRATION IN TELECOMMUNICATIONS	253
	TABLES OF ORGANIZATION	254
	CORE DEPARTMENTS IN TELECOMMUNICATIONS	258
	JOB DESCRIPTIONS IN TELECOMMUNICATIONS	259
	STAFFING: RECRUITMENT, TRAINING, AND EVALUATION POLICIES	264
	EMPLOYEE COMPENSATION, BENEFITS, AND ASSISTANCE PROGRAMS IN TELECOMMUNICATIONS	276
	INTERNAL COMMUNICATIONS	281
	TELECOMMUNICATIONS MANAGEMENT AND LABOR RELATIONS	284
	VALUES-DRIVEN TELECOMMUNICATIONS MANAGEMENT	292
	SUMMARY	295

PART 4 CORE DEPARTMENTS 299

11 Radio Program Management 301

THE FIVE FUNCTIONS OF PROGRAM MANAGEMENT 303
THE PROCESS OF RADIO PROGRAMMING 305
RADIO NETWORKS 307
RECORD COMPANIES AND OTHER MUSIC DISTRIBUTORS 308
THE FORMAT: DEVELOPING THE RADIO SCHEDULE 309
COMPETITION ANALYSIS 312
PROGRAM EVALUATION IN RADIO 318
SUMMARY 320

12 Television Program Management 320

PERSONNEL ADMINISTRATION 321
PROGRAM ACQUISITION 323
TELEVISION SYNDICATION 330
PROGRAM PRODUCTION 335
PROGRAM SCHEDULING 341
PROGRAM STRATEGIES 346
SUMMARY 349

13 Sales and Marketing Management 351

THE PRODUCT: SELLING TIME 351
THE MAJOR PLAYERS: ADVERTISERS, AGENCIES, REPS,
   AND STATIONS 353
RATE SETTING 359
SALES AND TRAFFIC 364
THE PROMOTION DEPARTMENT 366
COMMON PROMOTIONAL AND MARKETING ACTIVITIES 370
PROFESSIONAL SALES AND MARKETING ASSOCIATIONS 375
SUMMARY 376

14 Audience Analysis 378

FORMS OF MEDIA RESEARCH 379
SOURCES OF AUDIENCE RESEARCH 381
RESEARCH AND SALES DEVELOPMENT 389
RESEARCH AND PROGRAMMING 391
NEW DIRECTIONS IN AUDIENCE RESEARCH 395
ISSUES FOR MANAGEMENT CONCERNING AUDIENCE
   RESEARCH 397
SUMMARY 399

15	**The Future of Media Management**	**405**
	GLOBALIZATION AND VERTICAL INTEGRATION	406
	TECHNOLOGICAL CONVERGENCE	408
	ECONOMICS AND MEDIA MANAGEMENT	409
	TELECOMMUNICATIONS MANAGEMENT AND SOCIAL CHANGE	411
	THE BOTTOM LINE: SUCCESSFUL MANAGEMENT FOR THE NEW MILLENNIUM	414
	SUMMARY	415
	INDEX	419

PREFACE

The first edition of this text began prophetically with the statement "perhaps the only permanent feature of telecommunications management is change." Since those words were printed, the pace of political, technological, and economic development has produced profound change in broadcasting, cable, and related industries.

The major political event, of course, has been the demise of international communism, brought vividly into our living rooms with such potent images as the destruction of the Berlin Wall, the execution of Romanian dictator Ceaucescu, and the events surrounding the aborted coup of Mikhail Gorbachev and the ascension of Boris Yeltsin to power in Russia. That the revolutionary change is not complete was underscored by poignant coverage of the massacre of pro-democracy demonstrators in Peking's Tienanmen Square in 1989 and ongoing ethnic violence throughout Eastern Europe, especially in Bosnia (the region comprising the former Yugoslavia).

On the economic front, the first appearance of this text coincided with unprecedented change in telecommunications ownership, operations, and finance. In the late 1980s, a frenzy of mergers, leveraged buyouts, corporate raids, and takeovers reshaped the structure of the media industries. Two major consequences followed. The huge debt load created by massive refinancing left media companies comparatively conservative in their fiscal policies and increasingly preoccupied with cost controls. In addition, the emergence of a new breed of managers interested mainly in profit (at the expense, some argued, of community and public service) led to unprecedented concern about the ethical behavior of media companies and the social responsibility of media management. Computer, video, cable, publishing, and other companies began to jockey for position in a frenzied race to build "the new information superhighway." Traditional broadcasters felt they were being left at the starting gate and tried to make their point at the FCC and in the White House.

The implications of these political and economic developments for media industries cannot be overstated. All over the globe, press, radio, television, and newspapers once controlled by government bureaucracies, ministries of information, or puppet dictators are being turned over to new voices and private interests. Along with political freedom, the formerly repressed republics of the Eastern bloc are demanding western-style media: a free press, records, discs, and tapes (for the turntable, CD player, and VCR), even cable TV. This phenomenon has created unprecedented challenge and opportunity for media management. The industry, as well as this book, must look beyond the boundaries of the United States and take a truly international perspective.

Features of the Second Edition

- *Increased International Coverage*

While the previous edition was mostly concerned with the nature and structure of domestic commercial media, this edition extends coverage of management issues beyond our borders to include new audiences and markets throughout the world. In addition to domestic media owners and operators, in Chapter 7 and elsewhere, the text now profiles the major multinational media corporations and details their operations both in the United States and abroad.

- *Increased Economic and Financial Material*

The text now includes expanded coverage of financial and economic concerns, both at the macro and micro levels. As in the previous edition, issues of budget and finance are covered from the corporate and senior management perspectives (Chapter 9). In addition, Part Four (Chapters 11 through 14) now provides in-depth coverage of financial management in each of the core departments found in media firms: programming, sales, and promotion.

- *A New Section on Core Departments in Media Businesses*

The various tasks performed by management in four crucial areas—programming, sales, promotion, and audience research—are given extended treatment in Part Four, which is new to this edition. Separate chapters on radio programming, programming for broadcast and cable television, sales and promotion, and audience analysis provide important detail on these critical management functions.

- *More Coverage of Ethics in Media Businesses*

Throughout the book, increased attention is paid to issues of ethics in media management. Particularly, the performance of media companies in employee equity is given extended treatment, in Chapter 10, especially with regard to opportunities for women and ethnic minorities in ownership and senior management.

- *Extended Discussion of New Technologies*

Technological developments continue to reshape the media landscape. Unforeseen in the first edition was the gallop toward high definition and advanced television systems; first in the Far East, now on American soil. The proliferation of fiber cable and cellular service increasingly make the regional Bell operating companies a competitor for services previously identified with broadcasting and cable. New businesses, such as satellite teleports, interactive video, even computer data networks (such as CompuServe and Prodigy) now compete for a share of the leisure time and disposable income of consumers. While the main focus of the book is on the more "traditional" media businesses—radio, TV, and cable—there is increasing emphasis in this edition on new and emerging forms of mass communication, particularly in Part 2 (Chapters 4, 5, and 6).

- *Expanded Case Materials in a New Instructor's Case Manual*

Finally, the case studies, which proved so useful in the first edition, have been revised and expanded and moved to an accompanying Instructor's Case Manual. This will make the incorporation of case studies into the class more flexible for the instructor and more user-friendly to the student. In addition, the Instructor's Case Manual now includes review, examination, and other supplementary material.

The Plan of the Book

Three words characterize the approach of this text: pragmatism, professionalism, and personality. As stated previously, we are living in an era of unprecedented development and change. As we will see in Chapter 1, the television and radio landscape is dotted with new media and methods. Management of the existing and emerging media technologies will be viewed with a constant eye toward the overall economic outlook, consumer and advertiser demand, and hardware and program costs. Our concern is with the pragmatics of acquiring, financing, staffing, maintaining, and operating a media facility for profit.

But the pursuit of profit is not at cross-purposes with responsibility and public service. We hope to dispel two prevalent myths about media management. The first is that one can succeed in telecommunications with a minimum of skill and a carload of charisma. The second myth maintains that, as a class, media managers are by definition manipulative, uncaring capitalists who place the quest for increased profits ahead of the public trust. Successful telecommunications managers do require both professional training and a sense of civic responsibility. Media management is a profession, like medicine and law, one that is worthy of pride and professionalism (not to mention a university degree!).

We will also try not to lose sight of the fact that media management is above all a study of *people*. Perhaps more than any other industry, the development of American mass media is associated with individual initiative and achievement.

Our study of media ownership and management will also introduce us to the personalities who make the American telecommunications industries unique: from mavericks like "Captain Outrageous," Ted Turner; "wunderkinds," like Bob Pittman (who founded MTV); to female trailblazers like USA Network's Kay Koplovitz, and minority entrepreneurs like Robert Johnson of Black Entertainment Television.

The book is organized into five main parts. *Part One: Core Concepts* is an overview of electronic media management from a business and professional point of view. Chapter 1 surveys the many businesses, services, and products which constitute contemporary electronic media. Chapter 2 provides an overview of management theory and science and how that important and evolving discipline applies to modern media organizations. Chapter 3 describes the economic structure and corporate behavior of media companies.

Part Two: Core Businesses examines the major electronic media businesses in detail by reviewing consumer usage patterns, sales, revenue and expense trends, economic performance, and employment statistics. Chapter 4 examines the radio business. Chapter 5 looks at broadcast television, and Chapter 6 documents trends in the cable business, including a discussion of the role of the regional Bell operating companies in providing video as well as traditional telephone services.

Part Three: Core Processes centers on the steps necessary to enter the media marketplace. Chapter 7 examines patterns of media ownership, from small mom-and-pop operations to huge international communications conglomerates. Chapter 8 concentrates on the two methods of beginning a media venture—acquiring an existing facility or starting a new one from scratch, with particular emphasis paid to capitalization and acquisition financing. Chapter 9 examines financial management, including methods of financial reporting, analysis, and planning. Chapter 10 details personnel management, including station organization and staffing, compensation, training, and negotiations with guilds and unions.

Part Four: Core Departments describes how each of the main departments in electronic media firms is organized and managed. Chapter 11 discusses the role of the program manager in radio. Chapter 12 describes the programming process in broadcast and cable television. Chapter 13 examines the sales department: the main source of revenue for any media facility. Related to sales but increasingly important in its own right is audience analysis, which is the focus of Chapter 14. The text closes with Chapter 15, a look at the future of media management in an atmosphere of continued political, economic, and social change.

We have strived throughout to make this book look and read like a professional management publication. Figures, charts, and illustrations used are common to management reports and business communications. Our hope is that your career in media management will start as soon as you begin reading Chapter 1.

ACKNOWLEDGMENTS

Many people have helped this edition move from idea to reality. First, any professionally relevant text requires input from professionals. Especially helpful were: Val Carolin, CBS Radio Sales and WBBM, Chicago; Cecil Walker, Dick Mallary, Steve Smith, Marc Kaye, and Louise Willoughby, Gannett Broadcasting; Tom Johnson, Ed Turner, Bob Schuessler, Linda Fleisher, Steve Haworth, Juli Mortz, and Alyssa Levy, Cable News Network; Nick Trigony, Bill Killen, Tom McClendon, and Lynda Stewart, Cox Enterprises; Gene McHugh, Fox Broadcasting; Herman Ramsey, Tribune Broadcasting; Tom Dowden and Nancy Wood, Dowden Communication Investors; Paul Kagan, Paul Kagan and Associates; Dwight Douglas, Burkhart/Douglas and Associates; Roger Wimmer, Paragon and Eagle Research; Chris Porter and Mike Henry, Paragon Research; Don Grede, TVAnswer and Eon, Inc.; Jody Danneman, Atlanta Video Production Center; Dana Krug, Broadcast Investment Analysts; Jim Duncan, Duncan's American Radio; Joyce Tudryn, International Radio and Television Society; Amy Henry, Susquehanna Broadcasting; John Abel, Chuck (no relation) Sherman, Rick Ducey, and Mark Fratrik, National Association of Broadcasters; Louisa Nielsen, Broadcast Education Association; Paul Raymon, Raymon Media Group; Marc Doyle, Marc Doyle Associates; Chris Rapp, TeleCommunications, Inc.; Keith Randall, Canadian Forces Network; George Reed, Media Services Group; Marie Zimman, A.C. Nielsen; and Strat Smith, National Cable Center and Museum.

This book has benefited greatly from the input of colleagues who have used it and improved it over the years. Three deserve special mention. First, Dr. Joseph Buchman, who made many important contributions to the organization of the new edition. Professor Charles Warner of the University of Missouri gave freely of his time and expertise to the case materials in the Instructor's Case Manual. He is also a model of the successful interaction between the industry and the academic realm. Dr. Russ Mouritsen of Brigham Young University took on the critical but generally thankless job of preparing the accompanying Instructor's Case Manual. Welcome aboard, Russ! A special thanks goes to colleagues in the Management/Sales Division of the Broadcast Education Association, for their advocacy and insightful criticism of the first edition and their help in soliciting suggestions regarding the second. Special thanks to Paul Prince of Kansas State and Mike Wirth, University of Denver.

My colleagues at the University of Georgia continue to open their minds and libraries to me. Some whose books I have neglected to return include Tom Russell, Joe Dominick, Jim Fletcher, Bill Lee, Alison Alexander, Scott Shamp, Dean Krugman, Len Reid, Louise Benjamin, Conrad Fink, David Hazinski, and Allan MacLeod. Don Davis of Brenau College is still looking for things I took from him, too.

The reviewers for this edition provided invaluable feedback. They are Albert Auster, SUNY College at New Paltz; Joseph Butler, Syracuse University; Karen Buzzard, Northeastern University; Emory Johnson, California State University, Northridge; F. Leslie Smith, University of Florida; and George Whitehouse, University of South Dakota. Thank you for your many suggestions and improvements.

This book could not have been written without the research assistance provided by a small army of students through the years, particularly Al Moffett, Tom Berg, Doug Barthlow, Pat Priest, Larry Ftling, Patricia Ready, Penny Patterson, Penny Walls, Steve Strickland, Michael Baxter, and Brad Lohmeier. My students in the graduate and undergraduate seminars in Telecommunications Management have provided invaluable insight and information for this edition. While they are too numerous to mention individually, they will recognize their contributions in the pages ahead, and I am especially grateful.

Sonia Davis began the imposing task of processing the words; Leslie Hopkins finished it and kept the author reasonably sane throughout the project.

Hilary Jackson, Fran Marino, Dave Dunham, and Phil Butcher at McGraw-Hill are more than editors. They are true mentors and, happily, have become lifelong friends. My thanks for the patience and perseverance, especially to the Red Sox fan. On all other matters, her judgment has been flawless. There's always next year. And, for the Ranger fan: there's always XXXX 1994!

Finally, a special thanks to my family for their unwavering support. To my parents, who stressed the value of education as well as the joys of being a "media addict," my wife Candice, who indulges my passions (both expensive and free) and who shares an enthusiasm for life and love; and my terrific kids: Eric, whose quest for knowledge seems vaguely reminiscent, and Jessica, whose taste in clothes and toys made this edition necessary.

Barry L. Sherman

TELECOMMUNICATIONS MANAGEMENT
BROADCASTING/CABLE AND THE NEW TECHNOLOGIES

PART ONE

CORE CONCEPTS

Part One: CORE CONCEPTS orients the reader to the landscape of modern telecommunications. Chapter 1 defines key terms and identifies the major players in modern media: radio, TV, cable, and related technologies. Chapter 2 details the major functions of media managers and the nature of the relationship between management and labor. Chapter 3 analyzes the structure of media businesses and their behaviors relative to suppliers, competitors, employees, and consumers.

1

THE ELECTRONIC MEDIA LANDSCAPE

More than a half century ago, broadcast industry pioneer and visionary David Sarnoff wrote of an emerging breed of media manager:

> The needs of the times will bring forth perhaps a new type of executive, trained in a manner not always associated with the requirements of business management. He will have to reckon with constant changes in industry, the changes that scientific research is bringing. He will have to be able to approximate the value of technical development, to understand the significance of research.[1]

As usual, Sarnoff was on the mark. As we will see, media managers operate in a dynamic, competitive, and unique environment where the "rules of thumb" of successful management often do not apply. As a first step, we will survey the media landscape by identifying the elements which comprise contemporary communications. First, let's try to determine the common ground linking the wide range of goods and services which fall under the umbrella term *telecommunications*.

ELECTRONIC MEDIA: A DEFINITION

A simple, yet comprehensive definition of telecommunications is elusive. The widely used term can apply to an industry as vast and powerful as global satellite communications or to an operation as small as a neighborhood radio station. For the purposes of this book, telecommunications refers to the commercial and nonprofit organizations involved in the development, production, distribution, and exhibition of entertainment and information programming to the public via electronic (nonprint) means.

Let's break that rather inclusive definition down to its component parts.

The Bottom Line

Media firms run the gamut from the local storefront operation to the large multinational corporation. Whatever the size of the company, most media facilities share one characteristic: they are part of the private sector; that is, they are organized to deliver goods and services for a profit.

Historically, enthusiastic research and development specialists and idealistic educators and social critics have heralded the arrival of each new media marvel as a cultural panacea. For example, at the dawn of this century film was celebrated as a "universal language" with the potential for uniting the world with silent symbolism. After World War II, educational psychologists hailed television as the "great educator," soon to replace the teacher in every classroom. And in the 1960s, some groups viewed portable video and cable television as a manifestation of the triumph of grass roots democracy—a phenomenon that would end media control by the three major commercial networks and restore "power to the people."

Such blue-sky aspirations glossed over the importance of the bottom line. In the free enterprise system which undergirds media in the United States and increasingly throughout the world, media businesses operate to produce revenues and control expenses, and thereby to produce profit. Faced with an increasingly competitive environment, all media managers must continually address the following questions:

1 Is there *consumer demand* for a particular product or service?
2 Is there a *cost-effective* means for producing and distributing the product or service?
3 Does the product or service have *profit potential*?

The media landscape is littered with the refuse of unsuccessful ventures, marked by a failure to adequately assess these three important questions.

Some failures overstated the audience, expecting greater consumer interest than was forthcoming. For example, some years ago, the ABC television network marketed a box which would allow its local stations to send out scrambled movies late at night. By hooking up a special decoder box, viewers could unscramble the movies and record them on their VCRs. But, as ABC found out, too few people wanted such a service. (Too expensive? Too complex? Or maybe, who wants to stay up so late?) Early efforts to market in-home shopping, electronic newspapers, and other text ventures met a similar fate.

Other notable failures have been marked by excessive up-front costs with disappointing rates of return, question two of our big three. The first efforts toward direct broadcast satellite (DBS) service is a case in point. In the early 1980s, COMSAT Corporation sunk nearly $150 million into DBS, before abandoning the concept. A second DBS group lost nearly half that amount.

Sometimes, there is an apparent market for a service and a means of controlling costs, yet profit potential remains limited. This has been the case with low-power television stations and many new AM and FM stations. Some cities have proved too small, their advertising base too limited, to support additional television and radio outlets.

The emphasis on the bottom line in this book does not mean that the multitude of firms and people involved in nonprofit media, including public broadcasting and community-access cable television, are not a significant factor. Indeed, public, nonprofit media organizations are in an era of accountability. Shifts in government priorities, competition from new media services, past mismanagement, and other factors have combined to make managers of noncommercial media institutions increasingly involved with fund-raising, underwriting, cost control, and other fiscal concerns. Like their counterparts in the commercial media, they seek to operate efficiently by streamlining costs and eliminating waste. While turning a profit may not be the primary goal of such institutions, they are as concerned as private-sector organizations with issues of marketability and cost-effective production and distribution.

The Four Basic Activities of Media Companies

All media businesses are involved to some degree in one or more of four basic activities. These are the next components of our definition: development, production, distribution, and exhibition.

Development Innovation and invention, leading to the initial development of media technology, have always been crucial to industry growth. The inventive genius of Thomas A. Edison, among many others, led us into the industrial age. Early radio is associated with such electrical wizards as Guglielmo Marconi, Reginald Fessenden, and Lee De Forest. Television is the product of such electronic experts as Vladimir Zworykin and Philo T. Farnsworth. Each year, industry trade shows are marked by the introduction of new or improved telecommunication products, such as smaller, more light-sensitive cameras, monitors, and radio receivers, and more sophisticated graphics and animation programs for computers. As we will see, the industrywide commitment to innovation places great pressure on media managers in the areas of corporate planning, diversification, and growth.

As we shall also see, in recent years American telecommunications firms have largely abandoned their research and development (R&D) function, providing increasing opportunities to companies from overseas, most notably Japan.

Production Firms involved in the production function of electronic media are concerned with large-scale manufacturing of consumer electronic products, as well as the programming which fills them. On the consumer product side, there are television sets, radios, compact disc players, video recorders, telephone answering devices, and innumerable other items. Like R&D, this sector of the American economy has been a source of growing concern. Faced by escalating labor costs, foreign imports, and some executive-level shortsightedness, U.S. firms have largely abandoned the production function, which has been taken over by conglomerates based overseas. Today, Zenith Corporation is the last remaining U.S.-owned manufacturer of television and radio receivers.

Despite its losses in the area of equipment production, the United States remains the leader in program production. Our movies and recording artists are still the world's most popular; our television shows are seen on virtually all the world's television screens. The enduring popularity of our media programming abroad has generated a global market and, therefore, great profit potential for U.S. producers. This popularity has caused some foreign government leaders and educators to worry about "ideological invasion"—the social and political consequences of cultural domination by the United States.

Recognizing the role of American producers and performers in the creation of popular culture worldwide, foreign companies have increased their investment in U.S. domestic media. For example, Australian-born (but naturalized American) Rupert Murdoch is the financial force behind the Fox television network. Today, both Columbia Pictures and Columbia Records are owned by Japan's Sony Corporation. Similarly, the major entertainment conglomerate MCA was acquired by Japan's industrial giant Matsushita. We will assess the impact of foreign ownership in more detail in Chapter 7.

Distribution The distribution sector of the telecommunications industry includes those firms specializing in linking producers to consumers. Distribution is a primary function of the three major commercial television networks (ABC, CBS, and NBC) and the "upstart" Fox network; it is carried out through their system of affiliation with local stations around the country. Distribution firms also include program syndication services such as Viacom and LBS Entertainment; cable superstations such as WTBS, WGN, and WWOR; radio networks such as Mutual, Unistar, and Westwood One; and satellite radio services such as The Satellite Music Network.

Much of the change taking place in contemporary telecommunications is in distribution. Satellite, computer, cable, microwave, fiber-optic, and VCR technology have all introduced new means of taking a message from point A to point B. At the same time, one of the oldest distribution means—the telephone line—has been updated to allow for transmission of video and audio signals. As a result, the regional Bell operating companies (RBOCs) are now venturing into cable and related services, a controversial issue covered in the chapters ahead.

The question of which means of distribution is the most efficient, cost-effective, and above all popular with consumers is central to telecommunications management as the industry evolves in the 1990s and beyond.

Exhibition The fourth phase of telecommunications is exhibition, or consumption: the point at which the consumer "uses" the product. As a result of technological development, media exhibition is shifting more and more from the marketplace to the home. The "media room," a specially designed habitat for media consumption, has become the new challenge to interior designers. Large-screen sets, high-quality VCRs, and stereo TV sound (not to mention the cost of movie tickets and baby-sitters) are keeping increasing numbers of potential movie-goers at home.

While today feature films are first released for theatrical exhibition, theater owners are concerned with the prospects of first-run movies being made available directly for home viewing on videotape, via cable television, or via satellites. With more and more baseball games appearing on television, team owners are concerned that fans will eventually elect to remain at home, rather than attend the games in person.

While in-home consumption is undoubtedly escalating, the development of portable devices, particularly headset radios, miniaturized television sets, and cellular telephones, has permitted media consumption to occur virtually anywhere. Managers trained to think of audiences in terms of "homes" and "households" are grappling with the implications of the new technology for programming and advertising in the car, the workplace, and even the streets.

The point is that recent technological, political, and social changes have forced telecommunication managers to evaluate their business in light of each of the four functions. Many companies formerly involved in only one phase of the business have diversified into another. A few have eschewed diversification and continue to specialize in one function.

The resulting, unprecedented disequilibrium in the industry has led to buyouts, mergers, and corporate reorganizations and to the introduction of terms such as *joint venture, coventure,* and *convergence* into the media management vocabulary.

Information and Entertainment

The primary end goal of all media firms is the delivery of information and/or entertainment to the consumer.

Perhaps the two most salient characteristics of life in the United States in the late twentieth century are an increased need for information and a burgeoning appetite for entertainment. Scarcely a generation ago, more than 50% of the U.S. work force was involved in the production and distribution of manufactured goods. Now a growing majority of jobs in the United States are in service or information industries. The decreasing number of manufacturing jobs, coupled with increased opportunities for computer operators, programmers, data analysts, word processors, account executives, and restaurant workers in the classified sections of Sunday newspapers, is evidence of this trend.

At the same time, media usage has reached record levels. TV watching and radio listening are up. Despite growing competition, motion picture attendance is holding its own and, among young people, even increasing. Add to this the sales of home videocassette recorders, home video games and computers, sound systems, records, tapes, discs, and personal stereo headsets, and the picture of a media-hungry public emerges. Television usage figures exceed 7 hours per day, and a truism of media consumption is that the "typical" American household spends more time watching television than engaging in any other single waking activity. The number of hours we spend with our media devices may soon be greater than the number of hours we sleep. It is the social need and public desire for information and entertainment which drive the media machine in the United States, indeed throughout the world.

The Public

The next component of our definition of electronic media is that it is *public*: it is available for consumption by all members of society. The notion of public media is critical to this text, since it excludes private networks, business data services, and other point-to-point communications systems which have long described their businesses as telecommunications.

While management of such firms is as challenging and complex as management of the kinds of firms discussed here, the need to be immediately responsive to public trends and tastes is not as acute.

The seemingly minor distinction between private and public media is actually fundamental. Managers of public firms must be responsive to fads and whims, must endure constant scrutiny by the government, press, and consumer groups, and are more immediately affected by changing economic and social conditions than are private communications networks.

As noted earlier, private communications carriers, especially the regional Bell operating companies, are positioning themselves to become providers of public communications services (such as TV and radio programming). A snapshot of their operations is presented in Chapter 6.

Hardware and Software

The media industry is concerned with the distribution of both goods and services. Media durable goods, or *hardware*, include color television sets, radio receivers, satellite dishes, and stereo headphones. Media services, or *software*, are designed primarily to produce messages: movies, TV programs, advertising, video games, even computer data.

The development of hardware has typically predated the availability of software. Perhaps this problem has already vexed you. If you have a video game system, you may have complained that there just aren't enough challenging or interesting cartridges available. Or, as the owner of a state-of-the-art stereo system, you may have found yourself remarking that there isn't enough good new music to make you want to go to the record store and make a new purchase.

The lag between hardware and software availability has been a hallmark of media development—and a perennial problem for media management. The commercial viability of a media company depends on successfully addressing both concerns. First, to ensure economic viability in terms of advertising revenues

TABLE 1-1 HOME TELECOMMUNICATIONS DEVICES

Percentage of U.S. households with:	
Radio receivers	98%
Television sets	98
Telephones	96
Audio component systems	89
VCRs	80
Cable television	65
Video game systems	45
Pay cable	30
Answering machines	30
Cordless telephones	25
Home computers	25

Source: Compiled from various sources; updated by the author.

or subscriber fees, there must be adequate distribution of hardware, as measured by set sales, circulation, or subscriber households. Then, there must be a continuous supply of software to satisfy the large appetites of the users of the hardware. The evidence is that communications hardware has achieved more than adequate distribution, as shown in Table 1-1.

Radio and television receivers are available to more than 98 percent of American households. There are over 11,000 radio stations, 1500 television stations, and about 10,000 cable systems competing for a share of the public's leisure time. The challenge for media managers is to make available programming interesting enough to attract sufficient numbers of people to ensure the economic viability of the existing communications vessels.

Electronic vs. Print

The final component of our definition of telecommunications is that it comprises electronic, as opposed to print media. In describing and analyzing mass communication, some scholars have considered radio and television as a simple extension of the mass press: newspapers, magazines, and the like.

While the management of print media has much in common with that of broadcast and cable (both sell advertising, both cover news, and so on), there are many areas where the two types of mass communication do not overlap. Much of the impact of the electronic media derives from its technological base: bandwidth, frequency response, operating power, storage capacity, and so on. Additionally, the "corporate culture" of electronic media businesses is much different from that of the print press. Outside of their newsrooms, few broadcasters consider themselves "journalists." Print employees often treat their broadcast counterparts with distrust, even outright enmity. That is a topic for another book. Here, we limit our concern to the mainstays of electronic media: radio, TV, cable, and related firms.

MAPPING THE MEDIA LANDSCAPE

Having examined each element of our definition of electronic media, let's look at the various enterprises which will be surveyed in detail in the pages ahead. Figure 1-1 provides a partial map of the contemporary media landscape.

The products and services shown in italics are the focus of this text. The main emphasis is on the management and operations of *entertainment software* firms—those companies which provide mass- and special-appeal programming to the U.S. public. In particular, the text focuses on radio, television, and cable management. However, new hardware developments, such as satellite dishes, fiber-optics, and home video, are also discussed, with emphasis on their current and potential impact upon the core broadcast and cable businesses.

Until recently, the U.S. media industry was relatively stable. Basically, on the radio side, there were AM and FM. In television, there were VHF and UHF, with programming dominated by the three major networks, ABC, NBC, and CBS. Even as recently as the 1970s, cable was more promise than reality. In less than a generation, technological change and consumer demand have introduced a number of new players to the game. These new products and services have added a spate of acronyms—such as DAT, DAB, DBS, SMATV, MMDS, and HDTV—to the media manager's vocabulary. To get our bearings, let's briefly identify and review the major media businesses which form the core of our management analysis.

FIGURE 1-1 A map of the electronic media landscape. *Source*: Adapted from C. Fombrun and W. Astley, "The Telecommunications Community," *Journal of Communication*, Vol. 32, Autumn 1982, p. 61. Used with permission.

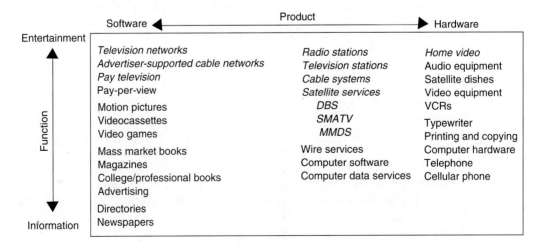

Radio Broadcasting

Conventional broadcasting is the backbone of our electronic communications system. In the last half of the nineteenth century, scientists and engineers learned how to use the natural phenomenon of radio waves for transmitting and receiving messages over long distances, without wires. The "wireless" became the first telecommunications service to deliver information and entertainment directly to home receivers.

AM Radio The first part of the radio spectrum to be set aside specifically for home receivers was the segment between 30 and 300 kilocycles per second (now kilohertz, or kHz). This corresponds to the area between 535 and 1605 kHz on the dial. Since amplitude modulation was the means used to propagate these signals, that bandwidth became known as the AM band. Technically, there are 107 channels available in the AM bandwidth. Of the 11,000 radio stations on air in the United States today, about 5000 (45 percent) operate in this band as commercial AM stations. How is it possible to make room for all these stations on just 107 channels? In the late 1920s and early 1930s, the Federal Communications Commission (FCC) adopted a system of channel classifications and power limitations. The 107 channels were divided into four broad categories: Class I, Class II, Class III, and Class IV. By varying operating power and signal transmission patterns, the FCC was able to accommodate thousands of stations.

Class I AM stations are dominant, clear channels. These powerhouse stations have the exclusive right to their assigned frequency at sunset and can be heard over wide geographic areas. Class I AM stations—for example, KDKA (Pittsburgh), KMOX (St. Louis), and WBZ (Boston)—are among broadcasting's oldest and best known. Many are still owned by their founding broadcast networks and corporations. These are the stations to which many of us tune in the late evening hours in search of out-of-town baseball games, country music broadcasts, or talk show personalities.

Secondary, or Class II, AM operations also occupy the powerful clear-channel positions. To avoid interfering with their Class I counterparts, they must limit and directionalize their transmissions and, in some cases, even shut down their operations at night. Recently, regulatory trends at the FCC and new developments in transmission equipment have made it possible for some of these so-called daytimers to extend their hours beyond local sunset.

Class III stations are designated as regional AM stations, and Class IV stations are known as local stations. Their service is more geographically concentrated and their operating power and efficiency are relatively confined. These channels represent the majority of local AM stations in this country. In fact, nearly three-fourths of the AM operations in the United States are Class III or Class IV stations.

To make AM more consistent with FM (below), the FCC recently changed classes from I, II, and III to classes A, B, and C. Class C AM stations today include the former Class III and Class IV stations.

FM Radio Frequency modulation, or FM, radio broadcasting occupies a higher-frequency portion of the radio band—the area between 88 and 108 megacycles per second (now megahertz, or mHz). Since FM broadcasting requires direct line of sight between transmitting tower and receiver, stations are limited in their geographic coverage by their power and the distance from the transmitting antenna to the horizon. In most cases this is an area from 10 to 70 miles. FM station classifications are therefore based on antenna construction heights and operating power. Class A FM stations transmit at low powers up to 3 kW, with a typical maximum range of 15 miles. Class B stations can operate at powers up to 50 kW, with a maximum range of about 30 miles. Class C stations may transmit at powers up to 100 kW, with coverage up to and sometimes exceeding 70 miles.

From a management point of view, Class B and Class C stations are the most attractive FM radio properties. Companies located in urban areas have built towers of impressive height to gain full coverage throughout their marketplace. For example, in support of its station in Atlanta, Shamrock Broadcasting (a division of the Disney Corporation) erected a 1794-foot tower, some 16 stories higher than the Empire State Building. Including the mound it sits on, the transmitting tower for WFOX-FM stands 2694 feet high.

In addition to being the preferred source for radio music, FM stations have been aided by FCC rules authorizing additional uses of their allocated bandwidth. FM stations may use their transmitters to beam specialty services (typically background music) to businesses or professional offices. Extended service creates additional revenue opportunities for FM managers.

About 5000 commercial FM stations are currently operating, representing 40 percent of all radio stations. Just over 1600 noncommercial FM stations are also on the air, many operated by educational institutions, religious groups, and other nonprofit organizations.

Developing Radio Technologies While new developments in video tend to get most of the attention in the trade and popular press, a number of new developments portend change in the radio landscape. On the AM side, there have been attempts to increase the number of stations, by reducing the amount of space allocated to each AM station (from 10 kHz to 9 kHz) and by adding more space at the top of the AM band (from 1605 kHz to 1705 kHz).

Since radio stations cross national boundaries, implementing these changes requires international treaties negotiated through the World Administrative Radio Conference (WARC). After some scrutiny, the FCC found that moving to 9 kHz spacing would be too expensive for implementation in the United States. Increasing the spectrum is a more likely development, with the possible allocation of up to 200 new AM stations in the next few years. However, even if these new stations are allocated, the uncertain financial picture for AM (see Chapter 4) makes expansion a risky venture for the radio entrepreneur.

Another development for AM has been the emergence of AM stereo. Two-channel AM was developed in the early 1980s to help AM stations compete with FM in the delivery of a high-fidelity signal. By the close of the decade, about one in three AM stations was broadasting in stereo. Thus far, the arrival of stereo has been less than a panacea for the problems of AM. For one thing, a single industry standard was late to emerge. Not all AM stations broadcast the same kind of stereo; more importantly, the stereo signal was not picked up by all AM stereo receivers. This points to another problem: apparently content with FM, few members of the public have replaced their radio sets with new receivers capable of picking up two-channel AM signals.

In 1990, a group of Canadian investors and government agencies began tests of digital audio broadcasting (DAB). Experimental broadcasts in the United States soon followed. DAB uses new digital transmission methods to beam as many as 16 or more simultaneous stereo signals in a given geographical area. The sound is just as good as that heard on compact disc (CD). In addition, experiments reveal that DAB suffers much less from problems that beset both AM and FM, like static and loss of stereo separation. It remains to be seen whether DAB will augment or supplant conventional AM and FM broadcasting.

Television Broadcasting

In the period immediately following World War II, the FCC began to issue licenses for television broadcasting. The first allocations went to stations operating in the very-high-frequency (VHF) band. We know these stations as channels 2 to 13. The first stations on the air were owned by the already powerful radio networks and by large manufacturing or business corporations. These stations began to offer popular vaudeville-type entertainment shows like *The Texaco Star Theater* with Milton Berle, and *Toast of the Town*, which most of us know as *The Ed Sullivan Show*. Many entrepreneurs wanted to cash in on the popularity of this new medium, and the FCC was soon deluged with applications for TV licenses. As a result, the commission issued a famous "freeze," which from 1948 to 1952 limited the number of stations on the air to 108. In April 1952, the FCC lifted the freeze and the expansion of television broadcasting began. An additional 70 channels were added in the ultra-high-frequency (UHF) band (channels 14 to 83). In late 1952, the first commercial UHF television station went on the air in Denver.

Today, there are approximately equal numbers of full-power commercial VHF and UHF stations operating: in the neighborhood of 575 stations each. The remaining stations (roughly 350) are noncommercial television stations, the majority of which operate in the UHF portion of the band.

Low-Power Television In 1979, the FCC authorized low-power television service (LPTV). The idea was to allow for the kind of diversity and range of choices in television which were available in radio, as well as to provide increasing ownership and programming opportunities for minorities. LPTV stations can operate on VHF and UHF channels unused by full-power stations in a given community. Power is restricted to 10 watts VHF and 1000 watts UHF, with an effective coverage area of 5 to 15 miles.

By the early 1980s, the FCC had been deluged with applications for LPTV licenses: as many as 30,000 applications were received for 4000 anticipated stations. After much maneuvering, the FCC began issuing licenses on a lottery basis and the backlog of applications began to dissipate.

Today, about 1300 LPTV stations are in operation, two-thirds of which are on UHF channels. Thus far, rural areas have seen the most benefit from LPTV service; over half of all LPTV stations are in the state of Alaska.

The fledgling LPTV industry has been beset by a number of problems. Most stations follow the advertiser-supported independent model developed by full-power stations, offering original and syndicated programming to their communities. However, the limited range of LPTV stations has generally yielded very small circulation figures for advertisers, and therefore disappointing revenue figures. In addition, LPTV operations have suffered from undercapitalization and poor management. Few LPTV entrepreneurs have obtained sufficient cash to ride out the difficult phases of licensing and construction.

The future for LPTV may be brighter. Many forthcoming channels will be assigned in major metropolitan areas, such as New York and Chicago. LPTV is trying to develop alternative programming, including foreign language programs, movies, music video, and financial and data services.

Cable Television Unlike over-the-air radio and television broadcasting, cable television is a closed communication system in which homes, or *drops*, are collectively wired by coaxial cable (via feeder and trunk lines) to a central originating source, known as the *head end*. The system is closed in that cable companies enter into a contractual arrangement with their audience (subscribers), and they typically negotiate private agreements with local municipalities for delivery of their services. Such contracts are known as franchise agreements.

Cable television service began humbly as community antenna television (CATV). During television's freeze period, many mountainous and rural areas were not within the limits of acceptable TV reception. Audiences felt left out of the new national craze; more important, local appliance sales and repair shops felt they were being denied a valuable source of income. Thus, in Pennsylvania, Oregon, and countless other places, radio repairers, hobbyists, and entrepreneurs began to erect large mountaintop antennas and to link them from house to house via coaxial cable.

In the early 1970s, some CATV operators began to realize that these antenna-and-receiver arrays could be used for the distribution of original programs. Cable began to spread to upper-income suburbs and municipalities, offering movie channels, sports, and other entertainment. By the mid-1970s, channel capacity had been expanded: converters were provided that increased cable service to 35 or 54 channels. Next, two-way interactive capability was explored. Cable subscribers could "talk back" to the head end by pushing buttons on a keypad, as in the Warner-Amex QUBE system. And cable could be used for home surveillance, shopping, and voting—and, at the height of the craze, even to play video games with unseen neighbors.

Cable television today is a broadband, multichannel service available to nearly 90 percent of the U.S. population. About 6 in 10 television homes now subscribe to some form of cable television service.

Pay Cable and Pay-per-View Television Pay cable refers to the extra-cost services which are available to the cable television subscriber. Media executives generally refer to these offerings as premium or tiered channels, because they are obtained by the cable subscriber for a fee in excess of basic cable costs.

The leading pay television channels, such as HBO, Showtime, Cinemax, and The Movie Channel, offer first-run and recent films, generally without commercial interruption. Other premium services are targeted for specific audience interests. These include channels specifically for children (Disney Channel) and for sports enthusiasts (Sports Channel).

Current estimates suggest that about one in three U.S. homes with television subscribes to at least one pay service (about half of all homes on the cable).

Pay-per-view (PPV), or per-program pay TV, is a potentially lucrative offshoot of cable television. Special-events programs are offered on a live, one-time-only basis to individual subscribers for a premium. Pay-per-view events have included Olympic sports coverage, heavyweight championship fights, rock concerts, and major wrestling promotions.

PPV faces some roadblocks. First, it requires special equipment, called an addressable converter, which allows the cable companies to isolate (and bill) each home that requests a special event. Second, and perhaps more crucially, it requires programs with sufficient appeal to convince an audience to pay for them! Over the years, viewers have come to expect television to be free. PPV entrepreneurs have to decide which programs people will purchase, how much they will be willing to pay for the programs, and when (or whether) they would pay again.

At the present time, about a third of all TV homes (and about half of all cable homes) have the equipment necessary to order and watch PPV events.

Both pay cable and pay-per-view change the goals of television programming from broadcasting—reaching as great an audience as possible—to narrowcasting—targeting specific programs and services to specific interests. The jury is still out on how narrow pay or pay-per-view targeting campaigns can be and remain commercially viable.

Satellite Master-Antenna Television Perhaps you have noted the appearance of a satellite receiving dish on the grounds of a nearby apartment complex. This is known as satellite master-antenna television (SMATV), also referred to as private cable, since it is cablelike service that operates on private property.

An SMATV company typically contracts with the builder or operator of a multiple-dwelling unit (MDU) to deliver programming to subscribing households from a central master antenna and satellite receiver. While SMATV operations do not offer the number of channels or range of programs provided by cable television systems, they have succeeded in wooing some upper-income urban and suburban subscribers away from local cable operations.

Multichannel, Multipoint Distribution Service Multichannel, multipoint distribution service (MMDS) consists of short-range microwave transmission from a centrally located transmitter to a number of microwave receiving dishes and tuners, called *downconverters*. Since cable is not needed to connect consumers to the head end, MMDS has also become known as *wireless cable*.

In 1983, the FCC adopted rules allowing for two 4-channel MMDS systems for each major U.S. market or community. In ensuing years, more than 16,000 applications were submitted to the FCC for this new telecommunications service. Despite the early interest, MMDS service has proved to be more promise than reality.

MMDS faces a technological drawback in that it requires line of sight between transmitting and receiving antennas. Thus, MMDS is susceptible to interference from trees, buildings, and other structures. Like LPTV, the range of MMDS service is typically limited to 10 miles or less. For these reasons and more, the primary users of MMDS thus far have been pay-movie operators targeting hotels, high-rise apartment buildings, and some businesses.

However, the relatively high frequency response and short transmission range suggest that MMDS has growth potential in the area of data delivery, or in the improved television transmission system known as high-definition television.

Direct-Broadcast Satellite Perhaps the most futuristic of the array of new media services is direct-broadcast satellite (DBS). With this arrangement, individual households replace their rooftop antennas or cable converters with small "napkin-sized" satellite receiving dishes so that programs can be downlinked directly from orbiting satellites.

DBS faces some handicaps. First, the signal transmitted to individual homes needs to be more powerful and concentrated than that currently downlinked to large dishes like those owned by broadcast stations and cable companies. More and higher-powered satellites are also needed. And, while there is promise of DBS providing 100 channels or more, some broadcast and cable programmers have vowed to keep their material off the satellite, lest the need for local television stations and cable systems be eliminated.

The development of DBS has also been hindered by enormous hardware costs. DBS is an expensive, capital-intensive operation. Estimates for DBS construction range in the hundreds of millions of dollars! As we have seen, COMSAT Corporation lost nearly $150 million in the 5 years before it abandoned its DBS service. Another company sank $68 million into a DBS venture before ultimately failing.

Despite these imposing obstacles, enthusiasm about DBS remains widespread. Limited-channel DBS systems are already operational in Europe and Japan. In the United States, long-time broadcaster and pioneer in satellite technology Stanley Hubbard is committed to the concept. His United States Satellite Broadcasting (USSB) has signed deals with major programmers Time Warner (HBO and Cinemax) and Viacom (Showtime, The Movie Channel, MTV, VH-1, and Nickelodeon) to distribute their shows via DBS. Another DBS venture, DirecTV, has also signed on with major programmers, including Paramount Pictures and The Disney Channel.

Two new technologies in television: direct-broadcast satellite (above) and high-definition television (below). (above, courtesy of DirecTV, Inc.; below, P. Gontier/The Image Works)

High-Definition Television A television picture consists of a series of parallel lines. How many lines make up the picture determines how large and clear the image can be. Our television system was standardized 50 years ago at 525 lines. This permits reasonable quality on small screens (like 19" or 21" TVs). But bigger screens suffer from grainy and fuzzy images (as you may have noticed in a bar or restaurant). Today, technological advances allow for a television picture as large and crisp as that of a 35mm motion picture. This technology is known as high-definition television (HDTV).

HDTV uses over 1000 scanning lines, making large-screen television much more practical, both for entertainment and for educational applications. Through much of the 1980s, HDTV was the focus of research and development in telecommunications laboratories. Regulatory concerns were also paramount, as the FCC struggled to adopt a standard for HDTV which would guarantee that existing television sets (and stations) would not be rendered obsolete by the new technology. Another concern was that American firms might not be able to compete with those in Japan and elsewhere in the manufacture and distribution of HDTV equipment.

To date, the issue of technical standards remains unresolved. However, the likelihood of American firms being assured a significant place in the coming HDTV arena was boosted by an announcement in mid-1993. A "grand alliance" between General Instruments, AT&T, Zenith, and the Advanced Television Research Laboratory was formed to pursue a single viable HDTV system. The proposal called for a digital system of 1050 lines for the United States, a system which could be used for television as well as computer applications.

Home Video

While DBS and even HDTV may seem very much like pie in the sky, the emergence of home video is truly down to earth. Unknown less than 20 years ago, the home video industry has exploded on the scene and created significant change in the media landscape.

Home video is the broad term which encompasses the various attachments available to viewers to increase the capabilities of their television sets. The first wave of home video began in the mid-1970s with the introduction of the video cassette recorder (VCR). In less than 10 percent of American households in 1979, the VCR now resides in 8 out of 10 homes. VCRs enable viewers to record television programs for later viewing (time shifting) and to make their own programs (with camera-VCR combinations known as camcorders, now in 1 in 5 households).

The primary use of the VCR—to play back prerecorded movies—has led to the creation of an entirely new (and largely unforeseen) media business. The video store is now as much a part of the consumer marketplace as the supermarket or post office. There are more than 50,000 retail video establishments in the United States. Videos are also available for sale or rent in supermarkets, convenience stores, pharmacies, and department stores.

While the VCR is the most common enhancement to the typical TV set, it is not the only one. The marriage of the television set to the computer began in the mid 1970s with the introduction of video games such as Pong and Pac-Man. Now the home television set is not only a source of game programs but also the input/output (I/O) device for an array of computational and data services. With keypad and keyboard attachments and coaxial cable or telephone (modem) connections, viewers can play games, compose and send correspondence (electronic mail), and arrange personal and business finances. By the early 1990s, it was estimated that 5 in 10 TV homes had at least one video game system; about half that number owned a personal computer.

Other new video components are proliferating rapidly, including large-screen television projection systems and video disc players. A device which made a splash in the 1980s, but whose growth has slowed because of increasing availability of cable, is the backyard satellite dish, known as a TVRO (for television receive-only earth station). There are about 3 million TVROs in use in the United States. TVROs permit households to receive dozens of satellite services, such as pay channels and cable networks, without need for a cable connection. As any trip down a rural road or to a wealthy suburb confirms, TVRO has found a place on the media landscape.

The continuing development of home video and other new technologies raises a number of issues which are of critical importance to media management. For example, network executives have expressed concern about whether VCR use erodes viewing of their programming. Cable managers are investigating how low-priced rentals of movies from video stores has altered their subscriber base. And advertisers, who foot the bill for much broadcast fare, are increasingly concerned that the new devices and the remote controls which operate them are enabling viewers to bypass (zap) or speed through (zip) their all-important commercial messages.

Some developments mentioned above (like DAB and DBS) have the potential to render conventional radio and television stations obsolete. In a nutshell, the technologies and businesses we sometimes label the "new media" call for the special kind of media manager envisioned by General Sarnoff many years ago. No longer can media management be complacent or smug. Broadcasting, once cynically called "a license to print money," is now a highly competitive and increasingly uncertain business. Once the darling of new media, cable is now a target of increasingly dissatisfied customers (many in the Congress) and of a potential new competitor from Bell, the RBOCs.

The nature of these and related businesses—their policies and behaviors (including their ability to anticipate and survive change)—is the topic to which we next turn.

SUMMARY

Electronic media refers to the commercial and nonprofit organizations involved in the development, production, distribution, and exhibition of products and services which provide entertainment and information to the public by electronic means.

Radio and television broadcasting are the backbone of the media business. Commercial radio broadcasting began in the medium-wave or AM band and has grown to include high-fidelity stereo broadcasting on FM. Commercial television began with network-owned and -affiliated stations located in the VHF band (channels 2 to 13). In recent years, independent stations, including many in the UHF band (channels 14 to 83), have proliferated and have shown some profitability. Many low-power television (LPTV) stations are now in operation in rural communities.

The contemporary media environment includes a variety of new programming and distribution systems. Cable television, which began as community antenna television (CATV), has grown to become a multichanneled source of a wide range of program and consumer services.

Pay cable and pay-per-view provide premium programming to cable viewers for a monthly fee. Still in the developing stages are multichannel, multipoint distribution service (MMDS) and direct-broadcast satellite (DBS). New distribution systems are developing simultaneously with the emergence of new modes of media consumption in the home. The VCR has spurred the growth of video retail. Video games, home computers, backyard dishes, and other devices have changed the television set from a passive box to an interactive input/output (I/O) terminal.

NOTES

1 David Sarnoff, *Looking Ahead: The Papers of David Sarnoff* (New York: McGraw-Hill, 1968), p. xiii.

FOR ADDITIONAL READING

Akwule, Raymond: *Global Telecommunications: The Technology, Administration and Policies* (Boston: Focal Press, 1992).

Baldwin, Thomas F., and D. Stevens McVoy: *Cable Communications* (Englewood Cliffs, N.J.: Prentice-Hall, 1988).

Bradley, Stephen P.: *Globalization, Technology and Competition: The fusion of computers and telecommunications in the 1990's* (Boston: Harvard Business School Press, 1993).

Dominick, Joseph R., Barry L. Sherman, and Gary Copeland: *Broadcasting/Cable and Beyond: An Introduction to Modern Electronic Media*, 2d ed. (New York: McGraw-Hill, 1993).

Luther, Sara Fletcher: *The United States and the Direct Broadcast Satellite* (New York: Oxford University Press, 1988).

Sapolsky, Harvey M. (Ed.): *The Telecommunications Revolution: Past, Present, Future* (London: Routledge, 1992).

2
THEORIES OF TELECOMMUNICATIONS MANAGEMENT

Theories of management and corporate behavior take two main forms. "Micro" theories focus in depth on small units of an industrial organization, such as the individual manager or employee. "Macro" theories take a long view; they attempt to encompass each of the various elements which comprise an industry. The next chapter, on telecommunications industry structure, is an example of macroeconomic theory.

This chapter takes a microview of the role of the manager in telecommunications and the relationship between media managers and their subordinates. These microexaminations are then fused into a comprehensive model of telecommunications management. Finally, as a prelude to the in-depth discussions of the media businesses which comprise Part Two (Chapters 4 to 6), persistent myths about the nature of media management are debunked when contrasted with contemporary economic and social reality.

The goals of this chapter are:
1 To define the process of management
2 To identify the tasks, functions, and roles of media management
3 To describe the relationship between management and labor in telecommunications industries
4 To provide a comprehensive model of media management
5 To compare the myths of media management with contemporary reality

MANAGEMENT: A DEFINITION

At a professional meeting, Clark Pollack, retired board chair of Nationwide Communications, wryly observed that media management is a simple task, since it involves only two resources: *people* and *money*.[1]

In brief, management consists of (1) the ability to supervise and motivate employees and (2) the ability to operate facilities and resources in a cost-effective (profitable) manner. Of course, mastery of these two basic skills is not as easy as it appears. The dynamics of human relations, communication, and motivation have been scrutinized by poets and psychologists as well as by countless management consulting services. And while a few media managers have enjoyed vast wealth as a result of successful operation of their enterprises, many find themselves drowning in a sea of red ink.

The root of the term *management* is the Latin *manus* (hand). Initially, the term was most widely used to refer to the care and training of horses, but gradually it came to be applied to a much broader area. People with management ability were those who could "handle" others: those who could train, govern, and influence other people and direct them in the performance of necessary tasks.

APPROACHES TO THE STUDY OF MANAGEMENT

Since its origins in the late nineteenth century, the formal study of management has followed three main schools of thought.[2] The first significant body of literature was formulated by the *functional* school. Its major goal is to carefully inventory the roles, tasks, and functions performed by managers in an organization. The attributes, skills, and experiences of successful managers are tracked through case studies, and from these case studies, "universal" or "ideal" management types and styles are developed.

The second main line of management research is the behavioral or *human-performance* approach. Here the stress is on the role of the individual in the organization. Human-performance approaches attempt to facilitate employee motivation, loyalty, and satisfaction, and try to achieve congruence between individual and organizational goals. Human-performance literature that compares U.S. enterprises with their counterparts in other areas of the world, particularly Japan, has received a great deal of attention in recent years.

The third major approach to the study of management is known as *management science*. Management science is concerned with the structure and behavior of large, complex organizations. Organizations are studied in light of their interrelationships with social, economic, political, and cultural institutions; their acquiescence or resistance to change; their internal networks of communication and control; and other variables. To the proponent of scientific management, the key to successful management is understanding the "mind" of the organization: the pressures, stimuli, and restraints which set the limits within which any individual in the organization must operate.[3]

This chapter takes a functional and human-performance approach. The tasks that media managers perform are inventoried and described. Then, the nature of the relationship between managers and employees in telecommunications businesses is critiqued and analyzed. Chapter 3 utilizes the management science approach to investigate the structure and behavior of telecommunications companies.

WHAT DO MANAGERS DO? COMPONENTS OF TELECOMMUNICATIONS MANAGEMENT

Management theorist Henri Fayol identified the basic management processes as planning, organizing, directing, coordinating, controlling, staffing, budgeting, and reporting.[4] Other authors have added other skills and roles to the list, including commanding, leading, actuating, supervising, negotiating, representing, innovating, delegating, evaluating, and spotting deviations. Acronyms for management tasks include POSDCORB (for planning, organizing, staffing, directing, coordinating, reporting, and budgeting, with an extra "o" included for pronounceability). A three-level managerial hierarchy has appeared that separates managers into three groups: upper, middle, and lower.[5]

Obviously, the attempt to catalog and codify the principles of management has itself become somewhat unmanageable. To establish the essential elements of telecommunications management, this section focuses on three distinct areas: (1) management skills, (2) management functions, and (3) management roles.

Management *skills* are the basic personal and professional competencies required of media management personnel. Management *functions* include the basic tasks performed by managers in telecommunications operations. Management *roles* consist of the many hats media managers wear in dealing with employees, regulators, and consumers. Figure 2-1 charts the elements of telecommunications management by skill, function, and role.

FIGURE 2-1 Telecommunications skills, functions, and roles.

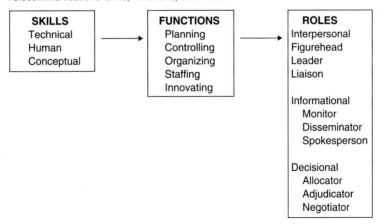

Telecommunications Management Skills

Technical Skills   Telecommunications is an industry characterized by technology and innovation. The industry is based on the "magic" of radio waves, futuristic satellites, high-powered transmitters, lightweight cameras, sophisticated computers, and other cutting-edge technology. Consequently, telecommunications managers must have some technical understanding of their businesses. Of course, it is impossible for media managers to keep abreast of all the developments in modern media technology. But a basic understanding of radio waves, signal propagation and distribution, computer operations, and recording systems enables the media manager to make informed purchase, expansion, and employee decisions in a complex, high-tech industry. This is why numerous media managers have risen to ranking positions from engineering, programming, and production, particularly in the industry's early days. Even today, some hands-on experience in and an understanding of the tools of media operations are important ingredients of successful telecommunications management.

Human Skills   Popular folklore and social stereotyping have contributed to an image of the media manager as a tyrannical mogul ruling over employees like a monarch over subjects. We associate Hollywood's golden age with benevolent despots like Samuel Goldwyn and Louis B. Mayer, and the rise of the television networks with take-charge executives like David Sarnoff, William Paley, and Leonard Goldenson.[6] More recently, mercurial personalities like Barry Diller, Ted Turner, Australian-born Rupert Murdoch, and the late Robert Maxwell have dominated the media landscape.

Behind-the-scenes biographies and case histories often paint a slightly different picture. Here media managers are described as slick negotiators and occasionally as magnanimous personalities with a strong personal interest in the well-being of their employees.[7]

What each of these views shares is that successful media management requires strong human skills in leadership and motivation. The history of media in the United States includes few entrepreneurial wallflowers or invisible executives. As a class, telecommunications managers are extroverted, dynamic, assertive, highly motivated, and gregarious. At a minimum, to succeed in telecommunications, the manager must have sufficient human skills to be able to work with others to build a cooperative effort within the organization he or she leads.

Conceptual Skills   Conceptual skills are the blend of intellectual abilities and sound judgment that enables successful media managers to get to the heart of an important personnel or marketing problem and make an appropriate decision. The telecommunications business requires leaders who respond quickly to changes in the environment: changes in consumer behavior, technology, employee patterns, government regulation, and so on.

Traditionally, conceptual skill in media management was thought to derive solely from experience: a mix of the work ethic and social Darwinism. Telecommunications managers who had the "right stuff" were the ones who simply survived: those who retained their leadership positions after economic recessions, format changes, and even station sales and corporate mergers. More recently, large media corporations have begun to realize that conceptual ability can be nurtured in a less threatening environment. Many media corporations have begun to recruit managers from other industries and from college undergraduate and graduate programs. Some industry leaders have established in-house management development programs, as examined in detail in Chapter 9. The point is, developing executives with conceptual ability is increasingly viewed as the key to the success of an operation.

It is difficult to quantify the technical, interpersonal, and conceptual skills that are required of telecommunications management personnel, or to say which of these areas is most important. It is equally difficult to identify how many of these qualities are innate and how many can be acquired through experience, job training, and personal improvement seminars. However, it is a fact that the changing telecommunications environment has put a premium on finding individuals with ability in all three areas.

Functions of Telecommunications Management

How are the basic telecommunications skills applied in the real world of media management? The following are the typical tasks performed by media managers, regardless of their position or the size of their facility.

Planning Planning is the process of determining in advance the objectives of the corporation and the means by which these objectives may be achieved. Simply stated, it is deciding in advance *what* to do, *how* to do it, and *who* is to do it. For the radio station manager, planning includes what music will be programmed, which disc jockeys are to be hired or fired, and how much to charge an advertiser for a 30-second spot. The cable television manager must decide, among other things, which neighborhoods will be "wired," which program services will be offered, and how the services will be priced.

The process of planning centers on forecasting. While few individuals are truly prescient, successful telecommunications managers have the ability to predict the future accurately and to plan for change. For example, it was young David Sarnoff who in 1916 saw the potential for the "radio music box" and who later parlayed the idea into the formation of RCA and NBC.[8] More recently, in 1973, an executive at Time-Life sensed that communications satellites could be used to send movies to local cable companies to distribute to their subscribers for a premium. The executive was Charles Dolan (now an influential media manager as head of Cablevision Systems, Inc.); the company, Home Box Office.[9] And in 1970, when the owner of a small advertising agency acquired an independent UHF television station, he decided to make his station available to cable systems for a nominal fee to help fill their channel capacity. The planner was Ted Turner;

the station, Atlanta channel 17, then WTCG, now WTBS, the original superstation.[10]

Telecommunications managers have five basic alternatives in the area of planning:

1 Retain the status quo.
2 Expand within the industry.
3 Diversify into a related industry.
4 Diversify into a new industry.
5 Divest or "sell out."

Deciding which of these alternatives the telecommunications company will take involves the coordinated effort of lawyers, accountants, market analysts and researchers, political pundits, and occasionally even intuitive relatives and friends. While others often provide the "intelligence" on the basis of which a decision is made, the actual responsibility of planning and decision-making rests with the media manager.

Organizing The second major management function of concern in telecommunications is organizing. Organizing is the assignment and coordination of duties within the corporation to ensure maximum efficiency in the attainment of predetermined planning objectives. In short, organizing consists of defining jobs and reporting relationships, determining who does what for whom.

Telecommunications firms require the disparate but equally important skills of programmers, engineers, accountants, researchers, salespeople, marketers, and others. Getting all these individuals working together for the common goal of the company is a difficult task. Not surprisingly, then, telecommunications is marked by structuralization, departmentalization, and the need for delegation.

Structuralization refers to the adoption of strict models of the organization, including management flowcharts, codified job descriptions, and clearly identified evaluation and grievance procedures. *Departmentalization* refers to the creation of separate departments for each function of the telecommunications enterprise. Television stations, for example, typically have separate programming, sales, and engineering departments. Some also have departments of news, promotions, public affairs, special programming, and research. The process of *delegation* allows each of the departments, with its own substructures, to operate quasi-independently. Each department in a telecommunications firm usually has its own separate planning goals, operating budget, and reporting relationships. The goal of delegation is to free upper-level management to concentrate on other functions (planning in particular) as well as to encourage employee creativity, initiative, and growth. However, delegation may lead to internal friction and a loss of the sense of the common corporate goal among employees.

The organizational challenges to telecommunications managers are to create a workable structure, to supervise departmentalization and delegation, and to reorganize as necessary as patterns of employee relations, technology, and market conditions change.

Staffing Like planning and organizing, staffing is an ongoing concern for the telecommunications manager. It consists of the recruitment, training, compensation, retention, and termination of employees. Simply put, it is the hiring and firing function of the telecommunications manager. Most broadcast educators and professionals agree that there is no shortage in the pool of potential telecommunications employees. Each year our colleges and universities produce thousands of people with degrees in journalism, telecommunications, broadcasting, or mass communications. And each day media managers are deluged with resumes from potential employees, many of whom volunteer to "do anything" to get their foot in the door. Paradoxically, many media managers complain about an employee shortage; they say that there are not enough "good" people around. This apparent contradiction is the result of the basic organizational structure of the telecommunications industry. The industry places a premium on people with both general and specialized knowledge. Yet some colleges provide people with broad liberal arts skills and general knowledge but give them little technical expertise or experience, and others emphasize the techniques of production but gloss over training in the humanities and social sciences.

The task of management is to find the middle ground between the two extremes: to place specialized individuals in specialized tasks, and at the same time encourage and develop potential department heads and managers with strong abilities to motivate and lead others.

As documented in Part Two, the telecommunications industry has a high rate of employee turnover. In television, for example, nearly 1 in 5 employees resigns or is terminated each year.[11] The turnover rate is not limited to entry-level positions; executive-level broadcasters frequently jump from one station or network to another, or form their own companies.

Different organizations take different approaches to the problem of staffing. As we will see, some organizations hire relatively unskilled workers to fill low-paying entry-level positions and encourage employees to work their way up to higher-paying management positions. Others conduct wide-ranging talent searches and offer extensive training and employee development programs. Some lure employees from other firms with significant monetary incentives and inducements.

Staffing is not merely an internal problem. Federal regulatory agencies and consumer groups have found that the media industry has traditionally underemployed women and ethnic minorities (see Chapter 3). To rectify this situation, telecommunications firms must meet federal equal employment opportunity and affirmative action guidelines. How well media companies live up to these guidelines leads to a number of annual reports and "report cards" from the National Black Media Coalition, Women in Communication, and other organizations.

Controlling Controlling is the communication function of management. The telecommunications manager must explain corporate and personal goals to his or her staff and must be able to determine how well those goals are being met on both an individual and a corporate basis. The management control process follows four main steps:

1 Establishing standards
2 Communicating standards
3 Measuring performance against standards
4 Correcting deviations

Organizational standards in telecommunication firms may be quantitative or qualitative. For a radio station, for example, quantitative standards might include projections of advertising revenue, anticipated audience shares in morning drive time, and the number of entrants in a mail-in contest. For the same station, qualitative standards might include having the best "sound" in town and producing a community image of the station as a loyal friend.

Once standards have been established, it is the task of management to communicate them to the staff. Many media managers have become masters of the memo; others attend monthly, weekly, or daily management meetings. Sometimes the means of communicating management goals matches the style and format of the telecommunications facility. Many small-market television stations and community radio stations communicate informally: in the casual atmosphere of an intimate studio or a neighborhood bar. At the other end of the spectrum are large communications conglomerates which feature plush boardrooms in which industry giants plot their strategies. Patterns of in-house communication are examined in more detail in Chapter 10.

The third step in management control is obtaining feedback; that is, measuring actual performance against established standards. Methods of comparing performance against standards may be formal or informal. For a cable television company, formal performance appraisals can include number of households wired, number of pay television subscribers signed up, number of orders for pay-per-view events, and quarterly billing figures. Informal performance measures can include public perception of the company's service department and the level of morale around the office.

Similarly, measures of the performance of individual employees may be formal or informal. Examples of formal measures are sales quotas met, days absent or late, and meetings attended. These can be listed on formal appraisal forms, prepared by managers on a quarterly, semiannual, or annual basis. Informal measures can include the manager's personal perceptions of an employee's motivation, dedication, commitment, and attitude.

After measuring corporate and employee performance against planning standards, the telecommunications manager then corrects deviations; that is, revises planning estimates and/or makes changes in staff or facilities. The changes made rest on the manager's ability to determine why preset standards were not met (or were exceeded). A decline in the ratings for a popular television show could be the result of poor writing or performing, a bad lead-in program, powerful competition, or the fickle nature of the audience. In this case, the television management executive might fire the writers or actors, change the show's time slot, or even cancel the program altogether. The point is that measuring standards against objectives rarely illuminates the cause of the deviation. Making that determination and charting the ensuing strategy are the tasks of the telecommunications manager.

Innovating Although it may seem that innovating is merely one element of the planning process, some analysts consider it to be the critical ingredient in the successful management of advanced technological corporations.[12] Innovation consists of developing new and better ways of conducting a telecommunications business.

Innovation can be both internal and external. Internal innovations may include automating program and accounting operations, building a new transmitting tower or receiving antenna at a radio station, and replacing coaxial cable with optical fiber in a cable system. External innovations may include developing a new program schedule, introducing a new line of radio receivers, and expanding into a new business enterprise altogether. The responsibility of the telecommunications manager is to keep the "forest" of innovative hardware and software in sight, separate from the "trees" of day-to-day management decisions. He or she must have the entrepreneurship to obtain or commit substantial funds for innovation and the resourcefulness to incorporate the innovation into the organization without compromising productivity. For these reasons, innovation is a hallmark function of the telecommunications executive, worthy of further discussion in the chapters ahead.

Roles for Telecommunications Management

In addition to performing the variety of tasks outlined above, the telecommunications manager is an actor, playing a variety of roles for staff, shareholders, and consumers or audiences. The term *actor* may suggest that telecommunications executives misrepresent themselves or are deliberately deceitful. Quite the contrary. By roles, we simply mean those organized sets of behaviors which are identified with a management position. These can be classified into three main groups. *Interpersonal roles* are those performed by the telecommunications manager as the leader of a media department or organization; *informational roles* are those involving the manager's function as corporate spokesperson or oracle; *decisional roles* are those of adjudicator and negotiator. Successful performance of these roles often equates with managerial and entrepreneurial success in the media marketplace.

Interpersonal Roles Telecommunications managers play three main interpersonal roles: leader, figurehead, and liaison. The leadership function is, of course, paramount. The telecommunications manager is responsible for the overall structure and functions of the media enterprise as well as for its ultimate growth and direction.

The telecommunications executive must represent, or personally embody, the goals and personality of the organization. This is the figurehead role, which consists mainly of ceremonial duties. The general manager of a television station will often represent the station at a community fund-raising event; the chief engineer might conduct a tour of the facilities for a group of foreign broadcast engineers; and the sales manager might take a potential client to lunch. Occasionally, a media manager's performance in the figurehead role can affect regulatory practice or enhance corporate image. For example, in the late 1950s, appearances by network television executives before a congressional investigation of allegedly fixed quiz programs helped restore the credibility of television in the eyes of the viewing public. More recently, impassioned speeches by Jack Valenti, head of the Motion Picture Association of America, contributed to massive publicity about a Supreme Court review of the impact of home video recorders on the movie industry. And criticism of the cable television industry has been deflected in part by the strong leadership presence before Congress and the courts of such leaders as Gerald Levin of Time Warner and John Malone of TCI.

As we have seen, the multifaceted nature of telecommunications firms has led to departmentalization and specialization within the firm. The stratified and specialized structure of the typical telecommunications enterprise requires the manager to act as chief liaison between individual departments or units or, in the case of group owners, between stations. As liaison, the manager creates and maintains contacts outside of the vertical units (departments) which make up the firm. Since the liaison function requires managers to spend much of their work time outside their own department, it is not surprising to find television sales managers who spend little time with their account executives or radio program directors who don't know or talk to their disc jockeys.

Studies have shown that managers split their time roughly equally between their subordinates and people outside the organization.[13] The tasks of the manager as liaison are to represent his or her staff and to obtain and interpret information from other units in the organization.

The liaison role leads into the next broad group of management roles: those of professional communicator.

Informational Roles The process of obtaining, synthesizing, and presenting information is a key aspect of telecommunications management. As monitors, media managers scan the communications environment for items of interest to their organizations and personnel. Telecommunications managers do this in a variety of ways. First, management personnel subscribe to general-interest newspapers and magazines as well as to specific trade publications such as *Electronic Media, Broadcasting and Cable*, and *Cablevision*. Second, managers belong to national organizations related to their specialty and attend and participate in the national meetings and regional seminars of these organizations. For example, most television program executives belong to the National Association of Television Program Executives (NATPE), sales managers attend Television Bureau of Advertising (TVB) sessions, and controllers and accountants belong to the Broadcast/Cable Financial Management Association (BCFM).

Once managers have obtained useful information, they must make it known to their subordinates. As a disseminator, the manager may conduct meetings and issue directives or staff memos. As suggested in the discussion of the figurehead role of media managers, the typical telecommunications executive is a professional communicator in his or her own right. As a spokesperson, the telecommunications manager makes personal appearances, gives speeches, and writes and delivers position statements or editorials.

Decisional Roles As discussed at the outset of this text, telecommunications is characterized by change. Therefore, telecommunications managers are the "change agents" in media organizations. They must respond to external conditions and must be the primary initiators of change within the firm. Planning for change involves the media manager in decision making. There are three main decisional roles performed by the media manager: allocator, adjudicator, and negotiator.

As an *allocator*, the media manager decides how much time, money, or equipment will be committed to pursue a project. At any given moment, the manager will be overseeing many ongoing projects. It might be helpful to view the manager in this case as a master chef, who enhances a promising recipe with extra ingredients, allows some dishes to cool and some to simmer, or puts a failed effort in the dog's bowl. The manager of a pay cable firm, for example, may be investigating the availability of a major feature film package and at the same time optioning a promising script for development, negotiating for a series of prize fights and searching for additional transponder space on a telecommunications satellite.

Successful management involves making dozens of entrepreneurial decisions on a daily basis and living with those decisions once they have been made. The media manager's decision-making role is not limited to resource and equipment allocation. There is also the pivotal task of making daily personnel decisions. Administration of the firm or department's staffing function leads to the manager's internal role as adjudicator and negotiator.

As an *adjudicator*, the manager makes decisions in areas of employee competition and conflict. Because of its competitive nature and the creative abilities of its employees, the field of telecommunications is often characterized by staff friction and rivalry. The manager is frequently called upon to mediate disputes. Some general managers have likened themselves to the referee in a boxing match; others to the chief psychiatrist in a mental institution. As adjudicator, the manager decides which of the parties in conflict is at fault and whether or not the dispute threatens the productivity of the department or company. In the latter case, the manager must make critical decisions involving disciplinary action or termination.

Since telecommunications operations involve an amalgam of technical and creative talents, the manager also performs a vital role as *negotiator*. Negotiations may be either formal or informal in nature. For example, managers of television stations in a large city will enter into formal negotiations with trade unions, such as the National Association of Broadcast Engineers and Technicians (NABET) and the International Alliance of Theatrical and Stage Employees (IATSE), to obtain a workable contract for their employees (see Chapter 10). In addition, they may be involved in informal negotiations with a manufacturer over the cost of a new videotape machine and with competitors on the establishment of a shared transmitter location. Skillful negotiation is a way of life for the telecommunications manager.

HOW DO MEDIA MANAGERS MANAGE? THEORIES X, Y, AND Z

Now that we understand what media managers do, we can turn to a set of concepts that probes how they do it—that is, the nature of the relationship between media managers and their subordinates.

Three main theories have arisen regarding the relationships between management and labor. These are Theory X, Theory Y, and Theory Z. The first two theories were described by Douglas McGregor in a widely read and influential text.[14] Theory Z emerges from the analyses of Japanese industrial organization undertaken by William Ouchi and others.[15]

Theory X

The management of personnel through physical force or power, intimidation or threat, and the fear of sanctions, penalties, or termination is labeled by McGregor as Theory X. This management style predominated in the early days of the industrial revolution, in a society structured along rigid class lines. The Chinese who built the American railroads and the immigrants who worked in the garment center sweatshops were perceived as inferior human beings. In management's view, the employees on the assembly line were barely distinguishable from the equipment they were operating. Over time (and with widespread reports in the press of management abuses), physical force and intimidation were replaced by moral authority coupled with control of monetary compensation. Managers were perceived as despots who rewarded good work with salary increments and who punished poor performance with postponed promotions, deferred increases, and, in the worst case, quick termination.

In McGregor's view, Theory X management is based on an absolute concept of authority: the goals of the corporation and the means by which they are achieved are a direct extension of the personality of its leading executive(s). For many years, telecommunications management followed a Theory X management pattern, and the high turnover and burnout rates that continue among media employees indicate that elements of Theory X persist. There are various reasons for this phenomenon.

First, telecommunications began contemporaneously with the rise of industrialization. Many of the same firms that developed the railroads, industrial machinery, and electrical components, including Westinghouse, AT&T, and General Electric, were active in the fledgling broadcasting industry. The same managers who supervised mechanical assembly lines participated in the wireless telegraph and phonograph industries.

In addition, as leading media historian Erik Barnouw has recounted, emerging media businesses have always had very close connections with the military, the bastion of Theory X.[16] Radio stations were seized by the Navy during World War

I, and the Navy Department maintained its interest in the medium following the war through the appointment of Rear Admiral W. H. G. Bullard to the board of the Radio Corporation of America. Throughout World War II and the cold war era of the 1950s and 1960s, David Sarnoff, head of RCA, maintained a close relationship with the Defense Department. Satellite and other telecommunications companies have had longstanding defense contracts. (An example is Raytheon, which built the famous Patriot missiles of the 1991 Gulf War.)

Many of broadcasting's first-generation managers learned about radio, television, and film while in military service. It is thus not surprising that they modeled their management style after that of their superior officers.

There are also practical reasons that the tenets of Theory X dominated for many years and persist in a number of media enterprises. Paramount among these is the pressure of time. Running a broadcast or cable system requires quick, immediate management decisions, and there is often little time to take into account the potential impact upon employees. The director of a television newscast cannot wait for her camera operator to "feel" ready for a shot; she must order him into position so that the shot is not missed. The radio disc jockey can't play a commercial when he feels like it, especially when the sponsor has paid a premium for a fixed commercial position. Daily media management requires dozens of snap, time-pressured decisions. For such decisions, a Theory X approach often functions best.

Theory Y

At the opposite extreme from Theory X is Theory Y, the human-needs approach to employee-management relations. In this approach, managers seek to influence their subordinates on the assumption that the best motivators involve satisfaction of basic human physical and psychological needs—most critically status, self-esteem, recognition, and appreciation. The goal of the manager is not to motivate by fear, threat, or intimidation, but to create a work environment in which the employee's personal goals match those of the organization. Employees are encouraged to view their jobs as an integral part of their total life experience. The task of management is to recognize the unique talents possessed by each employee, to match those talents to an appropriate position, and to provide sufficient tangible, personal rewards to the employee to encourage continued productivity and personal growth. These rewards may not merely be monetary: they may be emotional or spiritual as well. Theory Y has had its moments in the telecommunications business. Many of the influential programs of television's early days—for example, the famous *Your Show of Shows* starring Sid Caesar and Imogene Coca and written by Mel Brooks, Mel Tolkin, Neil Simon, Carl Reiner, and Lucille Kallen—were produced by committee. New ideas were shared at a roundtable that often included the producers, network executives, and even sponsors.

In a similar vein, in the late 1960s, the FM radio stations owned by ABC featured a "love rock" format, with a strong emphasis on poetry and folk-rock music. Management at these stations tended to be loose and informal; in the vernacular of the times, things were "laid back and mellow." A humorous episode from this period indicates how the tenets of Theory Y had taken hold. Elton Rule, then head of ABC, reports walking in unannounced to a management meeting. He was not recognized, and was invited to smoke a bowlful of marijuana; ultimately he was bitten by the German shepherd owned by the program director.[17]

More recently, the emergence of microcomputing in the 1970s and 1980s lent some credence to the Theory Y approach. Computer entrepreneurs, led by Apple's founder Steven Jobs, applied the Theory Y approach to the development of new hardware and software.[18] In their organizations, rigid bureaucratic structures (like those at "Theory X" competitor IBM) were to be avoided at all costs; employees were encouraged to be creative, self-motivated, and unstructured in their pursuit of new products and services.

Because of logistical and time pressures, very few media facilities can operate according to a Theory Y approach. However, in recent years, the declining productivity of U.S. manufacturers and the success of the Japanese industrial sector has led to the proliferation of a new model of management employee relations. Elements of Theory Z have made their way into telecommunications.

Theory Z

Theory Z is an amalgam of the X and Y models. A Theory Z organization strives to capture the best of both the Japanese and U.S. management approaches. The Theory Z organization is egalitarian, engages the full participation of employees in running the company, and emphasizes the development of interpersonal relationships between employees and managers. However, authority is not sacrificed: managers still make the influential planning and policy decisions and can rely on sanctions to achieve organizational goals.

The Theory Z approach is made clearer through a discussion of its main points.

Integrity In the Japanese industrial model, managers must have the trust of employees and their full belief in their integrity and in the integrity of the product. The manager with integrity acts consistently, treats secretaries and executives with equal respect, and approaches subordinates with the same understanding and values that characterize his or her family relationships.

Company Philosophy What is your business about? Its goals in the marketplace? The value placed on its people? According to Theory Z, the company policy may not be immediately visible, or it may not match the one printed in the employee handbook or the company's annual report. Managers must *audit*, or examine, the most recent key decisions made by the company to ferret out its "true" philosophy. Once the corporation's philosophy is understood, it can be communicated to employees and revised to reflect their inputs and needs.

Involvement at All Levels In a Theory Z organization, all levels of management must be involved: from the chief executive officer (CEO) on down. Theory Z cannot work unless openness and communication reach all levels, including the very top and bottom of the management ladder.

Structures, Incentives, and Rewards While proponents of Theory Z argue against strict organizational structures and reporting relationships, some elements of structure are needed so that employees and managers can be guided toward common long-term corporate goals. Incentives are needed to stimulate and maintain interest in the Z approach. The most common incentives are profit sharing, continued employee training, employee benefits, and personal enrichment plans. Such items as liberal maternity leave, on-site day care and exercise facilities, and other "quality of life" considerations are critical to the success of a Theory Z approach.

Interpersonal Skills Speaking and listening skills are essential if managers and employees are to profit from one another's input. Knowledge of group dynamics and problem-solving and discussion techniques helps to facilitate good employee-management relations. The success of the Japanese industrial sector has been credited in part to the concept of the *quality circle*. The quality circle is a round-table discussion in which all levels of an organization, from top management to assembly line, discuss a product in a free and open interchange.[19] This open communication helps keep management in touch with its labor force, and vice versa. Problems with the product as well as with hardware can be intercepted before they do irreparable harm to the productivity of the company.

Constant Evaluation The move to a Theory Z approach requires careful monitoring and evaluation. This can be informal, through self-evaluation, or formalized through objective tests. The key is that evaluation must be ongoing: it cannot be a simple function of annual profit or loss. The health of the organization must be monitored monthly, weekly, or even daily.

Union Involvement As the Theory Z approach is implemented, unions must be fully involved with company management. Management and unions must be convinced that there is nothing to be gained by merely protecting themselves and fighting each other. Seminars between managers and union officials should be held at a neutral location so that individuals can meet one another and replace negative stereotypes with more accurate, positive images. As we will see in Chapter 10, union involvement would likely produce a skeptical response from many media managers.

Employment Stability Employees should be retained throughout the process of conversion to Theory Z management. Companies are encouraged to limit layoffs and instead accept a short-term loss in profits. In the Japanese sector, during bad economic times, workers accept shorter workweeks or smaller checks. In exchange, a highly committed and experienced employment team repays management in better times with lower turnover, greater loyalty, and larger profits.

Long-Term Evaluation and Promotion   Many Theory X companies are typified by an "up or out" philosophy: the notion that early promotion is good and that delayed ascendancy in the corporate hierarchy is the mark of a poor employee. In a Theory Z operation, the promotion timetable must be slowed so that employees can gain a long-term sense of their jobs and the goals of the company. The productivity of too many companies (including telecommunications firms) is hampered by the continued presence of new managers who are not yet acclimated to their jobs or organizations. The recent wave of corporate mergers, reorganizations, and buyouts has made this situation especially acute.

Broadened Career Paths   In the Theory Z organization, managers are not lockstepped into one area or career path. The Japanese (as well as some German and Dutch firms) have recognized that managers who circulate across jobs, even without promotion, retain their enthusiasm, freshness, and satisfaction. Those who remain in the same position tend to lose enthusiasm and commitment. In the Z company, lateral job movement is encouraged.

There is growing evidence that elements of Theory Z have made their way, albeit slowly, into media management. These include a proliferation of employee training programs, management development plans, profit sharing and other financial incentives, and improved workplace conditions, along with more open lines of communication between managers and employees. However, high turnover and attrition, frequent labor and management conflicts, and occasional but highly publicized lawsuits brought against management by employees are evidence that telecommunications maintains elements of its Theory X tradition.

A MODEL OF TELECOMMUNICATIONS MANAGEMENT

By now, we've inventoried the tasks, functions, and roles performed by media managers and the various approaches that managers have taken to their jobs. The next section synthesizes these theories into a comprehensive model of media management.

Figure 2-2 provides a model of the social environment of media management. Note that the model places management in the center of a complex, dynamic structure of interlocking systems, all of which operate under constraints imposed by the political, social, and cultural conditions of society in the United States.

It is clear from the model that media managers are mainly intermediaries. Their daily decisions are determined by *external* (outside) forces, such as changing consumer tastes, the programming and sales policies of their competitors, regula-

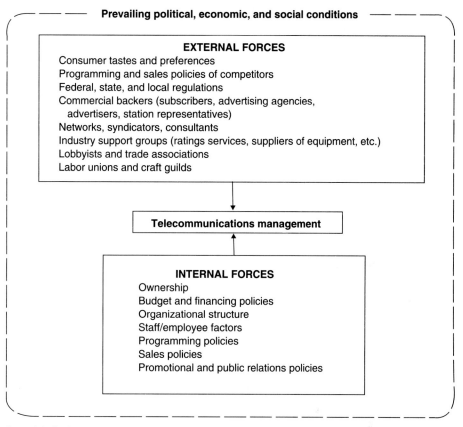

FIGURE 2-2   A model of telecommunications management.

tory agencies, investors, program suppliers and consultants, research services, industry associations, and labor groups.

In addition, media managers face daily *internal* pressures arising from ownership and organizational structure, budgets and other financial constraints, employee relations, programming and sales policies, and promotional and public relations strategies.

Management and organizational theorists have argued that the principal function of management is to maintain *stability*—that is, to keep the firm on a steady course no matter what changes take place in the internal and external forces depicted in Figure 2-2. Too much change in one of these forces might have a ripple effect that could drive the entire telecommunications system toward chaos.

Consider a change in ownership of a local radio station. Some of the implications for the station itself would be a change in format, news affiliation, announcing team, sales staff, and call letters. For the local market, it could mean a readjustment of audience listening patterns and, therefore, a change in advertising sales strategies and competitive policies. Even so-called minor changes can reverberate throughout the telecommunications system. Think back to recent fads and trends and to how quickly they were integrated into the world of telecommunications. The success of New Line's "Teenage Mutant Ninja Turtles" led to new cartoon programs, two feature-length films, the introduction of new video games and a line of toys.

In a more serious vein, growing public concern about driving while intoxicated led to concerted efforts by broadcasters to develop public service programs and announcements supporting higher drinking ages and tougher sentences for drunken drivers. To the dismay of many managers, it also nearly led to legislation aimed at banning beer and liquor advertisements from the airwaves.

More recently, increased public concern about the price of cable television led to more vigorous regulation of the cable industry. The reverberations of the Cable Act of 1992 have been enormous. The new laws governing rates, programming, competition, and new technology have had significant impact on cable systems, cable organizations, broadcasters, and members of Congress, not to mention the general public (for whom the legislation was originally intended)!

The present era is characterized by sweeping changes in the internal and external forces described in Figure 2-2, thus making the role of media management increasingly critical.

EXTERNAL CHANGES

Five related external developments are changing the telecommunications industry in the United States: technological change, economic change, population shifts, civil rights, and audience fragmentation.

Technological Change

Innovation has always been a characteristic of telecommunications, and in recent years the pace of technological development has intensified. In the radio business, for example, satellite dishes have replaced telephone lines for the delivery of network programs; microwave systems relay signals from the console to the transmitter; compact discs are supplanting turntables and cartridge machines; computers are printing logs and playlists and even running the audio board. Listeners are wearing miniature headset radios or carrying large "boom boxes." Even in the car, radio stations are competing for audiences with car phones, cassette machines, and CD players. And, as we saw in Chapter 1, digital audio broadcasting (DAB) is on the horizon.

Similarly, in television, video tape has replaced film, computer consoles have replaced flatbed editors, and satellite and microwave capabilities have allowed stations to broadcast more and more live news and entertainment programs. A significant number of viewers now have access to cable signals, video cassette recorders, remote control devices, large-screen televisions, and video game and computer attachments.

Technological change presents a serious challenge to contemporary media management. Which new equipment should be bought? Should the programming or sales departments be automated? How are new receivers changing the audience? Is high-definition television (HDTV) a threat? What about DAB? What about the VCR and remote control? Are viewers bypassing or editing out our commercials?

While the impact of technological change may take years to assess, media managers cannot afford to wait: they must make frequent, often daily, decisions and hope for the best.

Economic Change

Both the rise of radio in the 1920s and the arrival of television in the 1950s were accompanied by times of economic prosperity for media entrepreneurs and their audiences. Buying a radio or television set was a status symbol, and advertisers were eager to use the new media to promote their products and services. Today's new media developments face a much more uncertain economy, both nationally and globally. The traditional household, in which the father was the breadwinner and the mother stayed at home to raise children, is now the exception, rather than the rule. In 1960, nearly 70% of the work force was male; in 1980, that percentage had shrunk to less than 60%; by 1990, the work force was 54% male, 46% female.[20]

Today, 65% of married women with children under the age of 18 work outside the home; 40% with children under 6 years of age are reported to be working at least on a part-time basis.[21] In addition, in recent years unemployment levels have been at their highest since the Depression, factory production levels are down, and the number of Americans at or below the poverty line has been increasing.

Population Shifts

The rise of new media has typically coincided with a rising birth rate. For example, radio boomed during an era of large nuclear families, many of immigrant stock, creating a favorable climate for consumption of the new device. Radio took a hallowed place in the home, almost as a second hearth. Similarly, the rise of television paralleled the unprecedented baby boom following World War II, creating a willing audience for the new medium and for its programming and advertisers.

New electronic media are not so fortunate. Economic difficulties and a declining birth rate have resulted in significant demographic changes. In 1960, the birth rate was 23.7 per 1000 and the median age of the U.S. public was 27.5 years. By 1970, the figures were 18.4 and 28, respectively. By 1980, the annual birth rate had plunged to 15.9 per 1000 and the median age had risen to 30. In 1990, the birth rate had risen somewhat, to 16.0, but the median age of the population was the highest ever, at nearly 33 years.[22]

During the same generational interval, the rising divorce rate and other factors caused the number of households in the United States to grow but the number of persons per household to diminish.[23]

Taken together, these shifts have substantially changed the telecommunications environment. New media face an older, more fragmented population. Telecommunications managers can no longer count on the existence of a young, eager audience for their innovative products. Instead, they face a more mature and, therefore, more cynical and skeptical public.

Civil Rights

The late twentieth century has seen significant changes in the role of women and minorities in society. The volatile civil rights movement of the 1960s and early 1970s has given way to an era of more subtle yet perhaps more lasting gains for minorities and women. Each group has made substantial inroads into political, economic, and social institutions in the United States. Recall that women now comprise 46% of the work force. Both minorities and women have entered the upper echelons of business and management, albeit slowly. While the gains made by these groups may not have created true equity in our free enterprise system, there is no denying that a foundation has been laid.

The impact of the civil rights movement upon the telecommunications marketplace has been substantial. Marketers and advertisers have had to reanalyze their evaluation of minorities and women as audiences. Radio stations, for example, have found that ethnic minorities have more purchasing power than had previously been perceived. Television programmers have begun to realize that today's working woman is qualitatively different from the stereotypical housewife. Marketing and advertising targeted to these "new" groups have become hotly competitive.

Audience Fragmentation

Audience fragmentation is a direct consequence of the trends traced above, rather than a separate phenomenon. Economic difficulties, population shifts, and social reforms have merged to challenge the notion of a mass audience, the foundation upon which the telecommunications industry in the United States was built. The phenomenon has already substantially altered radio programming. Mass-appeal entertainment has been replaced by stations that offer narrow formats catering to specific population segments, such as all-news, rock, jazz, and ethnic stations. Similarly, cable television offers formats dictated by type of audience. There are movie channels for film buffs, sports channels for armchair quarterbacks, rock music channels for teens, and all-black, all-Hispanic, and all-Korean channels.

The reality of marketplace fragmentation has placed a great burden on telecommunications management. The field has become more competitive, more highly dependent on market and audience research, and more uncertain about audience acceptance of innovation.

INTERNAL CHANGES

Not only does telecommunications management face external technological, social, and economic changes. The nature of telecommunications management itself has changed. Major internal changes have necessitated a rethinking of the qualities or characteristics that were presumed to lead to success in media management.

The Media Management Myth

In the 75-year history of electronic media in the United States, a mythic view of management evolved. Like all myths, it was based on observation, folk wisdom, legend, and truth. According to the myth, to ascend to the top of the hierarchy of media management, individuals had to possess the combined talents of Horatio Alger, Thomas Edison, and Tom Swift.

Successful broadcast managers were a cadre of young men of immigrant or small-town stock. They had "discovered" broadcasting in their attic workshops, usually after a stint in the military, or purely by accident of fate or fortune. They started from scratch, often deeply in debt, and worked their way up from entry-level positions in engineering, programming, or sales to the top of the corporate ladder. Formal education was not a valued commodity; street smarts and business acumen were more important than a college degree. Consequently, media managers had a deserved reputation for toughness and guile. They became known for ruthlessly stifling competition from within through rigid administrative policies and corporate rules and for controlling outside forces through shrewd acquisition policies.

Once atop the media mountain, these first-generation moguls built communications empires, courted and befriended political leaders, and amassed great personal wealth. A few returned part of this wealth through philanthropic enterprises. Some established museums, others donated to medical research, and still others contributed to university journalism or business programs. Many became highly visible and active in civic, professional, and fraternal associations.

Aspects of this myth apply to media enterprises of all sizes: from the local radio stations to the large communications conglomerate. For example, John E. Fetzer began as a radio engineer in the medium's experimental days. In a career spanning 60 years, Fetzer parlayed his interest in radio into group ownership of eight radio and television stations in the Midwest. Upon retirement, he donated millions of dollars for the establishment of a modern business school at Western Michigan University. On a larger scale, there are the famous examples of William S. Paley, founder of CBS, and David Sarnoff of RCA. Paley, who was the heir to a cigar-manufacturing firm in Philadelphia, stumbled upon broadcasting, and just before his retirement became the primary benefactor of New York's Museum of Television and Radio. Sarnoff began his famous rise to power as a telegraph operator for the Marconi company that fateful night in 1912 when the *Titanic* sank in the North Atlantic; by his death in 1968, Sarnoff had become a leading figure in arts and cultural activities in the United States and abroad.

Of course, many media success stories are largely apocryphal. Their spread has been fueled by friendly biographers and eager image makers and press agents. However, the impact of the myth has been long-lasting. Until quite recently, few individuals ascended to the ranks of upper-level telecommunications management without having at least something in common with their mythic predecessors.

Indeed, research studies have documented that, historically, radio and television managers have shared backgrounds, interests, and professional preparation.[24] Let us take a closer look at the components of the myth.

A White, Male World   The history of telecommunications in the United States includes few ethnic minorities or women. Like the newspaper business, the world of radio and television was dominated by white males. While the household buying power of women made them the most sought-after audience group, and research showed minorities to be among the most frequent and loyal audience members, both groups were denied access to ownership, management, programming, and leading talent roles. Telecommunications firms have made strides in each of these areas, but the stereotype of the media manager as a white male is a persistent one.[25]

Archetypal American   According to tradition, broadcast leaders might have sprung directly from a Norman Rockwell painting or an Alfred Stieglitz photograph: from backwater towns in the heartland or the teeming streets of New York's Lower East Side at the height of the wave of immigration to the United States. But no matter what the background, the money used by these leaders to build their empires was "new money." Media empires were founded not on the family fortunes of the steel, oil, and banking magnates of the nineteenth century, but by persistent tinkerers who saw their preoccupation as a way out of the wheat fields or ghettos and who foresaw potential in what "established" executives felt was merely child's play or a passing fad.

The Organization Man   One aspect of the media management myth that has been repeatedly borne out by research is that few telecommunications executives came from other industries. People entered broadcasting because of ego needs, because show business was "in their blood," or because they simply loved radio and television and could not see themselves in any other profession. Some even claimed that they would work free of charge. Consequently, media organizations adopted an inverted-pyramid management structure which encouraged employees to work their way up from low-paying, often menial positions into the management hierarchy. Budding broadcasters were (and many still are) instructed to begin their careers in the smallest markets and to take any position they were offered. Perseverance, loyalty, and self-denial were perceived as the best predictors of management success.

The School of Hard Knocks A corollary to the inverted-pyramid approach was that experience was the greatest teacher. A formal education, particularly at the university level, was spurned. Media managers thought that a college education was a waste of time and money. Anyone interested in a career in telecommunications should be out there working in the field, not studying in the classroom! College graduates were perceived as spoiled troublemakers who lacked knowledge of the "real world" of broadcasting and mass communication, and their professors were seen as sequestered ivory-tower types who spent more time criticizing the media than training their students for professional careers.

A Challenge to the Myth

Three changes have taken place in recent years to challenge the persistent media management myth: attrition, education, and affirmative action.

Attrition Telecommunications is a relatively new industry. Scarcely 60 years elapsed between the first wireless radio demonstrations and the advent of satellite communications. Some of the media founding fathers are still living, but many of the legends of broadcasting have died relatively recently, including CBS founder William S. Paley, inventors Vladimir Zworykin and Peter Goldmark, and performers Jack Benny, Groucho Marx, Arthur Godfrey, and Lucille Ball. ABC's Leonard Goldenson has retired, and there are indications that other founding fathers will soon follow suit and pass the baton on to their heirs.

The new managers in the telecommunications industry are substantially different from their predecessors, and the infusion of new blood into the management hierarchy has substantially affected telecommunications products and services.

Education No longer is a college degree considered anathema to a media management career. A degree in media, marketing, or business is becoming a requirement for media managers.[26] In recent years, the number of colleges and universities offering degree programs in telecommunications, broadcasting, communications, and related fields has increased dramatically. *Broadcasting and Cable Yearbook* lists over 300 accredited programs in the field.[27] Clearly, today's media managers have had the formal educational training which sets them apart from their predecessors.

Today's managers have also had a very different informal education. They are "media babies," born and raised in an era of telecommunications. Leaders of many large media corporations, from Disney's Michael Eisner to Time Warner's Bob Pittman (who developed MTV), admit to being old-movie, rock-and-roll, or early-television buffs in their youth. This background as media consumers makes today's media managers different in many ways from the founding fathers. They probably have a more heightened sense of consumer and audience interests, as well as a keener appreciation of the history of their business and an awareness of its profound influence on social and cultural life in the United States. Some social scientists and psychologists hope that as a result they will have a more open mind to the possibility that violent and sexually explicit programming has deleterious effects on some viewers, particularly the young. In addition, it is hoped that these new managers will look more favorably on proposals to upgrade media programming and be less likely to consider telecommunications educators and researchers as "the enemy."

There is some evidence that an upgrading trend is indeed the case. The number of professional internships has increased dramatically, and joint academic-professional conferences and workshops are offered on a regular basis. Examples include the annual faculty and student seminars sponsored by the International Radio and Television Society and the research grants program funded by the National Association of Broadcasters. It is hoped that such programs will help today's media practitioners become more well informed and articulate than their predecessors, gain a clearer view of the purpose, character, and social effects of what they do, and maintain at least a partially open mind.

Affirmative Action and Equal Employment Opportunity The corporate boardrooms of the American mass media are no longer all-male or lily-white. Contemporary media management is marked by increased contributions from female and minority executives. One of the most innovative services, the Fox network, was formerly headed by Lucie Salhany. Minority ownership and management in cable are typified by people like Robert Johnson, president and co-owner of Black Entertainment Television; Montesuma Esparaza, head of Los Angeles' Buena Vista Cable Television; and Bonnie Sanders, a cable entrepreneur and member of the Metlakatla Indian community in Metlakatla, Alaska.

Challenging the media management myth—Robert W. Johnson of BET and Lucie Salhany, formerly of Fox Broadcasting Company. (left, Mike Carpenter; right, Fox Broadcasting Company)

CONTEMPORARY TELECOMMUNICATIONS MANAGEMENT: MYTH AND REALITY

It is clear that changes in our social, political, and economic life have "trickled up" to telecommunications management. At the same time that the electronic media are facing unprecedented technological and marketplace changes, a major internal reshuffling of personnel and management is occurring. This era of change has created a wide-open management arena. It is a time of challenge and opportunity in radio, television, and cable management.

Figure 2-3 sums up this chapter by comparing the media management myth with contemporary reality.

Unlike their predecessors, who tended to be hobbyists or merchants, today's media managers are trained specialists and professionals in specific fields such as finance, marketing, and research. Formal education in journalism, mass communication, business, and related disciplines is typical: there are even some Ph.D.s managing broadcast stations and cable systems! Announcers and engineers no longer have the inside track into management. Increasingly, radio and TV station and cable system executives are emerging from sales and marketing, and television executives from the news department.

As a new crop of young men and women from a variety of ethnic backgrounds makes its way up the corporate ladder, the stereotype of the media executive as a cigar-chomping, fiftyish white male is being retired. Some have followed the traditional route of remaining loyal to an organization and working their way up through the ranks, but today's trend is toward recruitment from outside: from other markets, other stations, and even other businesses.

Management decision making now relies less on intuition, street smarts, and other seat-of-the-pants methods. In today's competitive media environment, a stronger reliance is placed upon forecasting, projections, and research. Finally, the myth equates management success with such qualities as shrewdness, toughness, guile, loyalty, self-denial, perseverance, and drive. These qualities remain important, along with such "new age" qualities as a global outlook, the ability to manage change, and a commitment to the well-being of oneself and others.

SUMMARY

Successful telecommunications management requires basic technical, human, and conceptual skills. Technical skills are needed because broadcasting is built on a foundation of physics and electronics. While managers need not be engineers, some familiarity with basic technology and its operation facilitates decisions about equipment acquisition, operations, and innovation.

Managers need interpersonal, or human, skills to lead and motivate the creative and often competitive individuals who make up the media work force. In general, media managers are extroverted and gregarious and have charismatic personalities. Conceptual skills are the intellectual, analytic, and judgmental abilities which enable telecommunications managers to make difficult decisions

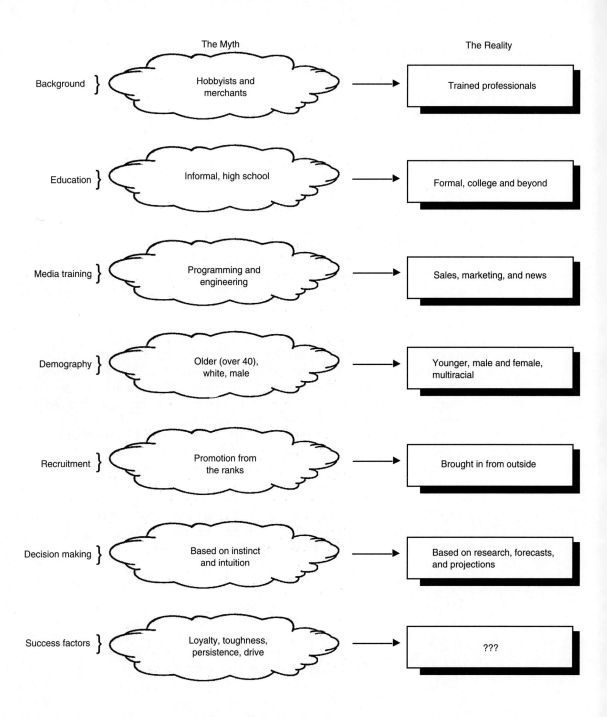

FIGURE 2-3 Telecommunications management: myth and reality.

and to react quickly to technological, social, and economic changes in the media marketplace.

Common management tasks or functions include planning, organizing, staffing, controlling, and innovating. Planning is crucial, since there is rarely a status quo in telecommunications: the industry is typically in flux. Organization—the coordination and assignment of duties—is also essential, since media facilities are characterized by internal structuralization, departmentalization, and the need for delegation. Staffing, the hiring and firing task of management, is a continual process in an industry characterized by employee competition, turnover, burnout, and attrition. The controlling function of management involves using preset planning objectives to assess corporate and employee performance. With formal or informal performance appraisals, managers communicate internal successes and failures to staff, make changes, and revise or restate corporate goals. Innovating—the development of new and better ways to conduct a telecommunications business—is a particularly important management function, since the industry is typically at the cutting edge of new technologies and managers must decide which new products and services are worthy of purchase or development.

Managers perform three main roles within and outside their facilities. In their interpersonal roles, managers lead their staff or company and serve as liaison between the organization and outside groups and institutions. As monitors of external conditions managers also obtain, synthesize, and present information. This is the media manager's informational role. Finally, the manager's desk is where the buck stops. In their decisional role, media managers allocate resources, adjudicate differences, supervise negotiations, and make the hard choices facing management in a dynamic, constantly changing industry.

Historically, three broad theories or styles have characterized the relationship between managers and their subordinates: Theory X, Theory Y, and Theory Z.

Theory X is the management of personnel through physical force, intimidation, threat, and/or fear of penalty or termination. This approach was popular in the early days of the industrial revolution as well as among the military. Because of the history of and traditions in the field of telecommunications, as well as time pressures, elements of Theory X persist today.

The Theory Y approach to management places a premium on developing interpersonal relationships between managers and subordinates. Theory Y had its heyday in broadcasting in the late 1960s and early 1970s, particularly in FM radio.

Theory Z, which is based on the organization of the Japanese industrial sector, is the most recent model. Theory Z combines aspects of both X and Y. It emphasizes the central role of management in achieving corporate goals as well as the personal goals of employees. Contemporary telecommunications management has incorporated elements of Theory Z into its structure.

Five related external social changes have had a direct impact on telecommunications management: technological development, economic changes, demographic changes, civil rights initiatives, and increased audience fragmentation. The myth of media management held that telecommunications managers were a cadre of small-town men or immigrants propelled by technical and engineering skills, guile, street smarts, and loyalty to the organization. Attrition, education, and affirmative action have changed the internal environment, creating opportunities for a new class of media managers. Contemporary telecommunications management is marked by an infusion of new blood: college-educated men and women from a variety of ethnic backgrounds who, as media babies, tend to be more familiar than their predecessors with the role of their businesses in the social, political, and cultural life of the United States.

NOTES

1 Clark Pollack, remarks before the Broadcast Education Association, Washington, D.C., November 19, 1982.
2 See, for example, James H. Donnely, Jr., James L. Gibson, and John Ivanovich, *Fundamentals of Management: Selected Readings*, 3d ed. (Dallas: Business Publications, 1978); Richard M. Hodgetts, *Management: Theory, Process, and Practice*, 2d ed. (Philadelphia: Saunders, 1979); Louis E. Boone and Donald D. Bowen, *Management and Organizational Behavior*, 2d ed. (New York: Random House, 1987).
3 Stephen D. Cohen, *Cowboys & Samurai* (New York: Harper Business, 1991); Yoshio Abe, *People-Oriented Management* (Tokyo: Diamond, 1987); Lester Thurow (ed.), *The Management Challenge: Japanese Views* (Cambridge, Mass.: MIT Press, 1985); Richard Pascale and Anthony G. Athos, *The Art of Japanese Management* (New York: Simon & Schuster, 1981).
4 See, for example, H. W. Lanford, *System Management: Planning and Control* (Port Washington, N.Y.: Kennikat, 1981); James H. Donnely, Jr., James L. Gibson, and John Ivanovich, *Fundamentals of Management: Selected Readings*, 3d ed. (Dallas: Business Publications, 1978); Stephen J. Carroll, Jr., Frank T. Paine, and John B. Miner, *The Management Process*, 2d ed. (New York: Macmillan, 1977).
5 Louis E. Boone, *The Great Writings in Management and Organizational Behavior* (Tulsa, Okla.: Petroleum Publishing Company, 1978).
6 See Edwin O. Palmer, *History of Hollywood* (New York: Garland, 1978); Erik Barnouw, *Tube of Plenty: The Evolution of American Television*, 2d rev. ed. (New York: Oxford University Press, 1990); Lawrence W. Lichty and Malachi C. Topping, *American Broadcasting: A Source Book on the History of Radio and Television* (New York: Hastings, 1975).
7 See, for example, Sally Bedel Smith, *In All His Glory: The Life of William S. Paley* (New York: Simon & Schuster, 1990); Kenneth Bilby, *The General: David Sarnoff and the Rise of the Communications Industry* (New York: Harper & Row, 1986); Hank Whittemore: *CNN: The Inside Story* (Boston: Little, Brown, 1990).
8 Lawrence W. Lichty and Malachi C. Topping, *American Broadcasting* (New York: Hastings House, 1975), pp. 30–31.
9 "The HBO Story: 10 Years That Changed the World of Telecommunications," *Broadcasting*, 15 November 1982, p. 48.
10 Whittemore, *CNN*, p. 7.
11 See Thomas R. Berg, *Predictors of Voluntary Turnover of Personnel at Commercial Television Stations*, unpublished Ph.D. dissertation (Athens: University of Georgia, 1988).

12 Simon Rano, *The Management of Innovative Technological Corporations* (New York: Wiley, 1980); J. E. S. Parker, *The Economics of Innovation*, 2d ed. (London: Longman, 1978).
13 See Henry Mintzberg, "The Manager's Job: Folklore and Fact," *Harvard Business Review*, July–August 1975, p. 55.
14 Douglas McGregor, *The Human Side of Enterprise* (New York: McGraw-Hill, 1960).
15 William G. Ouchi, *Theory Z: How American Business Can Meet the Japanese Challenge* (Reading, Mass.: Addison-Wesley, 1981); Yoshi Abe, *People-Oriented Management* (Tokyo: Diamond, 1987).
16 See Barnouw, *Tube of Plenty*.
17 Sterling Quinlan, *Inside ABC: American Broadcasting Company's Rise to Power* (New York: Hastings House, 1979), p. 181.
18 See, for example, Lee Butcher, *Accidental Millionaire: The Rise and Fall of Steve Jobs at Apple Computer* (New York: Penguin, 1988).
19 See Joel E. Ross, *Japanese Quality Circles and Productivity* (Reston, Va: Reston Publishing, 1982); David Hutchins, *Quality Circle Handbook* (New York: Nichols, 1985).
20 *Current Population Reports*, Series p-60, no. 142, (Washington, D.C.: U.S. Bureau of the Census, 1985); *Statistical Abstracts of the United States*, 1990 (Washington, D.C.: U.S. Department of Commerce, 1990), p. 384.
21 *Statistical Abstracts*, 1990, p. 385.
22 *Statistical Abstracts of the United States*, 1991. (Washington, D.C.: U.S. Department of Commerce, 1991), p. 14.
23 See *1990 Nielsen Report on Television* (Northbrook, Ill.: A. C. Nielsen, 1990), p. 7.
24 *People in Broadcasting* (Washington, D.C.: National Association of Broadcasters, 1962); Charles E. Winick, "The Television Station Manager," *Advanced Management Journal*, vol. 31, no. 1, January 1966, pp. 53–60; Thomas W. Bohn and Robert K. Clark, "Small Market Media Managers: A Profile," *Journal of Broadcasting*, vol. 16, no. 2, Spring 1972, pp. 205–215.
25 See, for example, "People Behind Programs: White Males Dominate Says Think Tank," *Broadcasting*, February 14, 1983, pp. 80–81; "Out of Focus—Out of Sync," *NAACP Report*, September 23, 1991.
26 See Jerry Hudson, "College Degree a Career Booster," *Feedback*, vol. 21, no. 3, Fall 1979, pp. 23–25; Vernon A. Stone, "RTNDA Report: Make the News Director GM?" *Broadcast Management/Engineering*, November 1982, pp. 77–82; Gary Kaplan, "Radio Should Go to College," *Radio and Records*, May 10, 1985, p. 7.
27 "Universities and Colleges Offering Degrees in Broadcasting," *Broadcasting and Cable Market Place* (New Providence, N.J.: Bowker, 1992), pp. H25–H29.

FOR ADDITIONAL READING

Dale, Ernest: *Management: Theory and Practice*, 4th ed. (New York: McGraw-Hill, 1978).
Dessler, Gary: *Management Fundamentals: A Framework*, 2d ed. (Reston, Va.: Reston Publishing, 1979).
Fletcher, Winston: *Creative People: How to Manage Them and How to Maximize Their Creativity* (London: Business Books, 1990).
Hodgetts, Richard M.: *Management: Theory, Process, and Practice*, 2d ed. (Philadelphia: Saunders, 1979).
Koontz, Harold: *Essentials of Management*, 5th ed. (New York: McGraw-Hill, 1990).

Naylor, Thomas H.: *Strategic Planning Management* (Oxford, Ohio: Planning Executives Institute, 1980).

Reinharth, Leon, H. Jack Shapiro, and Ernest A. Kullman: *The Practice of Planning* (New York: Van Nostrand, 1981).

Stacey, Ralph D.: *Dynamic Strategic Management for the 1990s* (London: Kogan Page, 1990).

Sweeney, Neil R.: *The Art of Managing Managers* (Reading, Mass.: Addison-Wesley, 1981).

3

TELECOMMUNICATIONS INDUSTRY STRUCTURE

Chapter 1 provided an overview of the range of businesses which comprise contemporary telecommunications. Chapter 2 inventoried the range of functions and roles performed by media managers and reviewed a variety of theoretical and practical approaches to the task. This chapter closes Part One by surveying the elements common to most media businesses—in particular, how they are structured and how they behave in the American corporate and cultural environment. Part Two will examine each main media business in detail, on its own terms.

The goals of this chapter are:
1 To describe U.S. enterprise in general and the telecommunications industry in particular
2 To list and describe the elements of telecommunication market structure, conduct, and performance
3 To describe the product life cycle and its applicability to telecommunications businesses

THE AMERICAN TELECOMMUNICATIONS INDUSTRY

In an influential text, Richard Caves argues that the management of any U.S. industry can be analyzed by examining its market structure, its market conduct, and its market performance.[1]

Market structure refers to the economic features of a market which affect the behavior of firms in the industry supplying that market. As is shown in greater detail in Chapter 5, since the market for broadcasting is advertising, and advertising effectiveness is built on the size and characteristics of audiences, the television industry evolved a specific structure for taking optimum advantage of the market: the network-affiliate relationship. Cable, on the other hand, earns its primary income directly from subscriber fees. Since its market is more direct, its product (its programming and delivery system) is more varied.

Market conduct refers to a firm's policies toward its product market and toward the strategies of its competitors for capturing that market. For example, over the years the competition between the three major commercial television networks (and more recently, the Fox network) has had a significant impact on the conduct of the broadcast industry. New program offerings are kept top secret; a network will often try to lure big-name stars and reporters from other networks by offering special incentives and inflated contracts.

Market performance refers to the economic output of an industrial organization—that is, its achievements in attaining maximum profitability, efficiency, employee satisfaction, and growth. In this sense, broadcasting in the United States can be viewed as a success story, since profits have traditionally been high and only Japan can match U.S. figures for saturation (number of sets in homes), hours spent viewing, hours broadcast, number of different channels and programs available, and so forth. But as we will see, the American media system can be criticized for its performance in two areas: employment practices (such as too few opportunities for women and minorities, and often-strained labor relations) and planning (such as misestimating the impact of cable and VCR on network television in the 1980s).

The first step in understanding telecommunications management is to examine telecommunications market structure, conduct, and performance in detail.

ELEMENTS OF MARKET STRUCTURE

There are six basic elements of market structure:

1 Seller concentration
2 Product differentiation
3 Barriers to the entry of new firms
4 Buyer concentration
5 Fixed costs
6 Demand growth

Seller Concentration

The degree of concentration in an industry ranges from monopoly, where a single seller controls the product, to oligopoly, where a few sellers dominate an industry, to pure competition, where a very large number of sellers compete in a wide-open market.

Usually, the degree of seller concentration can be specified in terms of *concentration ratios*, the percentage of the gross revenues of an industry accounted for by its largest companies. Thus, a monopoly would have a concentration ratio approaching 100%; a competitive industry would have ratios between 0% and 20%; and an oligopoly would fall somewhere between these extremes.

Figure 3-1 presents eight- and four-firm concentration ratios for the major electronic media businesses, plus ratios for some other industries to provide a basis for comparison.

There is universal agreement that the telecommunications industry is an oligopoly, and that the degree of concentration has grown precipitously in recent years.[2] Among all revenues for all segments of the media industry (not including telephone and computers), the top eight and top four media companies (Capital

FIGURE 3-1 Media concentration ratios: eight- and four-firm. *Source*: Prepared by the author from numerous sources.

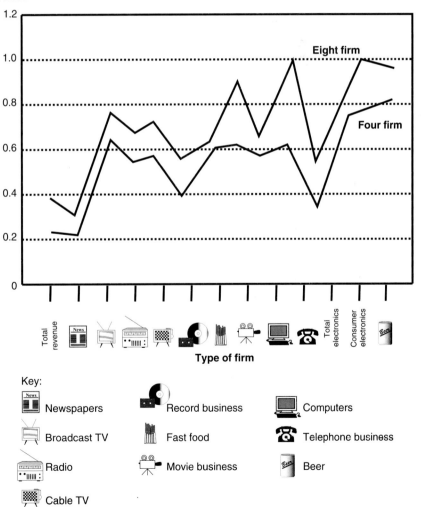

Cities/ABC, General Electric/NBC, Time Warner, and Tele-Communications, Inc.) account for about 40% and 20% of total industry revenues, respectively.

The regional Bell operating companies (RBOCs) are the most concentrated telecommunications segment, accounting for virtually all revenue in the telephone business. The top four telephone companies (AT&T, GTE, BellSouth, and NYNEX) account for about 6 in 10 revenue dollars alone!

Consumer electronics is similarly concentrated, with the eight leading manufacturers (among them Hitachi, Sony, Matsushita, and Philips) approaching 100% of total revenues.

Broadcast television is next most concentrated, with the networks and stations owned by the largest firms (including Capital Cities/ABC, NBC, CBS, and Fox) accounting for just under 75% of total industry advertising revenue. Cable concentration approaches 70%, led by the major multiple-system operators (MSOs): Tele-Communications, Inc. (TCI), Time Warner's systems, and Continental, Comcast, and Cox Cable.

Similar concentration characterizes the radio business, where the eight largest groups (including CBS, Cap Cities/ABC, Infinity, Viacom, and Cox among others) own enough stations to produce a concentration ratio exceeding 60%.

The oligopolistic structure of telecommunications is graphically depicted in Figure 3-1. Note that the electronic media firms are second only to the film industry in terms of degree of industrial concentration. Telecommunications is much more concentrated than the music business (50%) and the newspaper business (below 40%), though those industries tend to get more attention on the topic of concentration of ownership.

Within the industry itself, telephone, broadcast television, and cable television are the most concentrated telecommunications industries; radio is least.

Concentration of ownership is a controversial topic. Proponents of concentration (including the recently deregulatory-minded FCC) argue that only through concentration can sufficient resources be culled to develop new programs and services. Critics contend that concentration produces imitative programming and retards innovations such as the development of new distribution services or programs.

Despite criticism of concentration, the pace is increasing in telecommunications. As this edition goes to press, the regional Bell holding companies are in the process of coventuring with cable. More data on these trends are presented in Chapter 6.

Product Differentiation

Product differentiation is the degree to which an industry's products differ from one producer to another. In general, firms providing purely utilitarian products (pencils, toilet tissue, writing paper) or raw materials (steel, coal, electricity) tend to be undifferentiated. However, where products relate to personal lifestyle characteristics, needs, and preferences, the key to industry success lies in differentiation, the hallmarks of which are *product image* or *brand identification*. Product differentiation is critical in telecommunications.

Broadcast stations spend small fortunes developing a characteristic sound or image. The station's jingle, logo, or on-air identification is extensively researched and exhaustively promoted. Local television stations seek a unique identity for their programs, particularly their news shows. The news team is named *(Action News)* and the anchors' faces are plastered throughout the community, on billboards and bus cards.

The major television networks, through their choice of program offerings, try to project a unique image. CBS, for example, is known as the network with the strongest journalistic heritage (though less so lately), from Edward R. Murrow to *60 Minutes*. ABC's reputation has been built on its influential sports programming *(Monday Night Football* and *Wide World of Sports)*. For a time, NBC viewed itself as the home of wholesome family programs (from *Little House on the Prairie* to *The Cosby Show)* and TV talk (*Today, Tonight, Late Night*, and *Later)*. The upstart Fox network has built an image as the channel "with an attitude," by targeting youth generally (and urban youth in particular) with such offerings as *In Living Color, The Simpsons, Married with Children,* and *Beverly Hills 90210*.

Cable television networks are built on differentiation, with news and information channels (CNN, Headline, CNBC, C-SPAN); sports channels (ESPN, Sportschannel America); movie channels (Cinemax, American Movie Classics, Encore); music channels (MTV, VH-1); and on and on.

The debate over the degree of product differentiation in telecommunications centers on the distinction between program *choice* and program *diversity*. As we will see in Part Two, there is little doubt that the television and radio consumer has a growing range of choices: more sets, more channels receivable, more stations, more formats, and so on. What is less clear is whether such growth has led to program diversity or variety—that is, real product differentiation. Some critics maintain that media growth has not led to new program ideas, formats, or concepts. For example, in the pay-cable field, disconnects among households subscribing to more than one pay service have been high because viewers see little difference between programs offered on HBO, Showtime, The Movie Channel, Cinemax, and the other main pay services. Many cable services (e.g., *Nick at Night* on Nickelodeon; USA; The Family Channel) mainly rebroadcast programming originally aired on one of the TV networks.

Often, industries with an oligopolistic structure endeavor to provide an illusion of differentiation by varying small, nonessential aspects of a product. The automobile industry, for example, may stimulate interest and create demand for a new model by varying the trim or dashboard configurations without altering the engines or drive trains. Similar nonessential differentiation can be seen in the broadcast industry. The networks often "recycle" established situation comedy or adventure formulas in supposedly new series. How many situation comedies have featured a dim-witted husband and a "dizzy," accident-prone housewife, from the 1950s' *I Love Lucy* to the 1990s' *Roseanne*? Similarly, *Dynasty* can be viewed as a reworking of *Dallas*, and *Melrose Place* as more of *90210*. In the early 1990s, the success of *America's Most Wanted* produced a spate of crime-fighting programs, including *Rescue 911, Cops,* and *Unsolved Mysteries.*

Of course, the line between necessary and trivial product differentiation in an industry that caters directly to popular taste is a fine one. To succeed in the competitive media business, stations, networks, and cable systems must have a unique identity but must also deliver familiar products expected by audiences.

Barriers to Entry

Barriers to entry refer to the extent to which competition from new companies is controlled, blocked, or thwarted in an industry. There are three main classes of barriers to entry. The first is based on *absolute costs*. Industries which have a large commitment to sophisticated machinery and equipment, which need specially trained employees, and which require scarce resources to operate have high absolute costs and, therefore, are difficult for new firms to enter. Telecommunications falls into this class. Building a broadcasting station requires an available frequency in the electromagnetic spectrum. It also involves a considerable financial commitment to purchase studios, transmitters, and other expensive equipment. And it requires highly skilled people in management, engineering, sales, research, and other areas. The same is true of constructing a new telephone or cable service, entering the HDTV market, launching a satellite company, or developing other new communications technology.

Costly equipment, such as this satellite dish farm, are among the barriers to entry for telecommunications firms. (courtesy of Turner Broadcasting Systems)

A second, related class of entry barriers is based on *economies of scale.* Scale-economy barriers arise when new firms do not achieve low operating costs until they have grown enough to gain an adequate share of the market for a particular product or service. Firms that require high up-front capitalization expenses and anticipate a low return or payout for many years face such barriers. Many of the newer telecommunications services encounter high scale-economy barriers. Cable television is the best example. The cost of laying cable can be as high as $500,000 per mile. Consequently, cable companies have been big borrowers and generally do not anticipate profits for 5 to 7 years. Other new technologies facing high scale-economy barriers include MMDS, HDTV, DAB, and DBS.

The situation is somewhat better for traditional television and radio stations. Technological improvements have made it comparatively inexpensive for new stations to begin operations. This is particularly true in low-power television (LPTV), where new stations are being launched for a fraction of the cost of their full-power VHF and UHF counterparts. New transmitter and other operations technologies have made new radio stations (when frequencies are available) comparatively cheap, in the range of $100,000 to $250,000 for construction.

The third main class of barriers to entry is based on *product differentiation.* Highly successful firms often have an established name or image that allows them to dominate the market, making it difficult for new firms to compete, particularly in the early stages of development. Occasionally, a firm's brand name becomes a generic term for a product it makes: Coke can mean any cola drink; Band-Aid, any sheer bandage; Kleenex, any soft facial tissue; and Frigidaire, any refrigerator. This phenomenon is not limited to the consumer product industry. In telecommunications, for example, although Sony's brand of home video recorders (using Beta format) was not the largest seller in the market (with consumers preferring VHS), to many consumers, the home video recorder was a "Betamax." Similarly, to many people any small headset radio is a "Walkman," a registered Sony trademark. The phenomenon can occur among radio and television stations and cable networks as well. Often consumers refer to their pay-cable service as "HBO," even if it's Cinemax, The Movie Channel, or Showtime. This phenomenon has hindered competition in the long-distance telephone market. Many consumers fail to differentiate among Sprint, MCI, and AT&T, which has the benefit of a long-term identification as a long-distance carrier.

Buyer Concentration

Buyer concentration refers to the number of buyers available for a company's products. In a general sense, the fewer the potential number of buyers, the lower the profits of the industry and the greater the barriers to entry for new competitors. The situation is exacerbated in industries where intermediaries constitute the first market for a product, which is then resold to the consumer. The clothing industry is an example. Manufacturers sell their new lines of clothing to buyers representing major department stores. The buyers thus determine which clothes will make it into the retail marketplace.

On the surface, it would appear that telecommunications has little buyer concentration. Its products are sent directly into millions of homes; each consumer can make individual decisions with the flick of a switch or the rotation of a dial. However, in actuality, there is considerable buyer concentration in commercial broadcasting. The "products" being bartered are audiences. Television and radio stations "sell" listeners or viewers to advertisers. The individual audience member is basically insignificant: the goal of the station is to produce large, identifiable, and, therefore, salable audience segments. Program and advertising success are defined in terms of audience delivery, usually counted in hundreds, thousands, and even millions.

Taking this discussion further, if the product of commercial broadcasting is the audience itself, it soon becomes evident how much buyer concentration exists. The most popular stations have the most sought-after product: the largest audience. These stations are most likely to be pursued by the largest advertisers. It is difficult or impossible for less popular stations to gain a share of the advertising budget of agencies' main clients. That is why, year after year, the same small group of large companies are the leading network television advertisers (among them General Foods, Procter & Gamble, McDonald's, Burger King, Anheuser-Busch, and Sears).

If the audience ratings are there, the networks can set astronomical prices (such as $1 million for a spot on the Super Bowl telecast), which are willingly paid by the advertisers. The same situation exists in radio. Popular stations often turn advertisers away. New stations, lower-powered stations, or those with more focused and possibly more eclectic tastes (such as progressive jazz or modern rock), cannot realistically expect a share of the budget of the major advertisers and must vigorously compete for a piece of a much smaller advertising pie.

Whether the recent trend toward more stations and systems and more channels and receivers will lead to increased audience fragmentation, and therefore less buyer concentration in telecommunications, is an issue to be examined in the pages that follow.

Buyer concentration also characterizes contemporary cable television. Today, the largest cable system operators also own major portions of the leading cable networks. TCI has substantial interest in Turner Broadcasting and is a programming partner with Fox; Time Warner owns HBO and Cinemax. Smaller systems argue that they can't compete for programming with larger ones. Independently owned and operated, they often pay higher prices for programming than do similar-sized and larger systems co-owned by the programmers.

There is little doubt that the depressed economic climate faced by many media operations is due at least in part to increased concentration, both among buyers and among sellers.

Fixed Costs

Fixed costs refer to the continuing expense for equipment (plant) in an industry. Firms which must continually retool to remain competitive and profitable face high fixed costs. These costs eat directly into revenues and drive up expenses, since outmoded equipment has little or no resale value and must be considered a loss for the company. Since the telecommunications industry is characterized by innovation, its fixed costs can be quite high. The equipment supporting a local television newscast is a case in point. Thirty years ago, a local newsroom could be outfitted with a small set and supported by a few camera operators who provided news film in silent 16mm. Now, local news programs feature futuristic sets with sophisticated meteorological gear for weathercasters and state-of-the-art action vans which originate stories live from the field via satellite or microwave. The newscast of the 1960s might have cost as little as a few hundred dollars to produce. Today, a local newscast may represent millions of dollars invested in physical plant. Every year, the broadcast industry's annual conventions include acres of exhibits of new technology. And each year, to remain competitive, hundreds of stations retool.

Radio, TV, cable, and telephone companies face a daunting task of retooling for the 1990s and beyond. For radio, will DAB become a reality? Is broadcast TV headed for HDTV? Can cable rewire with optical fiber and digital compression (see Chapter 6) to squeeze more channels into its systems? Will the telephone companies provide fiber links to the home and replace the need for broadcast or cable? Above all, is there sufficient investment capital in each of these industries to make new technological developments proceed from drawing board to practical reality?

Demand Growth

The last element of market structure, demand growth, refers to the level of producer and consumer demand for an industry's goods and services. In general, growing demand leads to higher revenues and profits. It can also lead to increased competition, higher barriers to entry, greater product differentiation, and higher fixed costs. Most economists agree that demand growth is essential to the survival of an industry. Shrinking demand has a ripple effect that leads to corporate bankruptcies, unemployment, and higher prices—that is, the entire recessionary scenario.

Until recently, the telecommunications industry experienced virtually unabated demand growth, on both the producer and the consumer side. Despite high prices, advertisers flocked to radio and television, whose ability to sell products was unmatched. Television viewing levels grew steadily every year since introduction of the medium in the late 1940s. Radio listening and sales of receivers escalated at an unabated rate. New media devices, especially VCRs, achieved phenomenal rates of diffusion. Continued demand growth was the main reason that prospective entrepreneurs and managers flocked to radio and television stations, cable systems, satellite services, and other communications businesses in the go-go decade of the 1980s.

By the beginning of the 1990s, both producer and consumer demand began to fall. Marked by the industrial malaise affecting American industries in general, and their own round of mergers, acquisitions, and bankruptcies, advertising agencies began spending less money on radio, TV, and cable.[3] Nielsen ratings began to show a drop in the time spent watching TV in American homes.[4] The bottom fell out of the home video market, as fewer people bought VCRs or went to the video store to rent movies.[5]

The bottom line is that, perhaps for the first time in history, continued demand growth is no longer a certainty in the telecommunications business.

ELEMENTS OF MARKET CONDUCT

An industry is characterized not only by its market structure but also by its market conduct, which is defined as a company's policies toward its consumers and its competitors. There are three broad areas of market conduct:

1 Price-setting policies
2 Quality-control policies
3 Competitive policies

Price-Setting Policies

In theory, a highly competitive marketplace leads to the greatest control of prices by consumers. A monopoly, on the other hand, can fix its prices, since consumers cannot shop elsewhere for the goods or services offered by the monopolist. In the middle are the oligopolies, which have a more complex pricing structure. There is a degree of interdependence among oligopolies. If one firm raises or lowers its prices, the other producers are obligated to follow suit. A certain amount of price collusion or price fixing is common in oligopolies, as when major-league baseball owners agreed not to sign high-priced free agents in order to keep down payrolls. Rival firms in oligopolies can easily set price ceilings or minimums around which consumer prices will vary. This has been the case in recent years in the airline industry, as the number of competing airlines has diminished to a "big three" of United, Delta, and American.

As pointed out previously, the telecommunications industry is largely oligopolistic. Therefore, we would expect pricing to include elements of seller agreement, price controls, and price leadership by major firms. While antitrust laws make collusion between sellers strictly illegal, it is not uncommon for salespeople at different radio stations to be aware of the advertising rates of their competitors, or for advertisers to leak their budgets so that stations seeking to "get on the buy" can adjust their rates accordingly. It is also not uncommon for the television station with the highest rating to set its advertising costs for the next quarter first, and for the second- and third-ranked stations to follow suit.

For many years, advertisers have accused the major commercial television networks of price fixing.[6] The high share of television viewing during prime time has made it an essential advertising buy, and the agencies claim that the networks have willfully and knowingly conspired to raise the costs of spots during this time period to astronomical heights.

The role of regulators in price setting has been particularly strong in the cable industry. In past years, local governments had considerable input into setting the rates that cable systems could charge for basic and premium services. In addition, many systems were required to seek approval from a franchising or supervisory board to secure price increases, and requests had to be backed up by evidence of escalating plant or other costs. The cable industry fought long and hard to restrict rate regulation, and in 1984 Congress passed the Cable Telecommunications Act, which limited the role of the FCC and local governments in rate setting.[7] However, the years following rate deregulation were met by considerable protest over cable rates. Consumer prices for cable services were attacked as arbitrary and inflated, often dictated by what the large MSOs decided to charge, rather than the actual cost of providing households with the utility.[8] The result was the Cable Act of 1992, with its rollback of cable rates and more vigorous policing of cable offerings and charges.

Quality-Control Policies

The second component of market conduct in an industry is the control of product quality. In the case of a monopoly producer, the consumer has no alternatives, and theoretically product quality is likely to suffer (think of the quality of automobiles in Russia or of telephone service in China). In the case of pure competition, because of the large degree of differentiation and the high number of sellers and buyers, product quality is theoretically superior (the values to be found at an outlet mall, for example).

Again, the characteristics of an oligopolistic industry approach the middle ground between these extremes. If a single firm in an oligopoly increases the quality of its product, it may be considered a *product leader*, forcing other members of the industry to follow suit. On the other hand, a high degree of collusion among members of an oligopoly can lead to agreements to hold down product quality, to delay the appearance of new features, and even to build in or plan obsolescence in a product.

The evidence concerning the telecommunications industry's record with regard to quality control is mixed. Some critics have argued that the competition among TV networks for large audiences has led to low-quality programs, aimed at the lowest common denominator in the audience and offering the "least objectionable" fare.[9] At times, the same companies have been lauded for costly but high-quality offerings—for example, ABC for its miniseries *Roots, Winds of War,* and *War and Remembrance*.

Often, a media firm commits a great deal of money and devotes a great deal of effort to become a product leader in a given area. For example, in the late 1980s CBS committed over $1.2 billion to acquire the rights to major-league baseball, college basketball, the Olympics, and other sports events.[10] NBC committed millions to establishing its news leadership, including the formation by its parent General Electric of the Consumer News and Business Channel (CNBC) on cable, and its own newsfeed service to its affiliates (to compete with CNN).[11]

Of course, what constitutes "high quality" will always be difficult to determine in the telecommunications industry, since its output is programming. While most people can agree on reasonable criteria for quality in an automobile (gas economy, safety, drivability), the criteria for quality programming is far from absolute. And the fact remains that programs universally perceived to be high in quality (documentaries, operas, stage performances) are typically mired at the bottom of the program ratings, offering evidence that they are not watched in sufficient numbers to make them economically viable.

Competitive Policies

The relationship among competing firms in an industry has been alluded to throughout this section. An industry may be openly *competitive*; that is, all major companies in the industry actively encourage diversity, innovation, and differentiation as each struggles for market share. Or an industry may be *coercive*; that is, the major companies take collusive steps to weaken competition and raise barriers to entry. Broadcasting and cable have frequently been cited as examples of coercive industries. Histories of broadcasting document how the industry came to be dominated by the three major networks through shrewd patterns of patent acquisition and through aggressive lobbying, as well as lax enforcement of antitrust legislation.[12] Regulatory efforts to promote the development of new technologies and to encourage diversity of ownership and programming have been thwarted by initiatives in behalf of existing media powers to corner their share of the market. For example, the complex set of rules adopted by the FCC in 1972 (and largely abandoned in 1980 and 1984) is now recognized as a concerted effort by the broadcast industry to repel competition from the cable business.[13]

More recently, when the FCC tried to promote the use of new technologies such as DBS, MMDS, and LPTV by encouraging entrepreneurs to apply for licenses, many of the applications were made in behalf of existing networks and large communications conglomerates.[14]

The coercively competitive nature of telecommunications can be seen in the behavior of firms in small markets as well. To limit the potential for success of a new radio or television outlet, competitors often artificially adjust their advertising costs, discounting ad rates to help drive the competitor off the air. A successful salesperson or air personality on one station frequently becomes an immediate target for acquisition by its rivals. Programming and promotional ideas are imitated or stolen outright. The grapevine among competitors is always active (though often inaccurate).

If the "ideal," or grade A, industry welcomes competition, encourages diversity, and freely exchanges information and ideas, the evidence suggests that telecommunications deserves a grade of D or F in this element of market conduct.

ELEMENTS OF MARKET PERFORMANCE

Ultimately, the success of any industry (or economy) is gauged by comparing its potential with its performance in four main areas:

1 Efficiency in the use of resources
2 Stability of prices and employment
3 Fairness in the treatment of individuals
4 Progressiveness in enlarging and improving the flow of goods and services (planning and growth)

In the ideal case, an industry is maximally efficient if it achieves the highest possible income through the best allocation and use of its resources. It is successful if its prices remain stable and within the reach of the majority of its consumers, and if it approaches a condition of full employment, with jobs available for a range of qualified personnel. A firm is successful to the extent that it creates opportunities for all groups—including minorities and women. Finally, the ideal firm builds into its operations long-term research and planning functions, with the goal of expanding its goods and services in the future through a solid plan for growth.

The evidence generally indicates that telecommunications has been highly successful in the first area, has been moderately successful in the second, and has performed comparatively poorly in the final two areas.

Efficiency

The system of mass communications in the United States is arguably the most efficient in the world. Virtually all of our citizens have access to television and radio transmissions, and most use them daily. The available spectrum space for broadcast transmission is nearly exhausted, but it is being expanded to make room for still more "voices," from DAB to DBS to HDTV, and for personal communications services, from CBs to cellular telephones. Unlike communications in much of the rest of the world, our program services are broadcast around the clock and are available in all regions of the country.

Most media businesses are run efficiently, without waste from unused technical and human resources. An exception for many years was the three commercial TV networks, ABC, CBS, and NBC, which boasted enormous management and news staffs. This situation was rectified (painfully, to many employees) in round after round of layoffs following the network takeovers in the late 1980s (see Chapter 10).

In telecommunications, outmoded equipment is typically repaired or replaced with greater regularity than is the case in other industries. And the highly competitive nature of commercial broadcasting leaves little room for human inefficiency. Individuals who "can't cut it" are usually replaced in short order.

Stability

Historically, media businesses have been stable in their pricing and employment practices. The cost of a radio advertisement remains within reach of most advertisers, and the cost of a new television or radio receiver remains within reach of most consumers. However, in recent years, costs for equipment, programming, and personnel have been escalating rapidly. Broadcast and cable revenues have grown, but expenses are up dramatically. Program costs continue to rise. Negotiated contracts, especially for union or craft guilds, are at record high levels, as are the general and administrative costs for such items as energy, telephone service, and office supplies. Cost increases have, of course, been passed on to the consumer—in the case of commercial broadcasting, to the advertiser. The price of a 30-second television spot has risen at both the network and the local level. In general, radio remains something of a bargain, but powerful and popular radio stations may charge as much today for a commercial as did television stations just a few years ago.

The pattern of employment has been more stable than that of pricing. Despite the well-publicized layoffs at the major television networks cited above, the overall size of the media work force has been growing, and the rate of employee growth has generally kept pace with the increase in stations and cable systems.[15] In other words, many of the employees displaced by network and local television layoffs have found work in new areas, such as cable news (see Chapter 6). The range of remuneration has also been relatively stable. Although big-name news anchors, actors, and disc jockeys command astronomical salaries, as we will see in Part Two, most media employees are paid standard to low salaries.

Equity

The struggle by the FCC and consumer groups to gain entree for more minority groups and women into telecommunications management is discussed throughout the pages that follow. On the whole, telecommunications businesses have not had a particularly favorable record in the equitable treatment of individuals. The problem is evidenced by the high degree of industry turnover and burnout and the often adversarial relationship between media managers and the labor force (discussed in Chapter 10).

There are two major approaches to remedying such inequities. The first is government pressure, most commonly in the form of regulatory guidelines for equal employment opportunity (EEO) and affirmative action programs. As might be expected, media managers often view these requirements as misguided bureaucratic intrusions into the operations of their enterprises.

The second, and perhaps more promising, approach to addressing the grievances of women, minorities, and other personnel in the media is the industry-led initiative to facilitate employee participation in ownership and management. In recent years, internship programs, loan programs, and on-the-job training programs have all been increasing in the telecommunications industry. While many of the programs were begun to "get the FCC off our backs," they may bear fruit in the long term by increasing employee participation in media management (see Chapter 10).

Employee equity in telecommunications has been a recurring issue, particularly regarding women and ethnic minorities. (courtesy of Gannett Broadcasting)

Planning and Growth

The final measure of the performance of an industry is its planning and its prospects for growth. Planning is perhaps the most critical task facing the media manager in a business marked by technological change and subject to the whims of consumer demand.

Industrial growth can follow three main paths. The first is *horizontal integration*, where the firm begins to spread its influence across a range of geographic markets. This was the pattern followed by the three commercial television networks in spreading their influence through the network-affiliate relationship. This phase characterizes cable today, as the leading MSOs have acquired larger and larger franchises throughout the United States. Chapter 7 reveals how regulatory decisions, including rules limiting group ownership (the ownership by one company of media holdings in different markets) and cross-ownership (the ownership by one company of media firms in the same market), have tended to limit horizontal integration by media firms. However, recent relaxation of some ownership restrictions and elimination of others have led to a flurry of activity in horizontal integration.

The second pattern of industrial growth is *diversification*, in which a company expands into related and semirelated enterprises. Elements of this form of growth are plainly visible in the contemporary communications landscape. Over the years, the television networks, for example, have included magazines, record companies, and even amusement parks among their holdings. In addition to its television network, CBS owned and operated CBS Records, manufactured musical instruments, and published magazines like *Woman's Day, Field and Stream,* and *World Tennis.* Facing financial woe in the late 1980s, CBS spun off its diversified holdings (selling its record division to Sony), following a management decision by majority stockholder Lawrence Tisch to concentrate on its core network and owned stations.[16]

Capital Cities/ABC is the most diversified network today, with holdings in radio (including networks, stations, and the Satellite Music Network) and cable (majority ownership of ESPN and an interest in Lifetime, among others). CapCities/ABC also owns weekly newspapers and a number of consumer and special-interest magazines.

Major station groups are also active in a variety of enterprises. For example, Cox Enterprises maintains a profitable automobile auction division as well as interest in cattle ranches. Jefferson Pilot is a radio station owner, a sports producer, and was the promoter-agent for figure-skating phenomenon Katarina Witt.

The third type of industrial growth is *vertical integration,* in which a firm branches from one phase of operations into another—for example, from innovation to production, from production to distribution, or from distribution to exhibition. Vertical integration is a common form of growth, since it tends to streamline expenses and helps regulate costs. However, if the same small group of companies controls all phases of production, a monopolistic situation can ensue. Thus, regulatory agencies such as the FCC, the Federal Trade Commission (FTC), and the Securities and Exchange Commission (SEC) are charged with keeping a watchful eye on acquisitions and mergers which might have a monopolistic result.

Until recently, concern about the control of information by an admittedly oligopolistic industry tended to limit vertical integration in telecommunications. Television networks, for example, were forbidden to own the syndication rights to the programs they broadcast. They were likewise prohibited from owning cable systems and from acquiring more than seven each of radio and TV stations. Telephone companies were prohibited from offering video and audio services which emulate radio and television, and were enjoined from owning radio and television stations and cable systems in their service areas.

As this book goes to press, these regulations are being streamlined or eliminated altogether. Radio groups may now own up to 36 stations (18 each in AM and FM). Relaxed "financial interest and syndication rules" (see Chapter 12) enable networks to participate in program ownership and TV syndication to a degree restricted in prior years. Telephone companies are now permitted to offer video services on their phone lines.[17] Additional cross-ownership restrictions are expected to fall in the near term.

In areas in which regulations like these have not existed, a considerable amount of vertical integration has occurred. There is no limit, for example, on the number of systems a single cable company can purchase or on the extent to which cable system operators can own programming services. The growth of the Fox TV network was stimulated by FCC waiver of rules barring network ownership of production, distribution, and exhibition. In addition to running its own network, Fox owns its own studios (the remnants of 20th Century Fox), syndicator (Fox TV), and even stations (the former Metromedia properties, among them WNYW in New York, KTTV in Los Angeles, and WTTG in Washington).

As is the situation with horizontal integration, vertical ventures are likely to increase as the costs of telecommunications production escalate, federal regulations are weakened, and telecommunications seeks relief from growing competition from overseas, particularly Japan.

Despite the recent flurry of interest in horizontal and vertical integration, telecommunications managers have not had a distinguished record in the planning area. For example, in recent years many managers on the AM band were unsuccessful in countering the rise in FM listening, leading to plummeting AM shares and to the sale of many stations. Another example is the unaggressive response of major TV networks to the declining shares of audience for network programs, particularly in cable households. In fact, some observers have pointed out that the response was to become more conservative and to take fewer risks, and the result has been a lackluster menu of new programs from year to year.

A final example of poor planning lies in the cable industry. Many cable entrepreneurs initially felt that the mere wiring of communities would bring in cable revenues successfully. Few managers realized that consumer demand would be capricious, that disconnect rates would be high, that pay programming might not be as attractive as anticipated, and that, if audiences did not perceive programs to be unique or attractive, the promise of a large number of channels was meaningless. Similarly, the major pay-TV services were caught napping by the emergence of the retail video market. The availability of prerecorded movies on demand on VHS tape caused revenues to plummet at HBO, Showtime, and other pay-movie channels.

As a result of these and other failures, media companies have placed increased emphasis upon planning for the near and far term. Major groups, such as Bonneville, CapCities/ABC, and Gannett, have "think tanks" of senior executives preparing position papers for their companies over the next decade and beyond. Unprecedented alliances have been formed to position American media companies against their foreign competitors. One example is the consortium of companies allied in the development of HDTV, cited in Chapter 1.

The role of research, particularly in simulation and forecasting, has become more and more important. Media managers can no longer rely solely upon their instincts, or street smarts, in planning for the growth of their firms.

THE MEDIA ARE MATURE: THE PRODUCT LIFE-CYCLE CURVE

In recent years, a new approach to the study of American business has received a great deal of attention. *Product life-cycle theory* helps clarify the stages of growth of a business, from its introduction to its demise.[18] As we will see in detail in Part Two, the core businesses which comprise this text—radio, TV, cable, and related

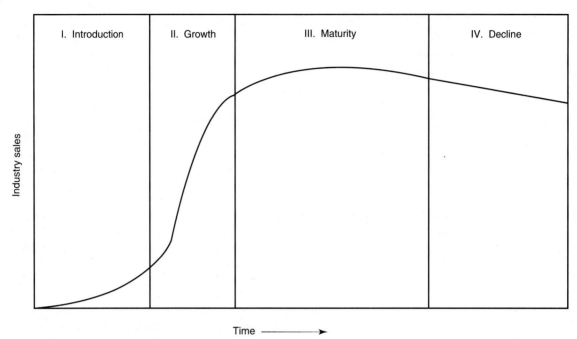

FIGURE 3-2 The product life-cycle curve.

industries—have followed the pattern of the product life cycle, as illustrated in Figure 3-2.

According to product life-cycle theory, many industries and their prime products evolve in the pattern of an S-shaped curve, through four distinct phases of development.

Phase I: Introduction

Phase I is the period of product introduction. In this phase, consumers tend to behave skeptically; they have to be induced to sample a product or be convinced of its value by others whose opinions they value and respect. Generally, the new product is comparatively poor in quality and undifferentiated at this point (think of early VCRs or video games like Pong).

However, for the producer lack of differentiation is offset by the presence of little or no competition in the marketplace (like the Atari company, before Nintendo, Sega, and others entered the scene). The major costs to producers in Phase I are in research and development (to make and refine the product) and in marketing (to convince consumers to try it).

As Part Two underscores in detail, media businesses follow this classic introductory pattern. Radio, TV, cable, VCRs, computers, and even telephones first faced a skeptical public. "Early adopters," particularly hobbyists, had to convince others of the value of their new acquisition.

The new technologies discussed in Chapter 1 remain in this phase, as firms are investing small fortunes in research, development, and market-testing of such innovations as DAB, DBS, interactive TV, and HDTV.

Phase II: Growth

The growth phase is the steepest part of the curve. Here, a product can literally take off. For consumers, the new product becomes a status symbol or a perceived necessity ("I've gotta have it!"). To get on the bandwagon, some consumers will buy products of uneven quality and will even accept inferior substitutes or imitations (like "fake" cellular phones in cars in the 1980s).

In order to capture market share, producers emphasize differentiation, choice, and quality ("accept no substitutes") and strive in their marketing to promote product leadership. Competition is most intense in this stage, but so are earnings and profits for the lucky few companies which survive. One result of the growth phase is that consumer prices tend to drop, placing the product within reach of the majority of potential buyers.

Again, media businesses have followed this classic pattern. The growth phases of radio, TV, cable, and most recently VCRs have been spectacular. Having a VCR in the 1980s was akin to owning a radio set in the 1920s, TV in the early 1950s, and cable in the early 1970s—a sure sign of having "made it" in consumer society. Media growth was accompanied by massive revenue and profit for those entrepreneurs who survived the intense and often bitter corporate struggle.

Phase III: Maturity

Mature firms have saturated their marketplace. Their products are priced within range of most consumers and are owned by a majority of potential customers. Competition has reduced the number of competing firms to an oligopoly, dominated generally by a core group of product leaders.

For a producer, maintaining market share is a function of encouraging repeat business: through promoting brand identification, loyalty, and the appeal of the product to the lifestyle of its major customer groups. Thus, the company places less emphasis on product quality (which has reached its highest level) and more on promotion and marketing. However, the need to spend on advertising and image promotion is accompanied by flattened profits and increased supplier and personnel costs. Many mature firms are overbuilt and overemployed, as a holdover from their heady days of product introduction and growth.

As the 1990s unfold, most major media are in the maturity phase. Most American households have TVs, radios, VCRs, cable, and, of course, telephones. The media are oligopolies, preoccupied with maintaining consumer loyalty in an atmosphere rife with variety and choice. Supplier and labor costs, plus debt from mergers and acquisitions left over from the growth phase, have led to rounds of budget cuts and layoffs. Many leading firms, including the major TV networks, face an uncertain future.

Phase IV: Decline

As the curve illustrates, maturity and decline tend to be the flattest phases: slow and steady, when compared with introduction and growth. A firm slips from maturity to decline when its quality falls below a level tolerated by its consumers and when its operations costs exceed its potential revenue.

Typically, a firm in decline loses market share to a smaller, more efficiently operated competitor whose "lean and mean" approach makes the older business seem like a plodding industrial dinosaur. Declining businesses lose touch with the marketplace; their products meet the perceived needs of the corporation and its suppliers, but not those of the consumer. Firms in decline face two alternatives; innovate into a related or completely different product line, or cease to exist altogether.

Like some American airlines before them, some media businesses seem to be in irreversible decline. As detailed in Chapter 5, network television may be one example. In news, CNN and others have shown the broadcast networks how to do it cheaper, faster, and better; in entertainment, they have been "outfoxed" by the Fox network, basic cable services, and others. As a result, their market share has slipped and their future remains in doubt.

John Abel, vice president of the National Association of Broadcasters and a former communications professor, is fond of pointing out that the great railroad firms of the nineteenth century disappeared when automobiles and airplanes emerged at the turn of the century. The magnates failed to realize that they were not in the railroad business per se, but in the transportation business. Few had invested in the newer technologies.

Extended to telecommunications, the lesson for today's mature and declining media companies is that they are not in the radio or network or telephone or television business. Rather, they are in the information and entertainment business, with the mode of delivery or transmission becoming increasingly irrelevant. Those firms which reposition for the data-driven economy of the next century will triumph; those which don't will surely disappear.[19]

A FINAL WORD

We have completed an overview of telecommunications management, having defined each term and traced the core concepts of modern media and management. We turn next to the core businesses of telecommunications, beginning with radio, then on to TV, cable, and related technologies.

SUMMARY

Telecommunications management involves (1) the ability to supervise and motivate employees and (2) the ability to operate facilities and manage resources in a cost-effective (profitable) manner.

The first step in understanding the task of the telecommunications manager is to examine the structure of U.S. industry in general and the telecommunications industry in particular. Any industry can be examined in terms of its market structure, conduct, and performance.

Market structure refers to the economic features of an industry. Elements of market structure include seller concentration, product differentiation, barriers to entry, buyer concentration, fixed costs, and demand growth. Data on the telecommunications industry reveal that while the industry ranges from small firms to multinational corporations, its structure is generally that of an oligopoly in which a few major buyers and sellers tend to dominate. The degree of product differentiation in telecommunications is open to debate. There is evidence that telecommunications offers great product choice, but little diversity. Barriers to entry and fixed costs tend to be higher in the telecommunications industry than in many other industries. Telecommunications has been a particularly attractive field for investors, since demand growth has been consistent through the years. However, there is evidence that demand growth is abating in the wake of increasing competition and a declining economy.

Market conduct refers to the policies of a company with respect to its market and its competitors. Elements of market conduct include price setting, quality control, and competitive policies. As in other oligopolistic industries, there is evidence in telecommunications of price fixing and price controls. Data on quality control are contradictory. While most program services tend to be imitative, or based on tried-and-true formulas, some media firms have shown improvement in product quality, particularly in the area of large-scale entertainment, news, and sports. Competitive policies in the electronic media have tended to be coercive; that is, the larger companies have attempted to thwart new competition through price controls and lobbying and by erecting barriers to entry.

Market performance is an industry's record with regard to use of resources, stability of prices and employment, fairness in the treatment of employees, and effectiveness in planning and growth. Telecommunications in the United States has been highly successful in terms of efficiency, use of resources, and economic stability. It has a somewhat more tainted record in the areas of employee equity, planning, and growth.

Product life-cycle theory plots the stages of industrial development, from innovation through growth, maturity, and decline. The major media businesses covered in this text are, by and large, in the mature stage, with a few (notably network television) apparently in decline. In order to stay in business, such firms must concentrate on marketing and innovation and must reposition themselves for an information-based economy.

NOTES

1 Richard Caves, *American Industry: Structure, Conduct, Performance*, 6th ed. (Englewood Cliffs, N.J.: Prentice-Hall, 1987).

2 The pioneering discussion is Harry J. Skornia, *Television and Society: An Inquest and Agenda for Improvement* (New York: McGraw-Hill, 1965). See also B. Litman, "The Television Networks, Competition and Program Diversity," *Journal of Broadcasting*, vol. 23, no. 4, Fall 1979, pp. 393, 409; Bruce M. Owen, Jack H. Beebe, and Willard G.

Manning, *Television Economics* (Lexington, Mass.: Heath, 1974); Benjamin M. Compaine, Christopher H. Sterling, Thomas Guback, and J. Kendrick Noble, Jr., *Who Owns the Media? Concentration of Ownership in the Mass Media Industry* (White Plains, N.Y.: Knowledge Industry Publications, 1979); Ben H. Bagdikian, *The Media Monopoly*, 3d ed. (Boston: Beacon Press, 1990); Herbert I. Schiller, *Culture, Inc.* (New York: Oxford University Press, 1989); Harold L. Vogel, *Entertainment Industry Economics*, 2d ed. (Cambridge, U.K.: Cambridge University Press, 1990); David Waterman, "A New Look at Media Chains and Groups," *Journal of Broadcasting and Electronic Media*, vol. 35, no. 2, Spring 1991, pp. 167–178.

3 See Robert Coen, "Little Ad Growth," *Advertising Age*, May 6, 1991, pp. 1, 16; Judann Pagnoli, "Advertisers Arm to Rebuild Beliefs," *Advertising Age*, November 4, 1991, pp. 1, 53.

4 See William Mahoney, "Big 3 See Erosion of Ratings Power," *Electronic Media*, January 1, 1991, pp. 26, 94.

5 See Jim McCullagh, "Video Generic Ad Campaign Hits Snag," *Billboard*, November 30, 1991, pp. 6, 92; Richard Natale, "Video SOS as B.O. Sinks," *Variety*, October 28, 1991, pp. 1, 68.

6 See, for example, "Alberto Culver Sues over TV Commercial Guideline," *Broadcasting*, November 21, 1983, p. 25.

7 See "A to Z on CTA," *Broadcasting*, October 15, 1984, p. 39.

8 See, for example, Kate Maddox, "TCI Improves, But Old Image Lingers," *Electronic Media*, November 4, 1991, p. 1; Doug Halonen, "Lawmakers Introduce Cable Bill," *Electronic Media*, October 21, 1991, p. 3.

9 Paul Klein, "The Men Who Run TV Aren't That Stupid . . . They Know Us Better Than You Think," *New York*, January 25, 1971, pp. 20–29.

10 See David A. Klatell and Norman Marcus, *Sports for Sale* (New York: Oxford, 1988); Joseph Lapoint, "Television Is Lavishing Money on Sports," *New York Times*, December 3, 1989, sec. 1, p. 25; "CBS Bites the Sports Bullet," *Broadcasting*, November 4, 1991, p. 27.

11 See, for example, Diane Mermigas, "NBC Plots Local Cable News Outlets," *Electronic Media*, August 14, 1989, pp. 1, 31; Wayne Wallye, "CNBC Wins FNN Bidding War," *Electronic Media*, May 13, 1991, pp. 1, 39.

12 See Eric Barnouw, *Tube of Plenty: The Evolution of American Television*, 2d rev. ed. (New York: Oxford University Press, 1975); Lawrence W. Lichty and Malachi C. Topping, *American Broadcasting: A Source Book on the History of Radio and Television* (New York: Hastings, 1975).

13 Stanley M. Besen and Robert W. Crandall, "The Deregulation of Cable Television," *Law and Contemporary Problems*, vol. 44, no. 1, Winter 1981, p. 122.

14 "Comes the Deluge in MMDS," *Broadcasting*, September 12, 1983, pp. 23–24; "CBS Puts DBS Costs at over $180 Million," *Broadcasting*, February 7, 1983, pp. 69–70; "UPI Owners Form New Low-Power TV Company," *Washington Post*, July 18, 1984, p. D3.

15 See *Equal Employment Opportunity Trend Report*, FCC-BEEO 18-01, January 13, 1989, p. 769.

16 See Diane Mermigas, "Three Executives Who Changed TV's Landscape," *Electronic Media*, January 1, 1990, pp. 34, 98; "CBS After Paley," *Electronic Media*, November 5, 1990, pp. 1, 55.

17 See Doug Halonen, "FCC Allows Telcos to Enter TV Business," *Electronic Media*, October 28, 1991, pp. 1, 40; Doug Halonen, "Justice Fights for Bell Firms," *Electronic Media*, September 16, 1991, p. 19; F. Dawson, "Telco Video," *Cablevision*, September 9, 1991, pp. 32–34.

18 See Sak Onkvisit and John J. Shaw, *Product Life Cycles and Product Management* (New York: Quorum, 1989); Gerrit Antonides, *The Lifetime of a Durable Good* (Boston: Kluwer Academic Publishing, 1990).

19 See Stan Davis and Bill Davidson, *20-20 Vision: Transform Your Business Today to Succeed in Tomorrow's Economy* (New York: Simon and Schuster, 1991).

FOR ADDITIONAL READING

Bagdikian, Ben H.: *The Media Monopoly*, 3d ed. (Boston: Beacon Press, 1990).

Caves, Richard: *American Industry: Structure, Conduct, Performance*, 6th ed. (Englewood Cliffs, N.J.: Prentice-Hall, 1987).

Compaine, Benjamin M., Christopher H. Sterling, Thomas Guback, and J. Kendrick Noble, Jr.: *Who Owns the Media? Concentration of Ownership in the Mass Communication Industry*, 2d ed. (White Plains, N.Y.: Knowledge Industry Publications, 1982).

Mosco, Vincent: *Broadcasting in the United States: Innovative Challenge and Organizational Control* (Norwood, N.J.: Ablex, 1979).

Onkvisit, Sak, and John J. Shaw: *Product Life Cycles and Product Management.* (New York: Quorum Books, 1989).

Owen, Bruce M., Jack H. Beebe, and Willard G. Manning: *Television Economics* (Lexington, Mass.: Heath, 1974).

Picard, Robert G.: *Media Economics: Concepts and Issues* (Newbury Park, Calif.: Sage, 1989).

Rano, Simon: *The Management of Innovative Technological Corporations* (New York: Wiley, 1980).

Schiller, Herbert I.: *Culture, Inc.: The Corporate Takeover of Public Expression* (New York: Oxford University Press, 1989).

Shepard, William G.: *The Economics of Industrial Organization* (Englewood Cliffs, N.J.: Prentice-Hall, 1979).

Vogel, Harold L.: *Entertainment Industry Economics*, 2d ed. (Cambridge, U.K.: Cambridge University Press, 1990).

PART TWO

CORE BUSINESSES

Part Two: CORE BUSINESSES describes the major players in contemporary telecommunications from a business management point of view. Included are current data on operations, revenues, expenses, profits, and consumer usage. Chapter 4 profiles the radio business. Chapter 5 looks at broadcast television, including networks, stations, and public broadcasting. Chapter 6 describes the cable television business in relation to conventional broadcasting and the regional Bell operating companies (RBOCs).

Each of the chapters in Part Two uses the product life-cycle curve as its organizing principle. Necessarily somewhat simplified, the historical discussion nevertheless serves the vital purpose of orienting the reader to the essentials of media management and how it evolved in each of the core electronic media businesses. A look back at the evolution of broadcasting and cable quickly demonstrates the central role of management in these dynamic, rapidly changing fields. As the major trends in radio, television, cable, and the new technologies are examined, contrasts and comparisons with contemporary management are stressed. Readers who want more complete discussions of media history can consult the reference lists at the end of each chapter.

4

THE RADIO BUSINESS

The contemporary communications explosion began quite modestly at the turn of the century when Guglielmo Marconi's wireless telegraph was adapted to transmit words and music instead of simple Morse code. Nine decades later, despite the advent of television, satellites, fiber optics, discs, and computers, the radio medium remains an integral component of the telecommunications environment. Since its beginnings, radio has experienced periods of steady and occasionally spectacular growth. The medium has also seen turbulent times of retrenchment, refinement, and reevaluation. This chapter presents a brief overview of the evolution and current status of the bedrock of U.S. broadcasting: the radio business.

The goals of this chapter are:

1 To trace the growth and present status of commercial and noncommercial radio by examining stations on the air, diffusion of sets, listening patterns, and programming trends
2 To identify sources of radio revenue, including advertising sales, and of radio expense, including program, administrative, and technical costs
3 To trace trends in network and local radio profits and profit margins
4 To examine the characteristics of radio personnel, including job classifications, salary levels, and trends in the makeup of the work force
5 To stress the implications of recent developments for contemporary radio management

RADIO: A MATURE MEDIA BUSINESS

We begin our discussion of the radio business with an examination of the diffusion of the medium, in terms of both the number of stations in operation and the availability of receivers to the American consumer. As we will see, on both counts, radio today is a mature enterprise. But first, a look back.

The number of radio stations licensed for broadcast by the FCC is charted in Figure 4-1. From the graph of station growth, some major trends can be noted. We can divide the medium's growth into four stages.

The Network Era

From 1925 to 1945, the era known as the golden age of radio, there were relatively few stations and all were on the AM band. With the exception of a single year during the Great Depression—1935, when there were only 583 stations—the number of stations in the United States stabilized at between 600 and 900.

Although the number of stations was relatively constant, the percentage of stations affiliated with a major network increased steadily through this period. In 1930, one-fifth of all stations were network-affiliated. By 1935, the figure had jumped to one-third. By 1940, 6 stations in 10 were programmed by a major network; at the end of World War II, only 1 station in 20 was *not* network-affiliated!

Thus, this era was the heyday of *network radio.* The National Broadcasting Company offered two network sources: NBC Red and NBC Blue. The other major

FIGURE 4-1 On-air radio stations, 1920–2000. *Source*: Christopher H. Sterling (New York: Praeger, 1984); *Broadcasting Yearbook* (Washington, D.C.: Broadcasting Publications, various years); projections to 2000 by the author.

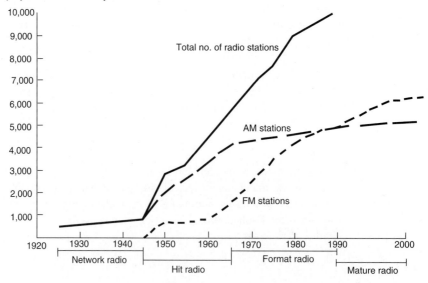

networks were the Columbia Broadcasting System (CBS) and the Mutual Broadcasting System. In 1942, NBC sold its blue network, the remnants of which later became the American Broadcasting Company (ABC).

During this period, radio occupied a central position in the household: one now usually reserved for the television set. Network radio was a mass-entertainment medium, featuring the biggest stars from the popular music, comedy, and variety fields. Highly rated programs and performers included Amos and Andy, Jack Benny, Bob Hope, Bing Crosby, and Kate Smith. In this era, sponsors often became synonymous with their programs. Major sponsors included Campbell soups (*The Burns and Allen Show*), Johnson's wax (*Fibber McGee and Molly*), Atlantic Oil (*Bob Hope*), Jello (*The Jack Benny Show*), and A&P (*Kate Smith*). For managers of local stations, revenue and profits stemmed mainly from payments by the national radio networks (known as *network compensation*). The networks in turn generated revenue from the national advertisers that purchased airtime on their most popular programs. Some of the larger stations in major cities created a revenue stream in local advertising, as leading car and furniture dealerships, among others, would sponsor local programs, from dramas to orchestra performances to very popular "barn dances."

Hit Radio

The pattern of radio broadcasting began a radical transformation in the post–World War II period. The arrival of television during this period did not inhibit the growth of radio, at least in terms of the number of facilities in operation. In fact, the number of stations on air began a spectacular rise, with most of the growth occurring in the AM bandwidth. This second wave of popularity marked the era of *hit radio* in the United States. Roughly spanning the 20 years between 1945 and 1965, it coincided with the spread of American consumerism, the growth of suburbia, the postwar baby boom, and the development of the 45-rpm record. Radio became a jukebox and personality medium, powered by the phenomenal success of star personalities such as Elvis Presley and celebrity disc jockeys such as Alan Freed (who, it is rumored, coined the term *rock and roll*), Dick Clark, and Murray the K.

During the hit radio period, the number of stations affiliating with a major network steadily declined, as locally played music programming replaced mass-appeal entertainment and dramatic shows. In 1950, the ratio of network-affiliated stations had dropped to about one-half; by 1960, network affiliation had bottomed out at one-third of all stations.

Control of programming and advertising shifted from networks to local station management. Stations maintained their network affiliation primarily for news and special-events coverage, although some daytime serials (known as "soap operas" because of the type of sponsors such programs attracted) remained in network schedules until the early 1960s.

During this phase, FM radio began to appear. Between 500 and 1000 FM stations emerged, most co-owned by AM stations (combos) or operated on an experimental basis by schools and colleges. Some began to program music pressed into Columbia's new 33⅓ high-fidelity disc format, and FM was promoted for its *fine music*.

Format Radio

Radio's third wave, its maturity period, began in the mid-1960s and continues to the present. This era has been marked by the phenomenal growth in the number of FM stations and the leveling of the AM bandwidth, such that by the late 1980s, the number of FM stations grew to comprise more than half of the total radio station universe.

Gradually, the FM stations emerged as the primary source of radio listening, especially music listening, with stations playing country, rock, classical, and a variety of other styles, and AM stations moving to talk, news, and information programs. In 1978, for the first time, the share of audience for FM stations exceeded that for AM. Today, FM accounts for 75% of total radio listening.

This era can be conveniently labeled *format radio*. In radio parlance, a format is a radio programming schedule. As is described in much detail below and in Chapter 11, the format is extensively researched and modified to attract a specific audience subgroup, such as teenagers, women 25–34, adults over 55, and so on.

With the number of radio stations leveling off between 10,000 and 12,000, radio management has largely abandoned the mass-appeal strategies of the previous stages in the medium's life cycle. As in all mature industries, marketing and promotion to these key audience segments have become critical to maintaining market share. As we will see, profit margins have narrowed, especially on the less targeted (and less listened to) AM band.

Decline

In the 1990s, radio faces the downside of the life-cycle curve: its decline and possible replacement by superior technology.

AM radio is most imperiled. The overwhelming majority of AM listeners are above 55 years of age (those who grew up in the network and hit radio phases). As these listeners literally die off, FM share is expected to rise to 90% or more. Already, many AM radio stations have begun to "go dark"—radio parlance for out of business.

In response, AM radio is attempting what some perceive as "last gasp" innovations. Enhanced transmission equipment, (including stereo), better receivers, an expanded bandwidth to reduce interference, and other technologies have been introduced to stem the tide of decline. It is too early to tell whether these innovations will succeed in luring audiences. However, it seems clear that by the year 2000, the number of FM stations will significantly eclipse those on the AM band.

FM is not immune to competition. Like most mature enterprises, it is vulnerable to the introduction of superior product, delivered more efficiently and at lower cost. As we saw in Chapter 1, digital audio broadcasting is on the horizon: offering CD-quality sound to radio listeners at home and in their cars. It is unclear at this time whether DAB will be a national/regional satellite or locally delivered broadcast service. Regardless, it holds potential to begin to erode FM listening by the turn of the century.

Radio's Life Cycle: Implications for Management

Many of today's radio owners and managers began their careers in the hit radio era and learned a number of key lessons about the radio business which they carry with them today. First, the concept of *localism* became paramount. Rather than relying on a major national network, successful stations had to develop their own programming formats and personalities. The local on-air personality became a valued commodity. Timely news reporting, civic awareness and involvement, and local public service became integral to station operations.

Second, the radio medium became permanently dependent on trends in the music business generally and the recording industry in particular. Music stations had to keep abreast of record sales to program the most popular new recordings. The songs chosen and the frequency with which they were played—known in the business as the music rotation—became the formula for survival and growth.

Managers came to view radio as a background, rather than a foreground, medium. They realized that radio was a companion for each member of the household—at home, at the beach, at work—rather than a medium enjoyed by the whole family at one central location. Thus, music and news elements had to be brief and to the point: directly targeted to a particular audience. The concepts of format and target audience took hold.

Finally, and perhaps most important, radio became a direct competitor with newspapers and outdoor advertising for the local advertising dollar. The biggest national advertisers migrated to the mass medium of television, leaving success in commercial radio directly in the hands of the local radio account executive (see Chapter 13).

The implications of these phenomena are hotly debated in contemporary radio management circles. Is AM truly "dead"? Has the number of stations reached a "critical mass"? That is, has the audience pie been divided into so many pieces that profitability for everyone is threatened? Can programmers, advertisers, and ratings companies effectively and accurately keep pace with an increasingly selective, often away-from-home listening audience? These issues are discussed below and throughout the remainder of the text.

NONCOMMERCIAL RADIO

Noncommercial radio is as old as the medium itself. In the first year of this century, much of the experimentation with wireless telegraphy, as radio was then called, was conducted in U.S. colleges and universities.

The frenzy to obtain broadcast licenses in the 1920s included applications from educational institutions and religious groups. By 1927, 98 noncommercial stations were on the air.

The Great Depression had a particularly devastating effect on noncommercial radio. By 1933, the number of noncommercial stations had shrunk to 43; those that remained generally operated for limited hours and at reduced power. In 1935, the FCC reported its opinion that commercial broadcasting provided sufficient opportunities for education and public service. In the view of the FCC, no special provision was necessary to provide noncommercial spectrum space for stations.

By the latter part of the decade, it had become apparent that commercial radio did not provide enough educational material to satisfy many educators and social critics. In 1938, under increasing public pressure, the FCC allocated 25 channels for in-school AM broadcasting. In 1945, 20 of the 100 new FM channel assignments were reserved for educational stations. This corresponds to the area between 88 and 92 on the FM dial.

Rapid and continuing growth in noncommercial educational radio followed. By the late 1950s, approximately 150 such stations were in operation. A decade later the number had doubled. By 1975, over 700 noncommercial stations were in operation; by 1993, there were over 1500.

The majority of noncommercial stations are licensed to schools and universities. Many of these provide students with an opportunity to learn about broadcasting by becoming fledgling disc jockeys and news reporters. Another large portion of these stations are operated by religious organizations.

One influential group of stations meet criteria established in 1967 and amended at various times since by the Corporation for Public Broadcasting (CPB). These criteria include minimum FM-operating power of 3000 watts, at least five full-time paid employees, an operating schedule of 18 hours per day, and an annual budget exceeding $150,000 per year. Such stations are eligible for federal financial and program support. They form the core of the National Public Radio network.

NPR was established in 1970 by CPB to produce and distribute programs. By 1980, NPR had established a national satellite networking system (ahead of the commercial networks) and was responsible for such widely acclaimed and popular programs as *All Things Considered* and *Morning Edition*. However, in the early 1980s, NPR was rocked by scandal. Mismanagement of federal funds and poor planning led to a budget deficit exceeding $9 million and, ultimately, to the resignation of the president of NPR, Frank Mankiewicz. Over the next few years, the NPR work force was pared from over 500 to under 300. By the early 1990s, a leaner public radio system boasted over 350 member stations and faced a tenuous economic future.

Like their commercial counterparts, public radio managers face a period of maturity and the possibility of decline. Public radio expanded in the 1960s and 1970s when government funds were plentiful and university enrollments and endowments were high. Today, many universities are questioning their investment in costly electronic "toys" for their students. Seeking to control excessive spending, federal and state governments are scrutinizing and trimming budgets for educational broadcasting outlets. Problems at NPR in the 1980s were symptomatic of a trend: noncommercial radio (and television) stations in the 1990s face an era of accountability—making effective management the key to their survival. Thus, the principles of sound personnel and financial management discussed in this book are applicable to public as well as to commercial broadcasting.

RADIO RECEIVERS

Not only have the number and types of radio stations changed over the years. Similar trends are apparent when we examine the history of the diffusion of radio receivers. As Figure 4-2 illustrates, the number of radio sets in use in the United States has far exceeded the saturation point of one receiver per person. In fact, there are nearly 600 million radios in use, more than two for each person in America. Radio is everywhere: three-fourths of all radios are in homes; the remainder are in cars, the workplace, and other locations.

FIGURE 4-2 Radio receivers in use, 1920–2000. *Source*: Christopher H. Sterling, *Electronic Media* (New York: Praeger, 1984); projections to 2000 by the author.

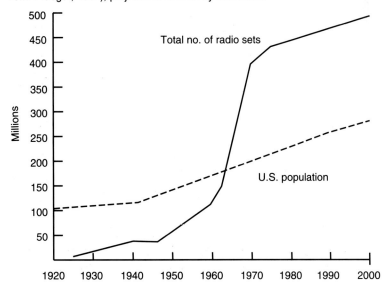

As Figure 4-2 demonstrates, the proliferation of receivers began in the mid-1950s and continues to the present time. The root of this rapid diffusion was the invention of the transistor and similar microelectronic technology, which made radio the most portable and personal of the electronic media. A radio receiver may now take any form; high-fidelity stereo components accompany us as we work, travel, and exercise.

By the dawn of the twenty-first century, there may be more than 700 million radio receivers in use—as many as seven in every household in the United States.

RADIO LISTENING: DAYPARTS AND DEMOGRAPHIC TRENDS

The rapid diffusion of stations and sets has, not surprisingly, paralleled a rise in radio listening. Radio listening peaked in the mid-1930s at almost 5 hours per household per day. The arrival of television caused listening rates to plummet, and in 1960 the average was just under 2 hours per day. In the 1960s, however, radio listening began a steady climb. During the 1970s, daily usage per household passed 2 hours; by the late 1980s, the 3-hour mark had been surpassed.

Audience research indicates that radio listening varies by time of day, known as *dayparts*, and by population characteristics, known as *demographics*. Listening patterns throughout the day are presented in Figure 4-3.

FIGURE 4-3 Radio listening in the United States. *Source*: Arbitron. Used with permission.

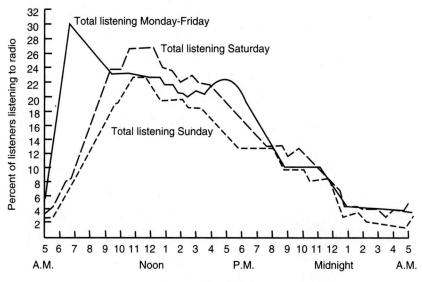

On weekdays, radio listening is greatest in the early morning hours (6 to 10 a.m.). This time period has become known as *morning drive*, since it corresponds to the time when most people are preparing for and commuting to their day's work. The second most popular time for radio listening is 3 to 7 p.m., a time slot known as *afternoon drive*. Next in aggregate listening are *daytime* (10 a.m. to 3 p.m.) and *evening* (7 p.m. to 12 midnight).

As we might expect, radio listening levels are smallest after midnight. Because most people are asleep and because listeners in this time period are occasionally cranky or strange, the slot from 12 to 6 a.m. is known in the industry as the *graveyard*. On weekends, radio listening is most concentrated in the late morning and early afternoon.

Over 96 percent of people in the United States use radio at one time or another throughout the week, and it is known that time spent listening is directly related to sex and age. In general, women listen to the radio more than men. Listening levels for women are highest during the daytime hours, not only for women at home but also for women in the work force, who tend to bring their radios with them to their places of business.

Generally, children between the ages of 2 and 11 spend comparatively little time with radio (as they're captivated by TV). However, adolescence (ages 12 to 17) spurs a sharp rise in radio use. In fact, in a typical week a teenager in the United States will listen to 20 hours of radio or more. The radio listening of teenagers is matched by their consumer behavior: this age group buys the most cassette tapes, hit records, and CD singles.

Like teens, the young-adult audience (ages 18 to 34) is characterized by high radio listening levels. This group also buys the most full-length record albums and CD releases, making young adults enormously popular among radio and recording executives.

As the median age of the average listener in the United States continues to rise, the adult sector of the population (ages 35 to 54) is becoming one of radio's most important targets. While this group doesn't purchase as much music as do teens and young adults, adult listeners make the majority of consumer purchases in the United States (from cars to clothes to cosmetics to groceries). Mature and senior radio audiences (over 55) tend to be small in number, but high in station loyalty and listening time. As the population of the country ages, the mature audience (and its greater purchasing power) is expected to become increasingly important to radio executives.

RADIO PROGRAMMING FORMATS

As we have seen, the combination of demographic and daypart variations in the radio business has led to the format approach to radio programming. Instead of seeking a huge aggregate audience, most stations fine-tune their program schedule to reach specific target-audience segments. Table 4-1 identifies the most popular radio formats by the percentage of total radio stations in each format and

TABLE 4-1 RADIO PROGRAM FORMATS

	Share of stations	Share of audience*		
		AM	FM	Combined
AC/oldies	18%	2.25%	19.43%	14.95%
AOR/classic	6	0.41	19.41	14.45
Country	20	3.73	17.24	13.72
News/talk	6	48.14	0.65	13.05
Top 40/CHR/contemporary	7	0.16	13.95	10.35
Black/urban	2	5.44	9.91	8.75
Beautiful/easy/soft	3	1.75	10.32	8.08
Spanish	3	8.46	2.52	4.07
MOR/variety/full service	4	13.36	0.04	3.52
Nostalgia	2	10.59	0.55	3.17
Religion/gospel	10	4.80	1.14	2.09
Classical	2	0.28	2.29	1.77
Jazz/new age	3	0.03	2.36	1.76
Unknown/others	14	0.60	0.17	0.28

* 12 years old and above, total broadcast week.
Source: *Broadcasting*, April 15, 1991, p. 75; *Billboard*, June 1, 1991, p. 15; *American Radio*, Fall 1992, p. A-19.

their share of the AM, FM, and total radio audience. (Format trends are discussed in detail in Chapter 11.)

Since the share of audience is crucial to the radio manager, let's look at that column first. Overall, the leading format in terms of audience size, is *adult contemporary* (or AC, in industry shorthand). Nationally, nearly one in six stations uses an AC format or its major variants, oldies and "soft rock." Adult contemporary stations target the female radio audience between the ages of 25 and 49, with an emphasis on current hit records, recent popular songs, bright personalities, and a sprinkling of news, weather, and traffic information. The oldies and soft-rock versions of the format feature enduring songs from the hit radio era, from the Beatles to the Supremes to the Beach Boys and the Box Tops, with an emphasis on love songs, ballads, and bright, up-tempo tunes. Familiarity and recognition are the keys to AC success: songs must be well known, well liked, and well remembered!

Album-oriented rock (AOR) stations are next most popular in terms of aggregate listening share. AOR stations have traditionally sought the teen and young-adult male audience. Their playlists include longer songs (cuts) by big-name groups—from AC-DC to Nirvana to Guns 'n' Roses—most with a hard-rock or heavy-metal sound. In addition to "supergroups," AOR stations feature solo artists like Bruce Springsteen, Tom Petty, and John Mellenkamp.

A major derivative of the format is included in Table 4-1. *Classic rock* stations appeal to older men (35 to 54) who were teenagers in the late 1960s and early

1970s. Classic rock leans heavily on the "big names" of this era, from Bob Dylan to the Doors, the Rolling Stones, and later Led Zeppelin, Supertramp, and Journey. Another AOR variation is *progressive, alternative,* or *new rock*. These stations move toward the younger edge of the demographic (12 to 34) and feature new music more prominently, such as that of R.E.M., Depeche Mode, Red Hot Chili Peppers, Morrissey, and Pearl Jam.

Country music stations are the most plentiful in America (about 1 in 5 stations nationwide), and account for almost 14% of radio listening. Country stations tend to be most successful in rural communities, and are particularly popular among women. However, the "country craze" that swept the nation in the 1980s brought high-powered country formats to stations in major markets such as Chicago, New York, and Los Angeles. The population growth of "sunbelt" cities in the 1980s, including San Diego, Phoenix, Nashville, Atlanta, Houston, and Dallas also fueled the format, leading to such derivatives as "contemporary country" for the new residents and "classic country" for the purists.

The staple of AM radio, especially in the nation's largest cities, is the *news/talk* format. This format includes a mixture of news, telephone call-ins, and discussions of topics from sex counseling to sports. While only 6 in 100 radio stations feature a news/talk format (almost none on FM), they account for 13 percent of all

Long-time radio personality Larry King gets the upper hand with guest Ted Koppel. After some lean times, network radio has rebounded in the 1990s. (courtesy of Westwood One)

radio listening. Most major cities have at least one news-only station; some cities in the top-10 markets (for example, New York and Los Angeles) are large enough to support two all-news operations. All-news and news/talk stations tend to be equally popular with men and women and tend to attract the most listeners during drive times.

Just over 10% of the radio audience chooses top-40 or *contemporary hit radio* (CHR) stations. These stations seek that large and loyal core of teens and young adults. Female and male soloists (from Madonna to Michael Bolton to Natalie Cole to Mariah Carey) are quite popular with this audience segment, which is largely female. This is the target audience that has made supergroups out of teenage heartthrobs throughout the years, from the Beatles in the 1960s, to the New Kids on the Block in the late 1980s, to Wilson Phillips and Taylor Dayne in the early 1990s. CHR stations emphasize current hits, played very often, almost like an audio jukebox. Oldies and unfamiliar songs are avoided at all costs.

The *black/urban contemporary* format emerged in the 1980s as a direct consequence of the collision and fusion of cultures in the nation's largest cities. "Black" is actually a misnomer for this format. Today's urban contemp sound is a mixture of black-based rhythm and blues, white and black rap, salsa of Hispanic origins, and dance music of all cultures—literally the "sound of the city." The key to urban contemp is the beat: an undeniable percussive pulse ideal for the big-city boom box.

Beautiful-music or *easy-listening* stations have had a long and profitable history on radio, especially on the FM side of the band. Sometimes sarcastically called "elevator music" or "audio wallpaper," this format has allowed many of the stations featuring it to make an enormous profit. The music, primarily orchestral arrangements with limited vocal accompaniment, succeeded in reaching women and older, established business executives. Station advertising successfully merchandised big-ticket items such as automobiles, real estate, and vacation packages. In recent years, the easy-listening format has declined in popularity and prestige. The primary reason is age. Today's homemakers and business executives no longer relax with Montovani or Tony Mottola; they'd rather hear Bonnie Raitt, Kenny Rogers, Linda Ronstadt, or the Beatles. The shift accounts for the arrival of the soft-rock or soft-AC format (which features mellow, laid-back songs by established performers intermixed with classic love songs by groups such as the Carpenters, Air Supply, and Bread). Many beautiful-music stations have relaxed their rules about using vocalized music, replacing 101 Strings with Neil Diamond, Paul Simon, Barbra Streisand, and Billy Joel.

The remaining formats in Table 4-1 each comprise less than 5% of all radio listening (though one, religion, accounts for 1 in 10 total radio stations). *Middle-of-the-road* (MOR) stations feature the music of vocal performers like Frank Sinatra, Vicki Carr, and Tony Bennett. A version of MOR called *nostalgia* concentrates on the big-band music of the 1940s. As might be expected, MOR tends to be directed toward an older (over 55) audience. MOR stations are also known as variety or full-service stations, since their programming features news, talk, weather, traffic, and other elements along with the music.

Hispanic stations, which are concentrated in the industrial Northeast, Florida, and the desert Southwest, make up about 4% of radio listening and more than 10% of all AM stations. As the mass immigration of Spanish-speaking people into the United States continues, the number of Hispanic stations will increase, along with the share of audience for the format.

Listeners to classical and jazz tend to be loyal, devoted, and highly knowledgeable about their favorite music. Unfortunately, there aren't enough aficionados to make these particularly popular (and profitable) commercial radio formats. Each of these formats averages about 2% of the listening audience at a given time—roughly 10% of the number of people attracted by AC, AOR, country, or news. As a result, classical and jazz tend to be the staples of noncommercial radio stations (which value motivated, educated, and informed listeners, especially when soliciting donations).

The radio dial is filled out by a number of other formats. Ethnic programming is common in urban areas: a Polish station in Chicago, a Vietnamese station in Los Angeles, and a station targeted at the Greek audience in New York. In recent years, there have been all-comedy stations, such as KMDY in Los Angeles and WJOK in Gaithersburg, Maryland, and there was an all-Beatles station near Houston. There are now all-sports stations on the air in New York, Chicago, Washington, Los Angeles, and other large cities. As we will see in Chapter 11, there are other new formats in development, particularly for financially strapped AM.

The evidence of station growth, diffusion of sets, and growing listening levels paints a rosy picture of the radio medium. On the surface at least, this is an era of plenty; there is an abundance of radios, stations, and program formats. What is unclear is whether more is necessarily better for the financial well-being of radio management. Data on radio revenues, expenses, and profit margins temper an optimistic view.

THE BROADCAST FINANCIAL PICTURE

In order to understand the contemporary financial status of the radio business (and the television business, examined in Chapter 5), it is necessary to identify and describe the major sources of revenues and expenses for broadcasting stations. These are illustrated in Figure 4-4.

Broadcast Revenues

There are three main sources of income for commercial broadcast stations: advertising revenue, network compensation, and barter (trade). The greatest revenue base is, of course, advertising sales: the sale of commercial time to advertisers.

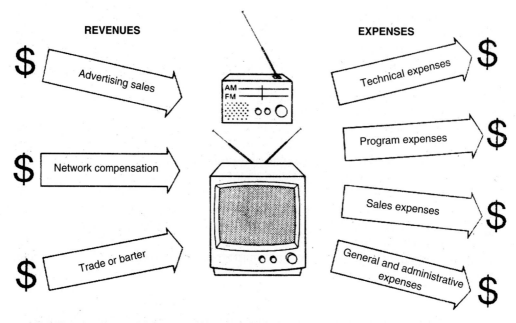

FIGURE 4-4 Broadcast revenue and expense items.

Broadcast advertising revenue falls into three broad classifications:

• *Network advertising*: the sale of advertising time by the major broadcast networks to leading national advertisers
• *National spot sales*: the sale of local station spots by radio and television station representatives to national and regional advertisers
• *Local spot sales*: the sale of local radio and television time by local station account executives to advertisers in their immediate service area

Network compensation consists of the payment made by major national networks to their affiliated stations in return for a share of the affiliates' airtime.

Stations also report income from barter, or trade, arrangements. These involve the exchange of advertising time for merchandise or other tangible assets. For example, many radio stations arrange to obtain transportation from local automobile dealers in return for advertising promotion. A second class of barter, known as barter syndication, concerns the trade of airtime in return for free or reduced-cost programming (see Chapter 11).

Broadcast Expenses

The bulk of the commercial broadcast dollar comes from network, national, and local advertising expenditures. Where does it go? To the Internal Revenue Service? To the news anchor or disc jockey? To maintaining the transmitter and paying the mortgage on the building site? Into the pockets of the owners and shareholders? The answer is, of course, all of the above. However, broadcast expenses can be classified into four main areas:

  • *Technical expenses*: transmitters, station equipment, spare parts and replacement materials, engineering salaries, and other items related to signal propagation, distribution, storage, and delivery
  • *Program expenses*: records and tapes, syndicated programs, salaries of announcers and newscasters, news services, music license fees, and other items related to programming content
  • *Sales expenses*: advertising commissions, payments to station representatives, sales training and development, salaries and commissions paid to account executives, and other items related to the sale of station inventory
  • *General and administrative expenses*: building and grounds, accounting, secretarial and clerical equipment and personnel, computers and automated business systems, and other items related to the internal operations of the station

Broadcast Financial Performance

Until recently, there were two standard sources for data on the broadcast industry's profit performance. The first source, the FCC's *Annual Report*, is based on data contained in required annual statements of income and expenses prepared by the stations themselves. In 1980, this requirement was eliminated as part of the commission's efforts to deregulate broadcasting, and recent efforts by the stations to sustain the practice with their own survey have achieved mixed results.

The other source of financial data is an annual poll of station management conducted jointly by the National Association of Broadcasters (NAB) and the Broadcast/Cable Financial Management Association. Through 1992, a separate financial report was prepared for radio and television. Citing a declining rate of cooperation from radio stations, the NAB discontinued its *Radio Financial Report* in 1993.[1] However, the U.S. Department of Commerce announced plans to publish annual reports on the radio industry to replace the NAB survey.

Over the years, the two financial yardsticks differed in their methodology and reporting standards. The FCC statistics were aggregate figures, for all stations, and generally included national network revenues as well as advertiser commissions. The NAB data were obtained from a sample of stations; they did not include network data or agency and representative commissions; and they were generally reported in terms of median station performance—that is, the profit and loss of the "typical" station.

It is useful to keep these differences in mind as we examine trends in radio and TV profits below.

THE FINANCIAL STATUS OF RADIO

Radio Revenue Trends

Like the number of stations, receivers, and formats, aggregate advertising revenues for radio have shown a steady increase, even when adjusted for inflation and increased expenses. The volume of radio advertising from 1965 to 1995 is shown in Table 4-2.

Some significant trends can be traced in light of the information presented in the table. First, aggregate radio station income from advertising has grown steadily, with total ad volume increasing fourfold in the period between 1975 and 1990, and it is clear that the rise of FM and the increase in radio listening have coincided with increasing advertiser interest in the medium of radio.

The second major trend is the rebound of radio network sales. Radio networking declined in the 1960s and 1970s, dipping to a low of 4% of radio revenues by 1975. However, the 1980s saw a resurgence of radio networking, with income from this source climbing back to 6% (and volume increasing from a mere $83 million in 1975 to over $400 million in 1990). The most recent upturn in network sales is a function of a resurgence in network affiliation, an increase in the number of radio networks available to local stations, and the popular programming provided by national networks and syndicators from *Paul Harvey News and Comment*, to *Rush Limbaugh*. The rebound of radio networking is covered in detail in Chapter 11.

TABLE 4-2 RADIO ADVERTISING VOLUME, 1965–1995
(Millions of Dollars)

Year		Net	Spot	Local	Total
1965	$	60	275	582	917
	%	7	30	63	
1970	$	56	371	881	1308
	%	4	28	67	
1975	$	83	436	1461	1980
	%	4	22	74	
1980	$	183	779	2740	3702
	%	5	21	74	
1985	$	365	1335	4790	6490
	%	6	20	75	
1990	$	433	1626	6780	8839
	%	5	18	77	
1995	$	553	1768	8729	11,050
	%	5	16	79	

Source: Radio Advertising Bureau reports (various); 1995 projection by the author.

A third trend in Table 4-2 is the decline of national and regional spot advertising as a percentage of total radio revenue. From the late 1960s to the present, there has been a consistent drop in the percentage of radio advertising generated by spot sales. The percentage of station revenue from this source shrank from 30% in 1965 to 18% by 1990, despite the fact that the dollar amounts increased more than fivefold. There are numerous reasons for this phenomenon. First is the number of stations. In the 1960s, national advertisers spent their money on three or four stations at most in a given market. Today, these advertisers can choose among 20 stations or more, resulting in greater competition for national spot dollars (and, of course, lower prices to the advertisers for radio spots).

The station representative business has become increasingly competitive and is now dominated by two large firms which effectively control most spot radio advertising. Interep, which owns Group W Radio Sales and Major Market Radio, is one "megarep"; the other is the Katz Radio Group, parent company of Banner, Republic, and numerous other rep firms. Through these powerful companies, which often represent numerous stations in the same market, national advertisers have been able to find considerable discounts in their radio buys.

The decline in the percentage of spot sales may also be seen as a result of the escalating profitability of television in this period. Advertisers can now buy television spots, particularly in cable television and on independent stations, at costs competitive with radio rates (see Chapters 5 and 6). In order to compete successfully with television, radio stations have had to be extremely flexible with the rates officially published for spot buyers in such industry guides as the monthly reports issued by Standard Rate and Data Service (SRDS).

Together, focused programming, greater competition, rep consolidation, high discounting, and a large inventory of spots (or *avails*) have led to remarkable growth in local sales. Indeed, radio has become primarily a local advertising medium. In the past 25 years, both the amount and the percentage of local sales revenue have increased dramatically.

In the mid-1960s, local radio advertising was a $500 million business; it now exceeds $7 billion in annual billing! In 1965, stations earned roughly 60% of their income from local spots; now the figure is nearly 80% for all stations and over 90% for the stations not owned by a major network or group.

With today's emphasis on local spot sales, it is not surprising that network compensation and barter comprise only a small fraction of radio station income. Network compensation has shrunk over the years to the point that it now accounts for under 2% of station revenue. In fact, the proliferation of satellite radio networks has made networking an expense for many stations. Some radio networks require their affiliates to pay for the earth station and the monthly programming it provides.

While barter arrangements provide essential services to most stations, they too comprise only a small portion of station income. Currently, the average cash value of barter to each radio station is only about 4% of station revenue.

Radio Expense Trends

General and administrative costs, from electrical power to office equipment, account for the largest share of the radio station's expense dollar. According to NAB estimates, about 40% of radio expenses falls into this category.

Sales expenses rank second in the cost hierarchy, accounting for about 30% of radio expenses. Programming and production, including news, account for about 25% of each expense dollar. Engineering costs account for the remaining radio expense: approximately 4% of the total.

One phenomenon is true of all four categories of expense: rapidly escalating increases. At one time, managing a radio station was a comparatively inexpensive operation. Once the transmitter was erected and a small studio built, a station could operate rather cheaply. In the early days of radio, stations were put on the air by amateurs for a few hundred dollars and a small investment in batteries and chicken wire. In the 1930s it was common for stations to budget their operation for under $5000. Unfortunately, the days of broadcasting as a low-overhead business have gone the way of the $5 pizza.

According to NAB figures, radio expenses increased an average of 8% to 9% a year during the 1980s. Most of the increases occurred at the end of the decade, and the upswing is continuing into the 1990s. As a media consumer, the reader has firsthand knowledge of these increasing costs. How much did a new tape or compact disc cost 5 years ago? And how much would the same recording cost today? Consider the radio program manager, who purchases records, tapes, and CDs for the station library. Have you replaced the cartridge on your turntable lately? Had to adjust the laser on your CD player? Consider the chief engineer, who must keep an inventory of spare parts on hand to sustain equipment at peak operation. And what is the price of your favorite newspaper or magazine, compared with its cost just a few years back? Consider the radio news director, who may subscribe to major dailies and newsmagazines and who pays to receive two or three newswire services.

While expenses in all four categories have escalated, programming costs have led the way. Increased benefits packages and other factors have caused salaries for on-air professionals to rise. Program expenses have also risen as a result of the royalties that stations must pay to music-licensing firms such as ASCAP and BMI for the use of prerecorded materials (see Chapter 11) and to satellite services and program syndicators for music formats. The cost of maintaining a local music library has shot skyward, and record companies, themselves facing financial uncertainty, have instituted increasingly restrictive policies on providing complimentary or demonstration copies of the music of their artists.

Next to programming, the highest rate of increase in station expenses has been in sales and promotion. With competition increasing, stations have had to mount elaborate advertising and image campaigns. Billboards, bus cards, even T-shirts are costly advertising items. Many stations now mount extensive TV advertising campaigns in an attempt to lure listeners.

Costs are up in the technical area as well. Transmitter tubes that cost $100 last year may cost $200 this year. A declining dollar, compounded by rises in the cost of vinyl, copper, iron, and steel (much of it imported from overseas), has resulted in increased prices for discs, tubes, tapes, and spare parts. Engineering salaries continue to rise. Many experienced and novice electrical engineers are abandoning broadcasting for lucrative career opportunities in the computer field. As a result, station executives must pay more than in the past to recruit, train, and retain engineers or face increasing costs for consulting engineering services.

As in other industries, general and administrative expenses have also increased. It simply takes more money to run a business now than it used to. Since most stations operate for most of the day and utilize sophisticated electronic equipment, utility bills have risen dramatically. Some radio stations lease special telephone lines from AT&T to link studios to transmitters and to establish networks for entertainment, sports, or special remote broadcasts. More than 8 in 10 radio stations today boast microwave or satellite technologies to link their studios to transmitters or to import programming. Utility costs conspire to create a constant financial headache for the media manager. At many stations, the monthly telephone bill alone can be $20,000 or more.

Other general and administrative expenses are on the rise. Since 1970, postal rates have quintupled. The prices of paper, typewriter ribbons, and even towels for the executive washroom have shown a similar dramatic increase. In an effort to streamline their operations, many managers have introduced office automation. Typewriters and adding machines have been replaced by word processors and microcomputers. Although this equipment should eventually result in cost reductions, for now the new devices have added a large expense item to the ledgers of many stations.

Sales expenses have also shown a marked rise. The $20 business lunch of a few years ago might exceed $50 today. And, as with programming and engineering, the cost of recruiting, training, compensating, and retaining sales personnel has shown a dramatic, ever-increasing climb.

In previous eras, radio was seemingly immune to downturns in the national and international economy. This is no longer the case, and rising expenses have cut considerably into the profits and profit margins of many (if not most) of the nation's radio stations.

Profits and Profit Margins

Expenses in radio and television has risen at a similar rate, but the impact of that rise on profits has been much more severe for the sound medium. Figure 4-5 charts trends in profits for the radio industry.

There is no doubt that aggregate and average radio station profits have been inconsistent in recent years. Riding the crest of the baby boom, the FM explosion, and some big years for the recording industry, radio operations increased their

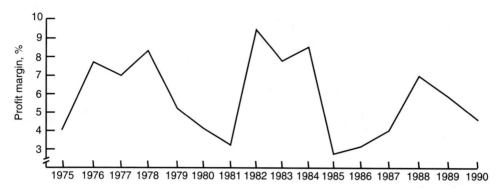

FIGURE 4-5   Radio industry profits 1975–1990. *Source: Radio Financial Report* (Washington, D.C.: National Association of Broadcasters, 1975–1990). Data are industrywide aggregates, subject to wide variation.

profits in the mid-1970s. From 1975 to 1978, total industry profit margins more than doubled, and the typical station returned from 7% to 8% on investment. Recall that this was the same period in which radio advertising revenues doubled, with most of the gain occurring at the local level. Listening levels at home and in the car were increasing, advertiser interest was keen, and the profit picture was good for the medium in general. By 1980, inflation in the United States was rampant, escalating at an annual rate of 7.1%. Utility, labor, and other costs were beginning their upturn. The number of on-air stations surpassed 9000. Then radio station profits began to nose-dive.

To understand the radio downturn, we must look at the industry by station type. Nationwide, FM stations did not report a profit at all until 1975. Together, independent FM stations that year reported a combined loss of $9.4 million. Three years later, independent FM stations accounted for $23.5 million in pretax profits! Similarly, in 1981, the NAB reported that the profit margin for the typical AM station had shrunk to a paltry 2.61%. In comparison, AM-FM combos reported a 4.24% return on investment and stand-alone FMs a comparatively healthy 5.69%. Further evidence of the generally difficult climate for the radio business during the recession of the early 1980s is provided by data contrasting the percentage of stations reporting profit with those reporting loss. In 1977 and 1978, 67% of all radio stations reported a profit. By 1980, the industrywide figure had shrunk to 59%.

The general economic recovery of the mid-1980s produced increased profits for radio, particularly in 1982 (not coincidentally, the first year following FCC deregulation). However, the next year found radio profits slipping. Indeed, despite a good year in 1984, by mid-decade more than one-third of all stations reported operating at a loss. While FM stations were averaging about 10% return on investment, AM station profits were less than 5 cents on the dollar.

Radio profit margins escalated in the late 1980s, fueled by the increased value of radio properties to investors (see Chapters 7 and 8) and the general boom at the time in the advertising business. The go-go 1980s, particularly from 1986 to 1988, saw aggregate radio profit margins rise again, to an average of 7% industrywide. However, virtually all profit was in the hands of the operators of full-time, large-market FM stations, which boasted an annual pretax margin exceeding 20%. At the other extreme, by the late 1980s, the majority of AM stations and AM/FM combo operations reported operating at a loss.

By the early 1990s, a severe recession exacted a toll on many businesses, including radio. The advertising business was hit especially hard: in 1991, radio ad spending shrank below 1990 levels, the first time in more than 30 years that radio ad volume had declined.[2] The trickle-down effect upon radio stations was deeply felt.

By the early 1990s, hundreds of AM stations had shut down. Thousands were for sale (but few buyers were forthcoming). FM suffered as well, with average profit margins decreasing each year from 1988 onward. Large-market FMs, which had returned 20 cents on the dollar in the mid-1980s, were considered "healthy" if their margins remained between 5% and 10%.

In fact, the gloomy profit performance of radio in the early 1990s, including the unprecedented number of station bankruptcies, led to the relaxation of radio ownership rules (including permissible duopolies) described in Chapter 7.

In the face of this gloomy economic picture, we might wonder why anyone would invest time and money in a radio station, even a profitable one. After all, it is possible to put funds in a variety of profitable investments without having to deal with irate listeners, cantankerous disc jockeys, and arrogant account executives. There are several possible answers. In the first place, accounting is an inexact science. It is possible that stations are simply not providing accurate profit reports to regulatory agencies or to the competition. Just as it is unlikely that stations receive the full amount reported on their rate cards for the sale of advertising time, so it is unlikely that their tangible losses exactly match the figures supplied to the NAB and other trade groups.

Second, NAB data include salaries paid to management staff as general expenses. Thus, while in aggregate the station may be losing money or reporting only a modest profit, it may still be paying handsome salaries to the general manager and to other station personnel. In addition, the history of radio is marked by turnarounds and success stories. Many managers are willing to endure short-term losses in expectation of building listener loyalty and advertiser revenue over the long term. Even if "loss leaders" are not turned around, group owners will often retain their radio interests as a convenient tax write-off, or they may eventually sell the station for a profit.

Ultimately, many owners and managers remain in radio because it is their medium of choice. There is still a great deal of ego involvement and satisfaction in running a radio station. For the former salesperson or disc jockey turned owner, business may occasionally be bad but it is rarely boring. Personal involvement with the radio medium sustains many financially unsuccessful managers and keeps them in the business.

THE BROADCAST EMPLOYMENT PICTURE

As is the case with revenue and profit trends, getting a fix on the status of broadcast employment is difficult. The FCC publishes data in its annual *Equal Employment Opportunity Trend Report*. The NAB provides data in a national survey of about 1000 member stations. Employment figures are also compiled by the U.S. Bureau of the Census, the Electronics Industries Association of America, and other trade groups, as well as by various academic sources. In general, there is consensus on at least three trends:

- The broadcast work force is comparatively small and is growing modestly (at best).
- Opportunities are greater in television than in radio.
- Gains are being made by women and minority groups.

Radio Employment

At 5:15 one April morning, Paul Young, a disc jockey at WVBK-AM in Herndon, Virginia, locked himself inside the station's control room. For the next 12 hours, he played the same popular country tune. An April Fool's prank? A special station promotion? Hardly. The record Young was playing nonstop was Johnny Paycheck's "Take This Job and Shove It," a protest of management's unwillingness to upgrade his $100-a-week salary. At 5:41 p.m., Young claimed victory: the station owner promised serious salary negotiations as soon as the college graduate, Vietnam veteran, and one-time male stripper got off the air!

While the above is an extreme case, it does illustrate the general employment picture in radio: jobs are scarce and pay is low, but opportunities are there for persistent, enterprising employees.

According to FCC estimates, some 75,000 people in the United States work in the radio business.[3] This figure includes personnel at radio networks as well as paid employees in public, religious, and other forms of noncommercial broadcasting. As for the commercial end of the business, the commission's most recent report indicates that about 65,000 people work at commercial AM and FM stations. An additional 3000 are employed in noncommercial radio. The remainder of radio employees toil in broadcast group headquarters and at the various radio networks.

Part of the appeal of a radio job may lie in its decidedly white-collar character. National employment figures estimate that 6 in 10 jobs in the United States fall into the Labor Department's blue-collar categories: clerical workers, craftspeople, operatives, laborers, and service personnel. In contrast, only 3 in 10 radio jobs fall into these categories. Of commercial radio jobs, 70% belong in the white-collar, or top-four, column: officials and managers, professionals, technicians, and sales personnel.

It may also be inferred that the radio business is typified by the presence of many "officers" and few "privates." In radio, 35% of the jobs fall into the professional category, which includes announcers, researchers, and production personnel. The next-largest category is officials and managers, accounting for 25% of the jobs. Sales personnel make up 20% of the radio work force; clerical and office personnel, 12%.

Throughout the 1980s, the number of jobs in radio grew at a steady rate of about 3% per year, though the recession saw a shrinkage of 1.4% of station employees in 1990 over the prior year.[4] Through the early 1990s, further attrition was noted in the radio job force. Total employment in commercial radio had dropped from over 75,000 in 1989 to about 67,000 at the close of 1992. In recent years, the largest increase in employment occurred in the professional and sales categories. This trend is not inconsistent with the growing competitiveness of the medium in programming and advertising.

The number of women and minorities in radio has been increasing. In 1977, women comprised 27.8% of the radio work force. In contrast, by 1990, nearly 40% of the radio work force was female (including the majority of employees in sales and promotion jobs). Minority employment (including blacks, Americans of Hispanic and Asian descent, and Native Americans) stood at nearly 20%, nearly double the work force percentages in 1975.

Gains are also being posted for minorities and women in the top-four categories. For example, the percentage of female managers rose from 19% in 1977 to about 30% in 1990. In 1977, 7% of radio managers were minorities; in 1990, the figure was approaching 12%.

On the negative side, many women and most minorities in radio are still performing clerical, secretarial, and janitorial tasks. In 1990, clerical and secretarial staffs in radio were 90% female; about half of the janitorial crews came from minority groups. These figures were virtually identical to 1977 figures.

Radio Salaries

Figures on radio station salaries and turnover rates indicate that the reader is studying the right subject matter: the money is in management and sales. Table 4-3 ranks selected radio job categories by average annual salaries.

As might be expected, sales managers make the most money in radio, with annual salaries in excess of $40,000 dollars per year. In large markets, the salaries of general managers and sales managers are in the six-figure category. Interestingly, the sales manager's salary is nearly double that of the station's program director.

Note too that air personalities and news reporters rank near the bottom of the salary scale, a scant notch above secretaries and the janitorial crew. While a few "name" disc jockeys command six-figure salaries, the average salary for people who speak into a microphone is only about $16,000.

There is some good news in radio employment. Job positions are on the rise for people with promotional savvy and knowledge of music. Nationally, promotion directors earn over $20,000 per year; nearly $40,000 in large markets. Similarly, music directors, many of whom double as air personalities, earn $21,000 nationwide; $30,000 or more in the larger markets.

TABLE 4-3 SALARIES FOR SELECTED RADIO PERSONNEL

Job title	National average*	Large market	Medium market	Small market
Managers				
General manager	$40,000	$120,000	$55,000	$29,750
Sales manager	43,500	112,793	50,000	27,000
Chief engineer	24,050	47,800	24,000	15,000
Program director	22,493	65,000	27,000	16,650
News director	18,500	43,000	20,100	15,000
Promotion director	23,320	38,000	16,000	17,625
Music director	21,000	30,000	18,960	13,430
Staff				
Salesperson	22,500	49,688	25,000	17,000
Air personality	16,000	45,000	18,000	12,265
News reporter	15,000	35,588	18,960	13,450

* Median.
Source: 1990 Radio Employee Compensation Report (Washington, D.C.: National Association of Broadcasters, 1990). Large market = population greater than 2.5 million; medium market = population 250,000 to 500,000; small market = population 25,000 to 50,000.

Howard Stern, Infinity Broadcasting's controversial morning personality. Stern's seven-figure salary is among the highest in radio today. (Rick Maiman/Sygma)

The low salaries in radio may be partially offset by fringe-benefits packages: 60% of all stations award Christmas bonuses, 40% pay medical and hospitalization benefits in full, and 15% offer tuition reimbursement for employees continuing their college education. On average, radio employees receive 2 weeks paid vacation per year, plus 7 paid holidays.

How comprehensive these benefits are and how they compare with trends in television and cable are topics examined in detail in Chapter 10.

SUMMARY

Since its inception in the early twentieth century, the radio business has been marked by steady growth in diffusion of sets, number and type of stations, revenues and profits, and size of labor force. Growth in commercial radio began in the AM bandwidth, among network-affiliated stations located in the nation's major markets. In the 1950s, the evolution of the car radio, the emergence of $33^1/_3$ and 45-rpm records, and a "youth culture" led to a rapid rise in the number of AM rock-and-roll stations and the development of the FM band.

Today, radio is a mature business, with evidence of decline, especially in AM. From the late 1960s to the present, the number of AM and FM stations has increased dramatically and radio listening has shifted from the AM to the FM band.

Noncommercial radio began simultaneously with the rise of the medium. The number of educational stations increased significantly in the 1960s and 1970s as a result of FCC policy, financial support from federal and local governments, and interest from colleges and religious organizations. At present, owing to regulatory changes, government cutbacks, and other factors, noncommercial radio faces financial problems which make sound management critical to its survival.

Commercial radio is programmed and advertising is sold on the basis of demographic categories and dayparts. Demographic categories are determined by age and sex: males and females ages 12 to 17, 18 to 34, 35 to 54, and over 55. Key dayparts are morning drive (6 to 10 a.m.), evening drive (3 to 7 p.m.), daytime (10 a.m. to 3 p.m.), evening (7 p.m. to 12 midnight), and graveyard (12 to 6 a.m.).

Radio programming formats are designed to appeal to target demographics. Popular formats include adult contemporary, contemporary hit radio, country, album-oriented rock, news/talk, middle of the road, easy listening, and all-news.

At present, over 13,000 radio stations fill the airwaves. Most FM stations are profitable, but increasing competition and rising expenses have led to decreasing profit margins for increasing numbers of stations. Most AM stations lose money; many have gone out of business in recent years. The radio work force is small (about 75,000 total employees), and has declined in recent years. Most radio jobs are in the white-collar categories: officials and managers, professionals, technicians, and sales workers. However, pay is comparatively low and is directly dependent upon station type and market size. Job opportunities are best today in sales, promotion, and music management.

NOTES

1 See "NAB Pulls Plug on Radio Station Financial Survey," *Broadcasting and Cable,* May 31, 1993, p. 26.
2 *Broadcasting,* August 23, 1991, p. 23.
3 *1990 Broadcast and Cable Employment Trend Report* (Washington, D.C.: Federal Communications Commission, June 11, 1991), p. 13.
4 *FCC EEO Trend Report,* BEE 18-01, May 24, 1991, p. 14.
5 Interpolated from data in *1990 Radio Employee Compensation and Fringe Benefits Report* (Washington, D.C.: National Association of Broadcasters, 1990), and *1990 Radio Financial Report* (Washington, D.C.: National Association of Broadcasters, 1990).

FOR ADDITIONAL READING

Collins, Mary: *National Public Radio: The Cast of Characters* (Washington, D.C.: Seven Locks Press, 1993).
Fornatale, Peter, and Joshua E. Mills: *Radio in the Television Age* (New York: Overlook, 1980).
Hilliard, Robert L.: *Radio Broadcasting,* 3d ed. (New York: Longman, 1985).
Keith, Michael C.: *Radio Programming: Consulting and Formatics* (Boston: Focal Press, 1987).
Lewis, Peter: *The Invisible Medium: Public, Commercial and Community Radio* (Basingstoke, U.K.: MacMillan Education, 1989).
Radio and Television Career Directory (Hawthorne, N.J.: Career Press, 1991).
Shane, Ed: *Programming Dynamics: Radio's Management Guide* (Overland Park, Kan.: Globecom, 1984).
Sklar, Rick: *Rocking America: How the All-Hit Radio Stations Took Over* (New York: St. Martin's Press, 1984).

5

THE TELEVISION BUSINESS

The rise of television in the United States has been rapid and complete. In little more than 40 years, the medium has spread into virtually every home. Its form has continually been refined: from small black-and-white screens to large color projection systems with full digital sound. Television's voracious appetite for programming has led to a broad range of content, from game shows and soap operas to documentaries and music videos. The centrality of television to political and social life in the United States, and its place as a fixture in the home, has created a business climate characterized by steady, often spectacular, growth in revenues and income. But recent recessionary trends have tempered the economic success of TV. This chapter examines the business of television from a managerial perspective. As with the previous chapter on radio, economic trends in television will be examined through an analysis of stations in operation, television receivers and sets in use, audience demographics, viewing trends, revenues and expenses, profits, and personnel.

The goals of this chapter are:

1 To trace the growth and current status of public and commercial television by examining stations on the air, household penetration, and viewing trends
2 To identify sources of television revenue, including network compensation; national, spot, and local advertising sales; and barter
3 To identify trends in station expenses, including program, administrative, technical, and sales costs
4 To trace trends in profit and loss for network-owned, network-affiliated, and independent stations
5 To examine television employment trends, including job classifications, salary levels, and opportunities for women and ethnic minorities

THE GROWTH OF COMMERCIAL TELEVISION

Figure 5-1 traces the rise in the number of commercial television stations on the air since 1950, with projections to the year 2000. Like the radio medium, commercial television has grown in four main stages.

The Golden Age

Television's *golden age*, which historian Erik Barnouw has called its "salad days," spanned the years 1948 to 1960, when the bulk of television programming was live, in living black and white. As noted in Chapter 1, the number of stations on the air was restricted by the FCC until 1952. From the late 1940s to the early 1950s, only about 100 stations were operating, virtually all of them in the nation's largest cities.

When the freeze ended, the number of applications to the FCC shot up. By 1955, there were 297 VHF and 114 UHF commercial stations on the air. In the second half of the 1950s, the number of VHF stations continued to expand, but the lack of receivers with UHF tuners caused these kinds of stations to shrink to just 75 by 1960.

During this period, more than 90% of the stations were affiliated with a major network. NBC boasted the most affiliates: 4 in 10 stations were in the NBC chain. CBS had slightly fewer affiliates. On average during the 1950s, 1 in 3 commercial

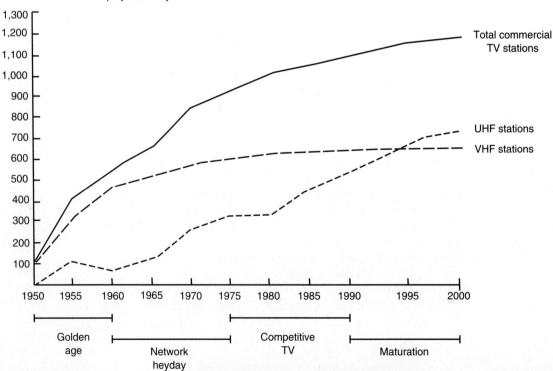

FIGURE 5-1 On-air commercial TV stations, 1950–2000. *Source:* Christopher H. Sterling and Timothy Haight, *The Mass Media* (New York: Praeger, 1978); *Broadcasting*, various issues; 1995 and 2000 projections by the author.

television stations was a CBS affiliate. The remainder of stations (fewer than 20%) were in the ABC chain. Many ABC affiliates were in the UHF part of the band, with a less favorable dial position. Thus, network television in the 1950s was basically a two-horse race between NBC and CBS. ABC struggled along, maintaining profits mainly through the strength of its group of popular AM radio stations, which led in youth-oriented rock-and-roll programming. A fourth network, DuMont, bowed out in 1955, after nearly a decade of unsuccessful competition with the industry giants.

Programming during this period was similar to that of radio's network era of the 1930s and early 1940s. There was a major emphasis on vaudeville-type comedy-variety programs featuring such personalities as Ed Sullivan, Milton Berle, and Red Skelton. Most prime-time evening programs were broadcast live, mainly from the East Coast, and high-quality anthology dramas provided exciting opportunities for young actors and playwrights. Such programs as *Philco Playhouse, Studio One*, and *Ford Theater* broadcast original dramas and comedies live each week. Unknown actors who got their start during this period included Charleton Heston, Jack Lemmon, Walter Matthau, and Grace Kelly. Budding playwrights included Reginald Rose, Paddy Chayefsky, Rod Serling, and Neil Simon.

Documentary and informational programs were also quite popular with prime-time audiences. On CBS, Fred W. Friendly and Edward R. Murrow reported, and frequently made news, on their *See It Now* newsmagazine program. At ABC, Bishop Fulton J. Sheen inspired national audiences with his philosophy that "life is worth living."

NBC's innovative president at the time was Sylvester L. "Pat" Weaver (whose daughter Sigourney would later become a major film star). In an effort to expand network influence into other time periods, Weaver devised the *Today* and *Tonight* concepts. He was also responsible for large-scale special programs featuring Hollywood-level production values and big-name stars. Such "spectaculars," as they were called, including Mary Martin's performance of Peter Pan, became ratings blockbusters for NBC.

When their airtime was not filled by network shows, local stations experimented with a variety of program concepts. Many bought the rights to the grade B movies of the 1930s and 1940s, which film distributors were eager to unload. Cartoons and children's serials, such as the *Three Stooges* and *Little Rascals*, also provided inexpensive programming. Many local stations dressed an announcer as a clown or police officer, bought hand puppets, and produced "instant" children's programs. News coverage was in its infancy. Most newscasts featured a stiff-looking radio-trained announcer reading the news, with only an occasional film clip provided by a government agency or newsreel company.

Network Heyday

The decade of the 1960s through the mid-1970s marked the rise to prominence of network television: a battle for viewers between powerhouses CBS and NBC, and the upstart ABC. This era is demarked as the *network heyday*. By the end of the

1950s network and local television, like its audience, had begun to change. The advent of video tape recording eliminated the need for costly and risky live production. Faced with the reality of declining motion picture attendance and the realization that television was not merely a passing fad, major film studios such as Disney and Warner Brothers signed contracts with the networks to produce films for television. Production moved from the East Coast to Hollywood; westerns, detective and medical series replaced comedy-variety and anthology drama as the staples of network programs.

The number of VHF stations continued to increase steadily, at a rate of about 10 new stations per year. As a result of the FCC's passage of the All Channel Receiver Bill, which required sets manufactured after 1964 to have both UHF and VHF tuners, commercial UHF stations became viable. Within 5 years, the number of such stations nearly doubled.

The network-affiliate relationship became the backbone of the television business. More than 9 in 10 stations were network-affiliated. Since the first stations on the air had signed up with either CBS or NBC, most new stations, including many new UHFs, became ABC affiliates. By the middle of the 1960s, ABC had succeeded in signing affiliation agreements with almost a third of the nation's television stations, bringing the network near parity with CBS and NBC.

With the influence of Hollywood and the proliferation of color sets, the primetime network schedule was dominated by genre shows (lawyers, doctors, cops and robbers, westerns) and situation comedies (*Beverly Hillbillies, Green Acres, My Mother the Car*). Spurred by significant historical events, such as the assassination of President Kennedy in 1963, urban unrest in the summers of 1964 and 1965, and the war in Vietnam from 1962 to 1975, news coverage became increasingly important. Each network expanded its national news from 15 minutes to half an hour. A network news ratings war soon erupted between Walter Cronkite on CBS and the formidable NBC team of Chet Huntley and David Brinkley. At the same time, each network established its own overseas bureaus in London, Paris, and, later, Saigon.

Gradually in the 1970s, ABC reached full parity with CBS and NBC, in large part by riding the crest of the popularity of its youth-oriented situation comedies (such as *Happy Days, Laverne and Shirley, Mork and Mindy*, and *Three's Company*). ABC's strength was also reflected in its quality coverage of sports programs (especially the Olympics), its renewed commitment to evening news, and the success of its major miniseries, particularly *Roots* in 1976. Many former NBC and CBS affiliates switched networks and signed agreements with ABC, and ABC also secured contracts with a number of new stations. By the late 1970s, each network was affiliated with about 200 stations.

While the period from 1960 to 1975 was a boon to the three commercial networks, significant changes were also occurring at local stations. The high ratings for national news led local management to expand efforts to cover early and late news. The availability of film and tape reruns of former network series led programmers to invest heavily in syndicated productions. A new class of stations with a menu of syndicated programs, movies, and local sports began to make inroads into local advertising. Although independent stations, or "indies," were still largely unprofitable, they began to loosen the stronghold of the network affiliates on the nation's major markets.

Competitive TV

The third period of TV's growth covers the period from 1975 to the late 1980s, and is labeled *competitive TV*. This era was marked by the emergence of cable TV, continued growth on the VHF band, further gains for independent stations, and a fledgling "fourth network," in short—an era of increasing competition for viewers.

The cable revolution, traced in detail in the next chapter, effectively began in 1975 with the national launch of Home Box Office (HBO). The period saw the construction of thousands of new cable systems throughout the country. The number of subscribing households grew from about 15% of the nation's total in 1975 to more than 60% by 1990.

Things were also changing inside TV stations. By the mid-1970s, lightweight portable cameras had replaced bulky studio equipment, allowing more comprehensive coverage of news and sports. The arrival of microwave and satellite equipment enabled news crews to report events live, ushering in electronic newsgathering, or ENG, as it became known. ENG benefited the networks, of course, especially in their coverage of large events like the major political conventions. But the revolution in newsgathering gave a major lift to local stations, whose news departments became enormous profit centers.

Smaller-format video tape (³/₄ inch and 1 inch wide, rather than the 2-inch format of 1950s and 1960s) made editing cheaper, faster, and easier, enabling local stations to assemble packaged reports for later broadcast, either on their news programs or on "magazine" shows. By the early 1980s, most larger television stations also had the capability to report live from the field, using microwave or satellite vans. Television news made money, increased audiences, and—through the celebrity status accorded some reporters, sportscasters, and weather personalities—provided seemingly limitless promotional opportunities for local stations, especially affiliates.

The competitive period was also marked by the rapid rise of independent television stations, a development that can be traced to a number of factors. Paramount among these were technological and regulatory developments which made reception of UHF stations easier. By the 1970s, many American households were replacing their first-generation television sets with new receivers capable of tuning in both VHF and UHF bands.

The growth of the cable television industry also contributed to the rise of the independent stations. Federal regulations required cable systems to carry all signals in their coverage area, including independents. The cable systems often assigned independent UHF stations to unused VHF frequencies; thus "channel 17" was actually seen on channel 10 or 5 on many sets. Newer multichannel cable systems removed tuning to a converter box, where all channels were "created equal." Powerful VHF network stations existed directly alongside previously second-class independents. In addition, the rise of digital tuners on new television receivers made UHF reception easier. On these sets tuning 18 or 48 became as simple as dialing 02.

The growth of independent television was also the result of changes in programming strategies and in audiences. Many independents (as well as some disenchanted affiliates) joined together to provide first-run films, miniseries, and sports events—a process known as ad hoc networking. Such miniseries as *The Bastard* and *Golda* were packaged in this fashion. Independents were also willing to take risks on more controversial programming than their comparatively conservative network competitors. In addition, there was some evidence that the public was beginning to reject imitative and bland network programs. Failure rates for network shows increased; there seemed to be fewer and fewer new hits each year.

By 1986, there was a new competitor for network audiences—the Fox Broadcasting Company, owned by Australian media magnate Rupert Murdoch. Fox succeeded in attracting audiences with its own attractive lineup of programs, including *The Tracey Ullman Show, 21 Jump Street* and more recently *The Simpsons, In Living Color, Married with Children*, and *Beverly Hills 90210*.

Increasingly, audiences tuned to Fox (and away from the three traditional networks) to watch television series, recent films, controversial documentaries, and music and variety shows. By 1993, Fox had grown to the point where it could offer 7 nights of programming each week.

Together, the competition from cable, independents, and some disenchanted affiliates led to an unprecedented decline in viewing of the three commercial networks, a phenomenon known as *network erosion*. The share of viewing at the three major networks fell from over 90% in the 1960s to just over 60% by 1990. The evidence was indisputable: like radio before it, broadcast television was now a mature business. Could it be in irreversible decline?

The Maturation of Television

Like radio, the TV business today has entered a maturity phase. However, it is too early to conclude that broadcast television is truly in decline.

By 1993, there were about 1150 commercial television stations on the air in the United States, including 560 commercial VHF stations and 590 commercial UHF facilities.[1] The percentage of stations affiliated with the three traditional networks was below 80%, the lowest level in television history. And those affiliates were increasingly disenchanted by the program offering coming from the networks. Cancellations, preemptions, and defections to other networks (including Fox) were becoming increasingly commonplace. Strapped for cash, the networks sought to reduce payments to their affiliated stations (see below), a move that placed further strain on the network-affiliate relationship.

As if those problems weren't enough, each of the major networks had been rocked by varying but unprecedented degrees of executive turnover, layoffs, and bitter employee infighting. The overseas news bureaus built proudly in the 1960s and 1970s began to close, rendered obsolete by international news coverage available from CNN, BBC, and other sources. The three networks became a "whipping post" for nearly every problem besetting the TV industry, from claims of excessive violence and an abhorrence of family values, to the high salaries paid to ungrateful superstar performers and athletes.

Things weren't much better at the local level. The flurry of station ownership changes which took place in the 1980s (traced in detail in Chapter 7) left management saddled with debt and high interest payments. The problem was soon compounded by shrinking advertising revenue caused by a deep recession. By the early 1990s, a number of stations and some television chains were facing bankruptcy.

As we will see below, the once impressive and unmatched profit margins of commercial TV stations began to erode. No longer was owning a TV station considered "a license to print money."

THE GROWTH OF NONCOMMERCIAL TELEVISION

Like its radio counterparts, noncommercial television, sometimes called educational television, public broadcasting, or PTV, has had a checkered financial and programming history. While commercial television was proliferating in the 1950s, particularly in the largest cities, public television was struggling.

For one thing, at first there were no federal monies available for the development of an alternative to commercial television. Thus, fledgling PTV stations had to rely exclusively on local donations, grants, or foundation support (largely from the Ford Foundation). While the FCC had reserved nearly 600 channels for educational noncommercial use, most were UHF, particularly in the nation's largest markets, where VHF stations had been granted to commercial broadcasters. Few viewers had tuning equipment capable of receiving the new signals. In addition, the few stations that were in operation faced considerable problems in the area of programming. Without a network system of distribution and with little money to purchase or produce programming, the public stations operated on a limited schedule, mainly offering video coverage of classroom-type activities (programming commonly referred to as "talking heads"). By the end of the 1950s, only 40 public stations were in operation and only a third of these were in the nation's largest 50 markets.

The 1960s marked an era of expansion of public television, owing mainly to increased availability of federal funding. The first federal funding bill for public television was passed in 1962. By 1965, the number of noncommercial television stations had reached 99; 6 of the nation's top-10 markets had public UHF facilities; half had noncommercial VHFs.

The development of noncommercial television reached a turning point in 1967. On the basis of a comprehensive report by the Carnegie Commission on Educational Television, Congress passed and President Lyndon Johnson signed a bill which fundamentally changed the way such stations were financed, organized, and programmed. The Public Broadcasting Act of 1967 called for creation

Jennifer Lawson, PBS programming chief. (Chad Wyatt)

of the Corporation for Public Broadcasting (CPB) to develop noncommercial broadcasting in the United States. Presumably independent from Congress, CPB was formed to oversee the funding and operations of the nation's noncommercial broadcasting system. However, to avoid the concentration of programming that characterized commercial television, CPB was to avoid establishing a central network to feed programs to member stations. Instead, local stations were to retain autonomy by developing their own programs with CPB grants. CPB was mandated to set up a distribution system that would allow member stations to select programs and broadcast them at their own discretion. In addition, it was to provide stations with research, training, and archival support.

Research and development support from CPB, as well as the increasing availability of seed money from its program development fund and a solid tax base at the local level, enabled public television to experience rapid growth in the late 1960s and early 1970s. In 6 years, the number of noncommercial television stations nearly doubled. By 1974, 244 were in operation, including 95 VHFs and 149 UHFs.

The period from the mid-1970s to the present has been marked by two factors which threaten to fundamentally alter noncommercial television: political infighting and a shrinking base of financial support. In 1970, CPB formed the Public Broadcasting System (PBS) as its program distribution arm. In ensuing years, CPB and PBS became adversaries. PBS became an advocate for its member stations, and was frequently at odds with its parent CPB, which was perceived by some to be a bureaucratic monolith. By 1973 the friction between CPB and PBS had become so pronounced that Congress threatened to suspend funding for CPB unless a truce was worked out. In response to congressional pressure, a partnership agreement was signed which emphasized CPB's funding role and which identified PBS as the major means of station interconnection.

Conflict between the two organizations continued throughout the 1970s, and in 1976 CPB called for a new Carnegie Commission to investigate the proper structure, function, and funding sources for noncommercial broadcasting in the United States. The resultant report, commonly referred to as *Carnegie II*, recommended the abolition of CPB and PBS. The former was to be replaced by a public telecommunications trust to guide the overall system and a program services endowment to concentrate on program development. The report also called for a major funding increase, to $1.2 billion by 1985. But in an election year, during a period marked by rising inflation and growing public antipathy toward tax-supported programs such as educational television, few of the recommendations of *Carnegie II* were enacted.

Today, the CPB remains the federal overseer of noncommercial broadcasting, but the rules of the game have changed somewhat. In 1990, Jennifer Lawson was named Vice President of National Programming and Promotion Services. She immediately suspended the station programming cooperative, in which PBS had funded programs by votes of committees of member stations. In its place came the new Public Television National Program Financing Plan. As with the commercial networks, whether a public TV program made it to air could now be traced to a single person—Lawson. Early on, she showed shrewd programming acumen, choosing *The Civil War* to lead off the fall programming season. This landmark series became the highest-rated programming in the history of public television.

Public TV needed a shot in the arm. Like noncommercial radio, public television in the 1990s faces not only the problems of bureaucracy and competition but also an eroding financial base. Shortfalls in the national budget and concerns about the supposed liberal ideology of public television has seen congressional appropriations trimmed (and even the threat of cancellation of CPB allocations altogether). The amount of money available from large foundations and from state and local governments has similarly decreased.

TELEVISION RECEIVERS

While television diffusion does not yet match radio's astonishing figure of over 2 sets per person and 5 per household, the "tube" is undeniably a permanent part of our environment. Figure 5-2 charts the continuing penetration of television sets to United States households since 1950.

As the graph shows, in 1950 fewer than 10% of U.S. households had television sets. In 30 short years that figure had reached 98%! Television advertising executives and public relations specialists are fond of pointing out that there are more working television sets in this country than there are refrigerators and indoor toilets.

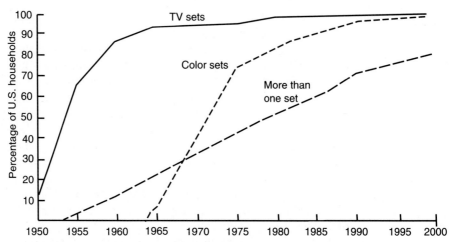

FIGURE 5-2 Television sets in use, 1950–2000. *Source: 1990 Nielsen Report on Television* (Northbrook, Ill.: A. C. Nielsen Co., 1990); 1995 and 2000 projections by the author.

The number of homes with more than one set (multiset homes) has increased at a steady rate throughout the medium's development. By 1965, about 1 in 5 homes had a second set. By 1970, 34% of American homes had two or more sets. In the mid-1970s, the figure rose to 40%. Currently, 2 of every 3 TV households has two or more sets, with the average number of sets per household at just about an even two.

The rise of color television has been just as spectacular. In the 10-year interval between 1965 and 1975, almost 7 out of 10 households acquired a color television receiver. At present, more than 98% of homes in the United States have color television. By the end of this decade, the figure will probably be close to 100%.

Another indicator of the growth of television is the number of channels which TV households can receive. In 1950, the typical TV household received only 3 channels. By 1960, the number had doubled to 6 (most likely three network affiliates, two independents, and one public station). By 1970, many cities had at least one additional independent or educational outlet, for an average of 7 channels of programming per household. The growth of cable (traced in the next chapter) brought the average number of channels up to 10 by 1980. By the early 1990s, the mean number of channels receivable approached 35.[2]

TELEVISION PROGRAMMING

Programming is the fuel which drives the engine of the television industry. There are at least 12 main categories of television programming:

- Newscasts (*CBS Evening News, Action News*)
- Documentaries (*60 Minutes, 20/20, Frontline*)

- Talk/interview programs (*Donahue, Oprah!, Geraldo,* and *Arsenio*)
- Quiz and game shows (*Wheel of Fortune, The Price Is Right*)
- Sports
- Movies (theatrical releases and made-for-TV movies)
- Situation comedies
- Variety shows (including musical variety and comedy variety, from the 1950s, *Ed Sullivan Show* through the 1990s' *In Living Color*)
- General dramas (from *Northern Exposure* to *Beverly Hills 90210*)
- Action-adventure dramas (with the emphasis on action and excitement, including police, war, detective, and western series)
- Music videos
- Animation (from *The Simpsons* to *Ren and Stimpy*)

TV Program Sources

Television managers obtain programming in three main ways: *network, syndication,* and *local origination.* As discussed above, 8 in 10 stations are currently affiliated with a major television network. Typically, stations affiliated with ABC, CBS, or NBC program two-thirds of their airtime with network shows, including the bulk of their afternoon, evening, and weekend schedules. Affiliates of the Fox network fill the bulk of their prime-time schedule with FBC programming. In addition, Fox has begun to expand its programming in the late-night block and on Sunday afternoons, with its addition of NFL football.

Syndication is the process whereby television stations obtain programming from program suppliers other than major networks. Syndication takes two main forms: *off-net* and *first-run.* Off-net syndication refers to programs which ran first on a major network. Examples include situation comedies such as *M*A*S*H, The Cosby Show, Designing Women,* and *Murphy Brown,* and dramas such as *Cagney and Lacey, Murder, She Wrote,* and *McGiver.*

First-run syndication refers to original series produced expressly for syndication to local television stations. Recent examples include the musical-variety program *Star Search* and the show-biz daily *Entertainment Tonight.* Most first-run syndicated programs are talk, quiz, or game shows, designed to run every weekday. Such programs are commonly called *strip* syndication programs because they are bought and programmed in units, or strips, of five (for example, Monday to Friday, 6:30 to 7:00 p.m.). Examples include *Wheel of Fortune, Family Feud, People's Court, Donahue,* and *Oprah*!

Local origination refers to the original programming produced by individual stations. The high cost of production and limited resources usually restrict local origination to news, talk, and sports programming. However, in recent years, a number of local stations have tried their hands at entertainment shows, including music video, comedy, and serious drama.

From a programming standpoint, managing a network affiliate is different from managing an independent station. With the lion's share of programming provided by the network, most ABC, CBS, and NBC affiliates expend the bulk of their programming efforts on local news. Of secondary importance is the acquisition of syndicated products to run near the prime-time schedule.

The situation is different for independents, especially those not aligned with Fox. Their paramount concern is to obtain sufficient syndicated products to fill their airtime and attract an audience. Film packages and off-network programming are the most sought-after commodities, followed by strip programs. With most of their program budget tied up in syndicated inventory, few independents have the financial and talent resources to originate local programs other than play-by-play sports coverage and low-budget discussion and children's shows. This situation is changing, however, as independents continue to gain clout among audiences and advertisers. In addition, two new networks (from Warner Brothers and Paramount) offer additional opportunities to obtain first-run programming.

VIEWING TRENDS

Like radio, television in the United States has achieved an astonishing rate of use. Figure 5-3 charts patterns of television usage throughout the day. All told, in the typical American household, the television set is on for more than 7 hours per day (compared with about 6 hours in 1970 and 5 hours in 1960).

Viewing levels rise throughout the day. Between the hours of 8 and 11 p.m. EST in the fall and winter months, between 60% and 70% of the homes in the United States are watching. For this reason, the period is known as *prime time*, and it represents the most important revenue source for television managers.

As might be expected, the time slots immediately preceding and following prime time offer the next largest audiences and, therefore, are next in importance to managers, sales personnel, and advertisers. The hour before prime time (7 to 8 p.m. EST) is often referred to as *access* time, since an FCC edict returned the period preceding prime time to affiliated stations to provide access for new and different types of programs. The large audience available (nearly 6 in 10 TV homes) and the absence of network programming mean that access time represents a significant profit center for commercial broadcasting. Many stations use this time period to air their most popular syndicated programs, such as *Jeopardy* and *Wheel of Fortune*. Another popular program in this slot is the previously mentioned *Entertainment Tonight*.

Early evening time is the period immediately preceding access time (6 to 7 p.m.). Following prime time is *late fringe* (11:00 to 11:30 p.m.). Fringe can be an extremely profitable time period for stations. Audiences remain comparatively large, and, equally important, network stations have *all* the commercial positions available for local sale. For this reason, in many markets, local news alone can account for as much as one-third of the affiliate's advertising revenue.

The importance of late fringe explains the intense competition between stations in the news area, as well as the high salaries paid news anchors, weather forecasters, and sports reporters (if they have sufficient ratings, of course).

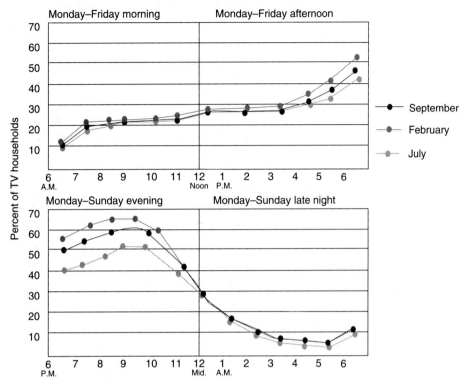

FIGURE 5-3 Television viewing throughout the day. *Source: 1990 Nielsen Report on Television* (Northbrook, Ill., A. C. Nielsen Co., 1990); used with permission.

The time slots attracting the next-largest aggregate audiences are *daytime* (9 a.m. to 4 p.m.) and *early fringe* (4 to 6 p.m.). On average, between 20% and 30% of U.S. households view television during these time periods. In recent years, the number of daytime viewers has increased. More employees are on split or night shifts, making larger audiences available during the day. Television sets have become more portable and, like radios, often accompany their owners to the workplace. In addition, daytime programming has become increasingly popular as soap operas have heightened the sophistication of their subject matter and their production values.

The number of television viewers is fewest in early morning (7 to 9 a.m.) and late night (11:30 p.m. to 1:00 a.m.), although sizable audiences are still available in both time periods. While morning news and entertainment shows such as *Today* and *Good Morning America* have attracted a loyal following, early morning still belongs primarily to radio. On average, only 10% to 20% of homes use television in the morning (compared with up to 35% using radio at this time).

There are more late-night than early-morning viewers. At midnight, the set is still on in as many as one-fourth of all homes, making the *Tonight* show, its many competitors, and the late movie U.S. institutions.

The competition in the late-night period stimulated by David Letterman's departure from NBC to CBS in 1993 illustrates management's increasing attention to this profitable daypart. While the aggregate size of the audience in late night is much less than that in prime time or fringe, late-night audiences are attractive to advertisers for their levels of education, attention, and disposable income.

TV Seasons and Sweeps

Television use varies according to season. Usage peaks in February, when it approaches 8 hours per day, and shrinks to just above 6 hours per day in July. Consequently, a critically important time for programmers and advertisers is the January–February *sweeps* period, when the national ratings companies are monitoring the viewing patterns of the nation's television viewers (some of the highest-rated programs in history have been aired during this period, including various Super Bowls and the famous ABC miniseries *Roots*).

Results of these ratings sweeps help determine the advertising rates for the remainder of the year. The down period in the summer enables the networks to experiment with untried performers and programs, and this is when *pilots*, episodes of prospective new series, are run. However, one of the more successful programming strategies employed by the Fox network has been to avoid summer reruns and counter ABC, CBS, and NBC with first-run episodes of prime-time series.

A TV pilot in production. Most pilots premiere in the summer months, outside of the key sweeps periods. (Jeffrey D. Smith/Woodfin Camp)

Household Viewing

Television usage varies according to household income, household composition, social class, and other factors.[3] The number of hours spent viewing is greatest in inner-city and minority households. Women and senior citizens watch the most television; men and teenagers the least. Interestingly, education is not a good predictor of television viewing habits; college graduates watch as much television as those who did not complete grade school!

As with radio, there are noticeable gender and age differences with respect to program preferences. Men prefer news and information shows, sports, and action-adventure shows. Women are more likely to enjoy general dramatic programs, movies, and situation comedies. The first choices among teenagers are situation comedies and music television, followed by dramatic series and movies. Not surprisingly, younger children enjoy cartoons and situation comedies.

Of concern to many parents and educators is the fact that some shows that rank high among children have been cited as the most violent programs on TV, including professional wrestling, cartoon series like *Teenage Mutant Ninja Turtles*, and syndicated fare like *American Gladiators*. The debate between the TV industry and child advocates has spurred legislation and other attempts at remediation.

Research on the audience for public television indicates that about half of the U.S. population watches at least one program per week. Public television audiences tend to be upscale in their education, income, lifestyle, and leisure pursuits. Of concern to public television programmers is the fact that the profile of the PTV audience matches many of the characteristics of the users of new technologies such as cable and pay cable, VCRs, and home computers. While the number of paid subscribers to public TV is increasing (to about 5 million), there has been some evidence of overall audience slippage in recent years, as cable networks like Discovery and Arts & Entertainment emulate PBS in the areas of nature documentaries, expanded talk shows, and international coproductions of dramatic series and films.

In sum, the data on set penetration and use indicate that television, like radio, is approaching saturation in diffusion, as measured by sets available, channels receivable and aggregate household viewing. And, like their counterparts in radio, television managers are not sure that the profit potential for the medium will keep pace with the continued proliferation of stations, receivers, program types, and audiences.

TV REVENUE TRENDS

Figure 5-4 summarizes revenue and expense trends in the television business today.

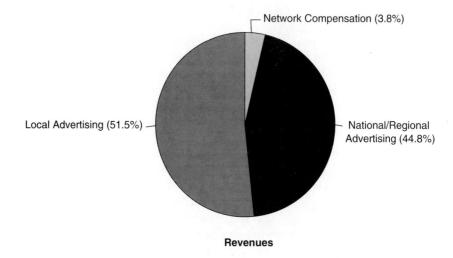

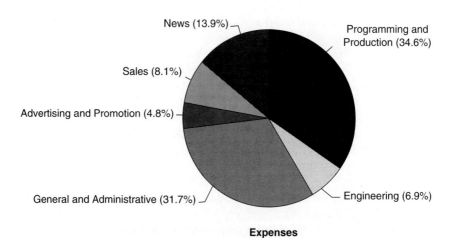

FIGURE 5-4 TV station revenues and expenses. *Source: NAB/BCFM 1990 Television Financial Report* (Washington, D.C.: National Association of Broadcasters, 1990); used with permission.

Network Compensation

The payments made by the national networks to their affiliated stations account for about 5% of the income of local affiliates. In the early 1990s, this network compensation averaged between $600,000 and $1 million per affiliated station nationwide. Airing their programs in the nation's largest cities is of primary importance to the networks. Therefore, the larger the market, the more money a network will pay to its affiliate in that market for airing its shows.

Affiliates in New York, Chicago, and Los Angeles can earn $15 million or more per year in network compensation from one of the "big three" networks: ABC, NBC, or CBS.

Network Formulas

The amount of network compensation paid to station affiliates depends on a number of variables, and the three major networks differ somewhat in their computations of the compensation rate. However, all computations are based on the following principles.

Network Station Rate (NSR) NSR is a flat fee which estimates the value of an hour of a station's air time. Since NSR is determined by a station's market size, coverage area, and popularity, it ranges from as little as $100 in the smallest markets to $10,000 or more in a major city like New York or Los Angeles.

Hourly Percentage, or Equivalent Hours As discussed above, certain parts of the broadcast day are more valuable than others to the broadcaster. Consequently, the network pays more to its affiliates for prime viewing times than it does for less valuable time periods. An *equivalent hour* is a clock hour multiplied by a percentage estimating the value of the time period in which it falls. For example, from 6 to 11 p.m. an equivalent hour is 100% of a clock hour, from 5 to 6 p.m. it may be 50% of a clock hour, and from 9 a.m. to 5 p.m. it may be computed as 35% of a clock hour.

Once the number of clock hours during which an affiliate carries network programming is adjusted to reflect equivalent hours, it is multiplied by the NSR to determine the station's gross compensation.

Overhead Expenses Each network deducts a portion of the station's compensation for its programming and operations expenses. In the traditional CBS formula, network overhead has been valued at between 8 and 10 equivalent hours per month; at ABC, the amount has been a multiple of the network's monthly station rate.

As profits at ABC, CBS, and NBC slipped in the late 1980s and early 1990s, affiliate compensation was one area targeted for cost cutting. Each network reduced payments to its affiliates, with more reductions anticipated in future years. For example, NBC introduced a "performance-based" sliding scale of compensation, which increased payments to stations earning higher ratings (and lowered compensation to underachieving stations). ABC followed suit with its own revised payment schedule, rewarding stations which carry, promote, and deliver high audience ratings for network programming (and offering less to those affiliates which frequently preempt network shows).

CBS caused the most consternation among its affiliates. The network proposed eliminating compensation altogether on Sunday night and on Monday from 9 to 11 p.m. (corresponding to the time slots of its most popular programs, including *60 Minutes, Murphy Brown,* and *Northern Exposure*). In addition, the network sought to increase its overhead deduction. Relations with disgruntled affiliates were strained as never before. Ultimately, in 1993, CBS backed down on its proposal to substantially reduce its affiliate compensation, opting instead for a more gradual reduction. The good news was that CBS prime-time ratings rebounded strongly, making carriage irresistible to most affiliates (and very attractive to their advertisers).

Fox Television

As a fledgling TV network, Fox has compensated its affiliates at a much lower rate than the big three. The Fox formula depends upon the financial performance of the parent company. Fox shares about 5% of its gross revenues with its nearly 130 affiliates. As with the major networks, market size and ratings are important. The Fox formula factors in the percentage of TV homes in the market plus the rating success of the station in that market. In 1990, total Fox compensation to its affiliates was about $10 million (roughly the amount CBS, NBC, and ABC paid to their stations in the top-10 TV markets alone). However, unlike the traditional networks, Fox plans to increase payments to its affiliates in coming years. The move will provide Fox's lineup of independent and UHF stations with a critical revenue stream, especially now that the network has expanded its schedule to a full 7 nights.

Why Affiliate?

Overall, network compensation is a small (and mostly shrinking) part of station income, of value mainly to TV stations in the nation's smaller markets. In 1985, network compensation accounted for about 10% of an affiliate's income. Today, affiliates receive some 5% of annual revenue from this source. Thus the value of an affiliation contract is more than monetary.

Why affiliate? As we have seen, network affiliation allows stations to fill up their airtime. It enables stations to present a variety of popular and special programs, and it generates strong lead-in and adjacent audiences for its local news and syndicated programs. Finally, network affiliation allows stations to "piggyback" on the considerable promotion and publicity value of the networks and their leading stars.

Barter

One perennial problem for stations not affiliated with a major network has been acquiring programming. Being newer, harder to find, and generally lower in stature than network affiliates, independent stations have always had cash-flow problems. The situation is compounded by the need of independents to fill the bulk of their own airtime.

Lacking money to purchase programming, independents and program suppliers have moved to a barter system. In barter, programs with national advertisements already included in the show are provided outright or at reduced cost to stations. Recent years have seen an unprecedented increase in barter syndication, particularly among independent stations. In the typical barter arrangement, strip newsmagazine and entertainment shows are provided by satellite to local stations, with a certain number of national advertisements presold. Today, bartered programming represents nearly half of the daily schedule at the typical independent station.

Barter is not unique to independent stations. The ownership changes, debt load, and depressed advertising market in recent years have left network affiliates short of cash to acquire syndicated product. By the early 1990s, many of the popular syndicated original and rerun programs most commonly seen on network affiliates, from *Entertainment Tonight* to *Oprah!* and *Donahue*, were being sold to stations on a barter basis.

As in radio, barter arrangements in TV may also involve trading advertising time to retailers in return for merchandise. In television, barter is frequently used to outfit newscasters, to provide vehicles for news crews, and to obtain meals or entertainment for sales and management personnel.

Like network compensation, barter for both programming and merchandise represents only a small percentage of station revenue. In the early 1990s, barter accounted for about $500,000 per station, or 3% of gross station revenue. As noted earlier, barter is of more value to independents than to affiliates, accounting on average for over $800,000, or 5% of total station revenue.

Advertising Revenue

With network compensation and barter comprising less than 10% of revenue, the major income source for commercial TV stations is the sale of commercial time to network and national, regional (spot), and local advertisers. The volume of television advertising expenditures from 1965 to 2000 in each of these categories is depicted in Table 5-1.

As in the radio business, trends in television advertising reveal a shifting pattern. All told, the volume of TV advertising has risen spectacularly in recent years. From 1975 to 1990, total TV ad revenues grew fivefold, from just over $5 billion to over $25 billion.

Network Advertising Sadly (at least to CBS, NBC, and ABC), the major networks have not been the primary beneficiaries of this largesse. As has been widely reported, network television is a mature advertising business. While network revenues increased by a factor greater than 3 between 1975 and 1985, there

TABLE 5-1 TELEVISION ADVERTISING VOLUME 1965–2000
(Millions of Dollars)

Year		Net	Spot	Local	Total
1965	$	1237	892	386	2,515
	%*	49	35	15	
1970	$	1658	1234	704	3,596
	%	46	34	20	
1975	$	2306	1623	1334	5,263
	%	43	31	25	
1980	$	5130	3269	2967	11,366
	%	45	29	26	
1985	$	8060	6004	5814	19,778
	%	41	30	29	
1990	$	9383	7780	7856	25,019
	%	38	31	32	
1995	$	9735	8555	11,210	29,500
	%	33	29	38	
2000	$	10,800	9450	13,500	33,750
	%	32	28	40	

*Percentages may not equal 100 because of rounding.
Source: McCann-Erickson, Inc., New York. Reported in the *Statistical Abstracts of the United States*, 1984, pp. 567–568; *TV Dimensions '92*, p. 18; 1995–2000 provided by the author.

was little change from year to year thereafter. By 1995, network advertising, representing nearly half of all TV revenues in 1970, may slip to a third or less of the total TV advertising marketplace.

Spot TV Sales National and regional spot sales have remained steady over a generation, representing about a third of all TV ad revenues from the 1970s to the 1990s. The volume of spot sales has increased in the period. In 1980, spot TV represented about $3.3 billion in annual advertising expenditures. A decade later, the spot business was billing nearly $8 billion annually. By 1995, the spot business is expected to exceed $8.5 billion, representing just under 30% of total TV ad spending.

Local TV Sales Where has the highest proportion of TV ad dollars gone? As in radio, to local sales. In 1970, at the height of the "network heyday," local advertising was a $700-million business, less than 20% of total TV ad volume. By 1980, local sales represented just under $3 billion (26% of total ad expenditures). In 1990, local sales eclipsed national/regional spot, becoming an annual business that grossed nearly $8 billion, almost a third of all TV ad dollars! With further erosion in network advertising expected and spot advertising expected to grow minimally at best, in the 1990s TV sales should become increasingly localized. At mid-decade, local sales may exceed $11 billion annually, representing nearly 4 in 10 total TV ad dollars.

Revenue by Type of Station   The amount of advertising revenue generated by television stations is largely dependent upon the type of facility. Affiliated stations obtain about 50% of their revenues from local sales. Another 45% of advertising revenues comes from national/regional spot sales made by their rep firm. The remaining 5% comes from network compensation.

The pattern differs slightly for independent stations. More than half (55%) of an independent's revenue comes from local spot sales. Almost the same proportion of revenue as at an affiliate (43% compared with 45%) emanates from national/regional spot advertising. Lower compensation payments from Fox TV and small regional networks (mostly sports) account for only 2% of the typical independent's gross revenue.

TV EXPENSE TRENDS

Like radio, television has become an increasingly costly enterprise, marked by spiraling expenses in all operating areas. (Refer again to Figure 5-4.)

Program and Production Expenses

In recent years, program and production costs, which account for nearly half of a station's expenses, have experienced a sharper rate of increase than any other area. The increasing number of stations, particularly independents, the proliferation of new delivery systems, including cable television, and the failure of many network shows to achieve lengthy network runs (and thereby generate sufficient reruns to sell in syndication) have created a shortage of programming product. As a result, the syndication business has become a seller's market, with stations paying premium prices for fewer and fewer titles.

Off-network programs with a solid ratings history are becoming increasingly rare, and when these are offered for syndication, they command record prices. In the late 1980s, *The Cosby Show* set syndication records in virtually all markets, garnering over $100,000 per week for reruns in the nation's largest cities. In the early 1990s, several other popular programs made their way to profitable off-net syndication, including *Northern Exposure, The Simpsons,* and *Murphy Brown.*

Local production costs also continue to escalate, particularly in the highly competitive news battleground. Though budget cuts have caused some salary slippage lately, news anchors, weather forecasters, and sportscasters with a strong track record in the ratings can command salaries normally associated with show-business celebrities and professional athletes. In addition, stations have invested "megabucks" in state-of-the-art news facilities: microwave vans, satellite uplink trucks, portable cameras, computers for generating graphics and meteorological data, and other hardware. In fact, among large-market affiliates, news budgets in excess of $10 million, representing as much as one-third of the total expense dollar, are typical.

Together, the high costs of syndicated and original programming (mainly news) have produced the sharp rise in programming costs experienced by local television stations.

General and Administrative Expenses

Typically accounting for one-third of a television station's expenses, general and administrative costs have been rising as a result of increases in utilities, postal rates, and other matters related to the conduct of business. A component of this increase has been the proliferation of office automation, including computers, fax machines, and photocopiers. The costs of employee benefits have skyrocketed, particularly in medical and related insurance coverages. Another spiraling administrative cost has been interest payments, left over at many stations from the frenzied flurry of station buying and selling in the 1980s. Many managers face daunting debt loads, as their collateralized loans (negotiated with little cash down) are coming due for repayment to increasingly eager banks and other lending institutions (see Chapter 9).

Technical/Engineering Costs

Engineering costs, which account for about 7% of a station's expenses, have also been rising rapidly. There is a serious shortage of competent television engineers, and with the explosion of new designs and formats for production equipment, the shortage may become critical. Today's television engineer must be familiar with satellite communications, microelectronics, laser optics, even computer programming. Consequently, stations are willing to pay generous salaries to their engineers (often to keep them from defecting to the computer and other microelectronics fields), and frequently retain the services of costly consulting engineering firms. And, as is true for the radio industry, the prices of equipment, spare parts, tubes, transistors, diodes, and even light bulbs continue to escalate, with no end to the rise in sight.

Sales and Promotion Expenses

The increase in competition in the television medium has caused a concurrent increase in sales expenses, which are now about 13% of a television station's expenses. Stations are finding it more costly to train and retain competent account executives. The process of matching potential advertisers to a station's programming formats has become extremely refined statistically, requiring the services of outside consultants and research services (see Chapter 14).

Ratings services alone can cost a station more than $1 million per year. Moreover, it is not uncommon for a television station in a major market to have four or five different computer systems in the sales department, each providing different services for use by account executives. As with radio, the costs of entertaining prospective clients and of traveling to sales representative firms, regional meetings, and training seminars are also on the rise.

Today's TV managers face daunting competition from other stations, cable channels, even VCRs. Thus, increasing dollars are being earmarked for marketing and promotion activity. As we will see in Chapter 13, annual promotion budgets in excess of $500,000 are commonplace in TV today. Major-market stations spend $2 million or more annually to keep their station "top of mind" with viewers.

PROFITS AND PROFIT MARGINS

For many years, television stations were considered "recession proof." In good times and bad, the volume of TV advertising climbed, and TV stations, especially affiliates, ranked at the top of American industry in profitability. Profit margins, normally between 5% and 10% for many successful businesses, were typically above 50% in TV. As stated earlier, managing a TV station was often seen as "a license to print money."

Financial reality has caught up with television. Recessionary periods in the late 1970s, early 1980s, and the first years of this decade have had a discernible impact on television's profitability. Still, the medium can be very profitable for some types of stations. Figure 5-5 summarizes profit trends in the industry.

Television profit margins began a precipitous slide in 1986, a trend that continues to this day. In 1985, the typical television station in the United States returned a sizable 26% return on investment. By 1990, the average television

FIGURE 5-5 Television station profit trends, 1975–2000. *Source: FCC Annual Report* 1981 (Washington, D.C.: U.S. Government Printing Office, 1981); *NAB/BCFM Television Financial Reports*, 1981 to 1990 (Washington, D.C.: National Association of Broadcasters). Data are aggregated for all types of stations; projections to 2000 by the author.

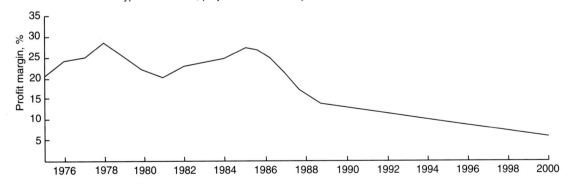

profit margin had slipped to about 10%. Still, industrywide that is a healthy rate of return, especially considering the poor performance of radio and other industries in the period.

On the whole, television stations remain profitable ventures. Industrywide, more than 7 in 10 stations annually report a profit. Unlike radio, television provides a greater return on investment than an entrepreneur might expect from more traditional sources such as securities and bonds.

However, not all types of stations share equally in the profit pie. The lion's share of TV profits goes to affiliated stations, especially those owned by the major networks and such leading broadcasting groups as Tribune, Gaylord, and Gannett. Their favorable dial positions, their location in the nation's largest markets, their strong connections to powerful parent corporations, and other factors make these stations cash cows. Annual profit margins for network-owned stations and for affiliates owned by the major broadcast groups typically exceed 40%.

According to the National Association of Broadcasters, the industrywide average profit margin for all affiliated stations in 1990 was a sizable 22%. In the top-10 TV markets, affiliates returned on average 28 cents of the revenue dollar in profit. Independent stations can also be profitable, especially those in a big market owned by a major media company. The leading stations of this type are owned by Fox (including WNYW in New York and KTTV in Los Angeles) and Tribune (including WGN in Chicago and KWGN in Denver). In 1990, major-market independents reported annual profit margins in the range of 14%.

UHF stations, especially those in smaller markets, remain second-class citizens in the television business. Nationwide, most UHF stations, both independent and network-affiliated, reported operating at a loss in the early 1990s. In fact, the majority were for sale (but there were few buyers). Only UHFs in the nation's top markets were operating in the black, and these were returning a mere 2 cents on the dollar.

The long-term future for television profitability is mixed. It can be expected that network affiliates in major cities will remain highly profitable. At the same time, major-market independents, especially those carrying Fox programming, should see their profit margins increase. In medium-sized to small cities, however, both affiliates and independents face diminishing returns, as competition for the attention of viewers and advertisers continues to increase.

THE TV EMPLOYMENT PICTURE

The television industry attracts the interest of thousands of college students who select mass communications, journalism, or broadcasting as their academic major. However, even though the television work force is considerably larger than that of radio, the TV business remains a difficult one for entry, advancement, and financial success.

The sheer number of TV stations and sets available may suggest that the industry is one of the nation's largest employers. Actually, TV is more of a "cottage" industry. The total television work force is only about 100,000 people.[4]

In contrast, over 2 million people fuel the publishing business, over 700,000 still work in the automobile industry, and over 250,000 are employed in the soft-drink business.[5] In 1990, commercial television stations employed about 70,000 people. Additionally, close to 25,000 people worked for commercial networks, production companies, and allied industries. Reflecting the lean economic times in the industry at large, there was evidence of work force shrinkage in the early 1990s.

Like radio, television is largely a white-collar profession. There are few smokestacks and assembly lines in TV facilities! However, the technical complexity of TV has traditionally made crew and engineering jobs the most common and available professions. Today, there are over 21,000 engineers and technicians at work in television stations. Performers, news anchors, producers, and other so-called professionals hold about 20,000 positions. There are about 10,000 people in managerial positions, 9000 in clerical positions, and nearly 6000 in the TV sales force.

The employee-management picture for television is quite different from that for radio. First, there is a much greater gap between white-collar workers and blue-collar workers in television. In the larger cities, some television blue-collar workers are represented by unions and crafts groups (see Chapter 10), and managers must be highly skilled in collective bargaining and labor relations. Second, while radio generally pays less than television, it does give more titles and power to its employees. In radio, 1 in 4 employees is a manager; in television only 1 in 7 employees can list "manager" on a business card. The power hierarchy is much more rigid in television management, and widespread departmentalization leads to fewer and more powerful (and better-paid) managers. Finally, the aggregate sales force in television is considerably smaller than in radio (8% compared with 22%), and TV's sales teams therefore divide a much larger advertising revenue pie (see Table 5-1). It is therefore not surprising that TV account executives often make extremely good salaries, usually above six figures.

Women and ethnic minorities have made affirmative action and equal opportunity gains in television. Women hold nearly 25,000 jobs in the commercial television industry, representing almost 37% of the total work force. Minorities hold down about 13,000 jobs, or 19% of the total. Gains have been greater for women than for minorities. In the years between 1985 and 1989, for example, female employment was up about 2%, while minority gains represented an increase of less than 1%. For both minorities and women, the greatest gains were in the sales and professional categories. However, both groups are still most visible in their traditional roles: 9 in 10 secretaries are women, and 4 in 10 television laborers are from minority groups.

Not surprisingly, employment tends to follow the order of profitability discussed earlier in this chapter. Network-owned and -affiliated stations in the nation's largest markets have the largest staffs; independents, particularly UHF independents in small markets, tend to "run lean," in a manner similar to small-market radio.

For example, on average, top-10 market affiliates employ over 250 people, with more than 75 in the news department alone. Affiliated stations in midsize markets employ from 75 to 100 full-timers; small-market affiliates usually operate with less than 50 full-time staffers. Independent stations maintain about half as many employees as do network affiliates. Major-market independents report full-time staffs of about 125; "indies" in midsize markets, about 40. Television stations in the smallest markets often have 40 or fewer employees.

Table 5-2 presents figures for the annual salaries of typical television station personnel. The main salary trend is obvious: television sales is a lucrative profession. The salary of an account executive in a small television station exceeds that of the production manager or music director at the largest radio stations! The next most lucrative salaries are in engineering and news. But note that the high salaries available to news directors and personalities are a direct function of market size. The news anchor who shifts from a medium market to a top-10 market can expect to double or triple her salary. A small-market weather forecaster who lands a position in a big city may increase his pay fourfold.

The greatly sought-after crew positions favored by many broadcasting students remain, like radio disc jockey positions, at the bottom of the salary ladder. Across all markets, the range of compensation paid producers, directors, and floor managers remains comparatively paltry—conspicuously close to the salaries paid secretaries, janitors, and general laborers.

It should be clear from Table 5-2 that the competition to land a job in a large market is fierce. Most television employees, especially on-air personalities, keep one eye on their copy and one eye on the trade listings of positions in larger markets. It is therefore no surprise that the turnover rates are high among both supervisory and nonsupervisory employees in the TV industry. Among supervisors

TABLE 5-2 TELEVISION EMPLOYEE COMPENSATION

| | | Market | | |
Job	National average[a]	Large*	Medium†	Small‡
General manager	113,347	193,801	113,365	74,655
Sales manager	91,848	144,995	84,309	60,861
Program director	43,922	82,686	41,873	26,149
News director	58,152	130,324	58,156	33,963
Research director	38,702	52,541	37,342	—
News anchor	64,763	283,583	52,529	29,211
News reporter	29,035	108,768	25,833	16,623
Producer/director	26,410	57,662	25,077	17,706
Floor director	22,594	36,162	20,220	13,886

[a] Median.
* Markets 1–10, affiliate stations.
† Markets 51–75, affiliate stations.
‡ Markets 151 and above, affiliate stations.

Source: *1991 Television Employee Compensation and Fringe Benefits Report* (Washington, D.C.: National Association of Broadcasters, 1991); used with permission.

and managers annual turnover exceeds 15% in major markets, and 1 in 5 department heads leaves or is fired each year.[6] In talent, sales, and technical positions, turnover averages 13%. Staff turnover is greatest in the smaller markets, where nearly 20% of employees change jobs annually.

Turnover rates present a "good news–bad news" situation for people seeking to embark on a career in television. The good news is that at a sizable number of stations, as many as one-third of all employees will lose their jobs in a given year. The bad news is that once you are hired, your job may be the next one that turns over!

Finally, the benefits package provided by most television stations tends to be more comprehensive than that provided by radio stations: 60% of television stations pay full medical and surgical benefits for their employees; 25% provide dental services as well.[7] One-third of all stations make tuition reimbursements, and 6 stations in 10 pay employees to attend clinics, workshops, and other staff development programs.

As in the radio business, TV employee benefits are correlated with market size. For example, only 15% of TV stations in small markets pay for an employee retirement plan, compared with 70% in the top-10 markets. Similarly, dental coverage is virtually unheard of in small-market TV, whereas half of the nation's largest TV stations pay dental benefits in full. Clearly, the historical trend of employees moving from small stations in small markets to larger stations in bigger markets in search of improved salaries and benefits remains.

SUMMARY

From the post–World War II period to the present, the television business has experienced steady growth in stations, sets in use, audiences, advertisers, profits, personnel, and payroll.

The television business has evolved in four distinct stages. The first period, from 1948 to 1960, was its "live" era. Relatively few stations were operating, and those were mainly in the nation's largest cities. Virtually all stations maintained a network affiliation with the industry leaders, NBC and CBS. During this golden age of TV, ABC affiliates, independents, and educational stations, many located on the UHF band, struggled for survival.

The second period of television growth was from 1960 to the early 1970s. This was the heyday of network television. The major networks moved production from the East to the West Coast, from live to filmed and taped production. ABC became competitive, mostly on the heels of successful youth-oriented programs. The nation's system of public television burgeoned as more federal and private monies were allocated and as the baby boomers entered college.

The third period of television growth, which began in the 1970s, can be characterized by a single word: competition. On the commercial side, the number of on-air stations grew to exceed 1100. Network affiliates were no longer the sole profit centers in television. Independents, in both the VHF and the UHF bands, became increasingly popular with viewers and advertisers.

6

THE CABLE BUSINESS

We are living through an era of unprecedented development and change in communications. At the vanguard of that change has been the wiring of the United States: the growth of cable service from an adjunct of over-the-air TV to a major player in the telecommunications business. This chapter traces the economic and business climate for cable management and examines the new participants in the cable business: the regional Bell operating companies (RBOCs).

The goals of this chapter are:

1 To trace the growth of cable television by examining number and types of operating systems, subscriber rates, and household viewing trends
2 To examine revenue, expense, and profit trends in the cable and pay television businesses
3 To examine the makeup of the cable television work force
4 To describe the regional Bell operating companies and examine the potential impact of RBOC entry into the field of home delivery of television and related programming

GROWTH OF CABLE

From its origins in the early 1950s to its maturity in the 1990s, cable television experienced rapid growth. The rate of cable growth can be tracked by number and location of operating cable systems, percentage of television homes in cabled areas, basic and premium subscriber rates, and viewing trends in cable households.

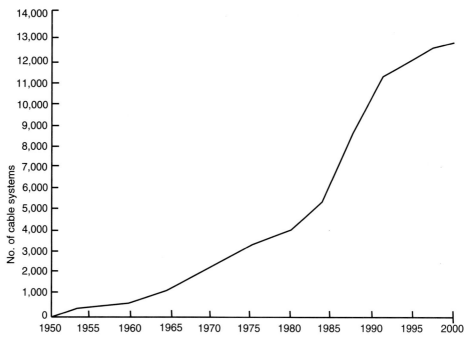

FIGURE 6-1 Cable systems in operation, 1950–2000. *Source*: National Cable Center and Museum *Marketing New Media*, October 19, 1992, p. 4; projections to 2000 by the author.

Operating Systems

Figure 6-1 documents the rise in the number of operating cable systems between 1950 and 2000. The growth of cable television systems, like the growth of broadcasting systems, occurred in three main stages, followed by a maturation phase. The first, spanning the early 1950s to about 1970, was the *community-antenna* phase. Cable systems existed predominantly to improve reception in mountainous areas and to relay big-city television signals to rural communities not reached by over-the-air broadcasting. Management in this era was largely a turnkey operation, involving comparatively little overhead and few personnel. Most cable managers were technicians and engineers. Few of the less than 2000 systems derived their income solely from the cable business; most cable operators also owned television and appliance dealerships, movie theaters, television stations, or public utilities.

The second phase of cable development was an era of spectacular growth, or *buildup*. Between 1970 and 1990, the number of systems quadrupled. More than two-thirds of the nation's households were in a cable area, and a large percentage elected to subscribe. The emphasis on cable programming shifted from improved reception and imported signals to exclusive movie and sports presentations.

Cable management shifted its focus from engineering and technical expertise to administrative and financial skills. Managers needed these skills to steer the growth of their systems—from planning capitalization for building and expansion to administering the increasingly complex subscriber billing process. Cable managers had to be skillful negotiators to deal with equipment and program suppliers, local governments, and consumers. During the buildup phase, a new class of cable managers emerged. Industry heads such as Ted Turner and Trygve Myhren, Ralph Baruch of Viacom, Jack Clifford of Colony Communications, Kay Koplovitz of the USA network, and Gerald Levin of Time Warner were, like the top broadcast executives, innovative entrepreneurs, shrewd negotiators, and charismatic leaders.

From the early 1990s to the present, cable has been in its third phase, a period marked chiefly by *retention* tactics. The industry faces a slowdown in construction, competition from other technologies such as home video, and rising consumer dissatisfaction. The emphasis is no longer on growth, but on consolidation and marketing. Retaining subscribers is the key to cable survival.

Like radio, cable is being increasingly packaged for specific demographic groups, such as young professionals, kids, and senior citizens. It can be argued that the cable business, like television, is now a mature enterprise: vulnerable to increased competition from new suppliers. More on this later.

More than 11,000 cable systems currently provide service to homes in the United States. Many of these systems are owned by large corporations with substantial cable interests. Companies which operate more than one system are known as multiple-system operators (MSOs). The leading MSOs run over 200 systems each and serve between 1 and 10 million subscribers. (For a listing of the top-20 MSOs, see Chapter 7.)

Interestingly, the growth of cable television proceeded in a direction quite different from that of over-the-air broadcasting. Radio and television stations first sprang up in large metropolitan areas. Then, with increased power and better transmission equipment, service was expanded to include suburban and rural areas. Cable, on the other hand, began in the most inaccessible rural areas, then spread to affluent suburbs. Only recently was cable introduced into major metropolitan areas.

Penetration and Subscriber Rates

The key to cable's growth is its subscriber base. As Figure 6-2 shows, there is a clear hierarchy involved in cable's quest for audiences. The bottom rung of the audience ladder is homes passed, the total number of television households living in cabled areas. This represents cable's potential audience and thus the limit of available income from subscriber fees.

The second level of cable audience growth is the number (or percentage) of basic subscribers. These are the families in cabled areas that elect to pay the monthly fee for basic cable service. Basic service typically includes local chan-

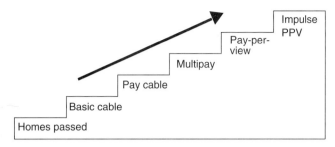

FIGURE 6-2   The cable subscriber ladder: Homes passed to impulse pay per view.

nels, regional television stations, a community bulletin board, satellite-delivered superstations, and advertiser-supported cable networks.

The third level of cable audience is the premium household—the pay-TV user. These are households which subscribe to a pay service for an extra monthly charge. The most common pay services are movie channels such as HBO and Showtime.

The next level is the elite cable audience known as multitiered, or multipay, households: those that subscribe to more than one pay service. Next come the pay-per-view households: subscribers that select special premium offerings for an extra fee. Pay-per-view events have included recent films, theatrical and musical performances, and boxing matches.

At the top of the ladder are the most lucrative households for the cable operator: multievent pay-per-view. These homes make more than one PPV pur-

The cable installer on the road. (courtesy of TCI)

chase per month, often on an impulse basis. ("That looks interesting; I'll order it!") As we will see, cable management is going to great efforts to capitalize on impulse PPV, by installing newer-model addressable converters, by improving automated billing processes, and by using new digital-compression technologies to add dozens of additional pay-per-view channels.

Cable's growth in the 1970s occurred mainly at the first two audience levels. Systems strove to pass more homes and to encourage households to obtain at least basic cable service for improved reception and increased numbers of channels. Once households subscribed, systems directed their efforts toward marketing the expensive, tiered services to provide movies and home-team sports.

For many years, cable managers operated on what seems to have been a false assumption: that "everyone" wanted cable and that an increase in homes passed would automatically lead to a proportional increase in basic subscribers. It was also felt that most subscribers would eventually desire a pay service and that they would remain with that service indefinitely.

As Figure 6-3 demonstrates, increasing homes passed has not necessarily been associated with basic and premium subscription. In 1982, for example, 60% of the homes passed by cable elected to subscribe. By 1985, more than 70% of the nation's TV homes had been passed by cable, and thus the potential cable audience was over 60 million homes. However, just over half of those homes subscribed to basic cable (about 36 million households), representing only about 4 in 10 TV homes.

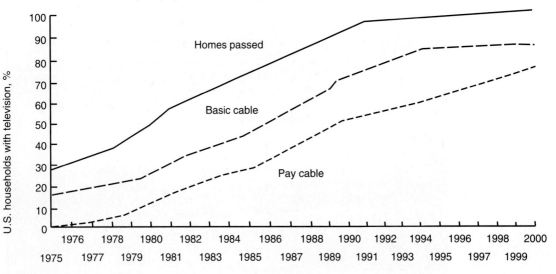

FIGURE 6-3 Percentage of U.S. TV households passed by cable, basic cable, and pay cable subscribers, 1975–2000. *Source*: NCTA; Paul Kagan and Associates; *Marketing New Media*, October 19, 1992, p. 4; projections to 2000 by the author.

Today, the cable reaches over 98% of America's TV households, or about 94 million homes. Thanks to growth in major urban areas and the popularity of the leading basic cable channels (such as CNN and ESPN), today more than 7 in 10 homes passed choose to subscribe. This represents roughly 2 in 3 TV homes in the United States.

While the increase in homes passed has been associated with increased basic penetration, the percentage of pay homes has declined in recent years. For example, in 1983, about 34 million homes subscribed to cable, with 26 million electing at least one pay service. This led to a "pay-to-basic ratio" (a key cable economic indicator) of nearly 77%. By 1990, about 50 million homes subscribed to cable, but only 38 million moved up to pay, for a ratio of 76%.

By 1993, of the 56 million cable homes, an estimated 42 million elected pay, for a ratio of 75%. The news was even worse for the major MSOs, as leading cable analyst Paul Kagan estimated their pay-to-basic ratio to be 72%, down from an all-time high of 86% in 1988.[1] Industry observers point to a number of factors behind this decline.

- Difficult economic conditions make premium television a luxury many families can't afford.
- The quality of films and original programs had not been high enough to sustain subscriber interest.
- Pay services face competition from retail video stores, which rent movies for as little as $1 each.
- Premium services can be redundant, offering many of the same films and special programs.

Cable Subscriber Churn

The scourge of the cable operator is the pay household which downgrades to basic, and even worse, the basic subscriber household which disconnects the cable altogether. In the business, this phenomenon is known as *churn*.

Industrywide, the rate of basic churn is just under 3% per month, or about 1 in 3 households per year. Most churn is caused by household relocation, a consequence of our mobile society. Two-thirds of the households which drop basic cable do so because of moving or other "noncable" issues (divorce, roommate changes, and so on). Of course, most of the churn from moving is made up by new residents.

Since pay revenues are crucial to cable management, pay churn rates are examined closely. Pay churn today approaches 7%, up from 4.9% in 1991.[2] Factoring out the percentage caused by moving, the numbers suggest that nearly 1 in 2 pay homes downgrades to basic each year. For the cable operator, this amounts to hundreds of millions of dollars of lost revenue, some of which (it is hoped) can be made up by a growing PPV universe.

The Growth of Pay-per-View

With 98% of homes passed by cable and with pay cable facing increasingly fickle audiences, cable operators are turning to an alternative revenue stream: the pay-per-view household.

Pay-per-view has been growing while the basic and pay curves illustrated in Figure 6-3 have flattened out. In 1980, for example, only 300,000 homes could receive pay-per-view. By 1985, 2.4 million homes, about 6% of cable households, were wired for PPV with addressable converters. By 1990, PPV capability was in 15 million homes, about 1 in 3 cable households. Today, it is estimated that PPV reaches over 25 million households, representing about 4 in 10 cable homes. Predictions for the year 2000 are that addressable PPV equipment and a multiplicity of channels for ordering movies and sports events will be in as many as 3 out of 4 cable homes.[3]

Audience and Viewing Trends

How do cable homes differ from nonsubscriber households? How do pay homes differ from basic homes? What is the profile of the PPV subscriber? While audience research in cable is occasionally contradictory, and increasing penetration has begun to blur distinctions, cable homes do have some specific characteristics.[4]

Basic Cable Households As might be expected, basic cable subscribers are somewhat dissatisfied with conventional broadcasting. They desire additional program choice and variety and welcome the opportunity to obtain a pay service. They watch more television overall than noncable households, but are less interested in programming from local sources. Much of the additional viewing comes in the fringe and late-night time periods.

Increased viewing in basic cable homes seems to subtract time from other leisure activities. Viewers in cable homes read less, listen to the radio less, attend fewer movies, and engage in fewer hobbies than nonsubscribers. However, they spend more time with the family and are more receptive to such technologies as video games and home computers.

Pay-Cable Households Pay-cable homes generally have a more upscale profile than broadcast and basic television households. Pay families are younger and have higher incomes, and there are more people, especially children, in the home. Pay subscribers tend to watch more TV than either basic-only or conventional broadcast subscribers, and television seems to be more central to their informational and entertainment needs. They seek programs of quality, and rate TV as a more important source of entertainment than such other media as radio and newspapers.

Pay-per-View Households A 1992 study completed for Showtime found some interesting characteristics of PPV users.[5] People who buy PPV events tend to be younger than regular subscribers, and many have small children. Like pay homes, PPV homes tend to have disposable income, with 51% reporting household income above $40,000 per year. Higher education is also prevalent; more than 4 in 10 homes ordering PPV events were headed by a college graduate. Black and Hispanic households represent a higher percentage of PPV homes than of conventional pay homes. However, as PPV technology spreads and the number of program offerings increases, these distinctions are likely to diminish, as they have with respect to basic and pay homes.

There is little doubt that the presence of cable and pay cable has permanently altered household viewing. The best illustration of this trend is the decline in network share, the percentage of television households viewing ABC, CBS, and NBC programming. Network share, once in excess of 80% throughout the day and 90% in prime time, is now under 50% overall and 65% in prime time. In pay-cable households, network share drops below 50%.[6]

CABLE PROGRAMMING

The range of programming offered by cable television falls into five broad categories: local broadcast signals, distant-signal stations, advertiser-supported basic cable networks, pay-cable services, and local-origination programming.

Local Broadcast Signals

In the early years of the development of cable, FCC rules required cable operators to carry all stations within or nearby the system's service area, provided the stations were "significantly viewed." Typically, significant viewing meant that the station was watched by at least 2 percent of the available television audience in the community, as measured by ratings services such as A. C. Nielsen. This rule became known as "must carry," since the cable system had no choice but to offer these local broadcast stations to their cable subscribers. In addition, systems were mandated to carry only the "most local" network affiliates in the area, a related rule known as "nonduplication" or "network exclusivity."

In 1987, however, the FCC's must-carry rules were declared unconstitutional. Cable systems could choose to carry local stations in their area, or they could elect not to. This situation infuriated broadcasters. On the one hand, failure to assure carriage meant that viewers with cable would have to disconnect the cable or throw a switch to see a local station. On the other hand, if local stations were carried by the cable system, broadcasters received no payment from the cable company. Their signals could be "cherrypicked" and retransmitted free of charge.

The Cable Act of 1992 offered a compromise. The law enables broadcasters to choose must-carry service or to negotiate some form of compensation from the cable system in return for signal carriage. As this edition went to press, the details of the compromise were still being worked out. However, carrying the signals of most of the over-the-air television stations in a community remained the backbone of basic cable service.

Distant-Signal Stations

Distant-signal stations, or superstations, are independent television stations from major markets which make their programs available to microwave and satellite distribution companies for lease to local cable companies. The leading superstations include Turner Broadcasting's WTBS (Atlanta), Tribune Broadcasting's WGN (Chicago), WPIX and WWOR (New York), KTLA (Los Angeles), and KTVT (Fort Worth, Texas).

Turner Broadcasting's WTBS, or "TBS Superstation," to use its most recent advertising slogan, is by far the most popular distant-signal station imported by cable systems in the nation. By 1994, the service was offered by more than 11,000 systems and reached over 60 million homes.[7] This might explain the surge in popularity of the Atlanta Braves, even despite their improved performance on the field.

Advertiser-Supported Cable Networks

Ad-supported cable networks are those channels which are provided at a cost ranging from a nickel to about a quarter per subscriber to cable operators. These channels have national sales offices which sell advertising to network and regional advertisers; local systems are provided with commercial windows (typically 1 to 5 minutes per hour) in which to insert their own local spots. With a mix of sports events, movies, and off-network syndication, some ad-supported cable networks, such as USA Network and The Nashville Network (TNN), emulate network and independent television stations. Others specialize in focused or targeted formats similar to those of radio stations. Examples include MTV and VH-1, the all-music channels; ESPN, the all-sports channel; Lifetime, a service which targets women; Turner Broadcasting's all-news operations, CNN and Headline News; and Black Entertainment Television (BET).

Pay-Cable Services

As discussed above, the major pay services provide what cable audiences seem to desire most: movies. To achieve product differentiation, they also carry original films, sports, concerts, and comedy-variety programs. The industry leaders include Time Warner's HBO and Cinemax services, Viacom's co-owned Showtime and The Movie Channel, and The Disney Channel. By 1994, HBO was available on over 9000 cable systems, with about 20 million subscribers.[8] Showtime was on about 6500 systems, with about 7 million paying households. There were about 7500 cable affiliates for Disney, with about 6 million subscriber households.

Local-Origination Programming

Local-origination programming emanates from the facilities of the local cable system. Some systems, particularly older or smaller ones, simply run a "crawl" or "scroll" of text which provides program information, news, weather, and classified advertising. As part of their franchise agreement, many systems maintain fully equipped television studios for locally produced programs. Such programming is frequently labeled public-access, or community-access, television. These channels are typically made available to community groups either on a paid (leased-access) or free (first-come, first-served) basis. As might be expected, the quality of access programming can vary considerably, depending on the ideology and production skills of the groups involved.

Local origination has been an area of growth for some large cable systems located in major urban areas. In 1992, for example, Time Warner Cable of New York launched its own 24-hour cable news service, called New York 1. In part, the development of NY1 was spurred by the success of News 12/Long Island, a service started some years earlier by Cablevision Systems to serve its many subscribers in the prosperous New York suburbs. As this book went to press, cable news channels were on air or in development in Los Angeles, Chicago, and Washington, D.C., among other communities.

Cable provides a range of additional services among its many channels, most of which cater to specialized interests. Included in this list are cable audio services, computer data services, public safety and security channels, and institutional connections for colleges, hospitals, and other private users.

Of course, like broadcast television and radio, the cable business is driven by the public's appetite for programming. As discussed above, people subscribe to cable to improve their television reception and to get more channels: primarily for movies, but to a lesser extent for additional sports and information options.

Programming is thus a central task of the cable operator. Which channels will be offered? What kinds of packages and options will be available to consumers? What local services will be provided? At what cost? The answers to these questions are of considerable managerial significance and form the basis of the in-depth analysis of the programming process provided in Chapter 12.

CABLE REVENUES

Running a cable television system is quite different from running a radio or television station. Cable faces much greater capitalization expenses for the construction and maintenance of equipment. Unlike broadcasting revenues, the bulk of cable revenues comes directly from subscriber fees, not indirectly from advertiser or network support. And unlike their open-circuit competitors, which are licensed to use the airwaves free of charge, many cable operators must return some of their precious revenues, in the form of franchise fees, to local governments.

The sources of cable revenue include:

- Monthly subscriber charges (for basic service and/or premium service)
- National, regional, and local advertising sales
- Pay-per-view charges
- Miscellaneous sources, including installation charges, additional outlet fees, rentals of remote controls, commissions from home shopping transactions, and leasing of studio space and production equipment

Subscriber Revenues

Traditionally, the backbone of cable revenues has been the monthly fees paid by subscriber households for basic and premium services. Figure 6-4 tracks revenue from those income sources from 1975 to 2000. A number of trends are evident. The first is that cable television revenues exploded in the 1980s. While their counterparts in radio and television broadcasting were laying off employees, becoming burdened by debt and dealing with ownership and management turnover, most cable systems were raking in record revenues. This alone may explain the enmity which arose between broadcasting and cable in the period—as well as the jousting between Congress and the industry which culminated in the controversial Cable Act of 1992.

Between 1980 and 1990, the number of cable systems doubled (from about 5000 to just over 10,000). During the same period, cable operating revenues increased more than tenfold, from about $2 billion to over $20 billion. For comparison, review the relatively flat revenue growth curves for radio and television stations and networks detailed in the preceding two chapters. To mix a metaphor, clearly cable was booming while broadcasting burned.

The revenue news has not been all good for cable. Note in Figure 6-4 the disparity between revenues from basic subscriptions and that from pay cable. Pay cable is indisputably a mature business, growing modestly in aggregate revenues from about $3 billion in 1985 to about twice that figure by 1995.

Projections to the end of the century suggest that competition from new delivery systems (including telephone companies, as described below), the shrinking general economy, and the chilling effect of new rate controls will lead to a flattening of overall cable revenues. However, by the year 2000, cable will approach $40 billion in annual revenues, making it a significantly larger and more powerful industry than the $2 billion enterprise of 1980.

With pay cable in maturity, if not decline, clearly cable's increased revenue is coming from other sources—most notably advertising and pay-per-view.

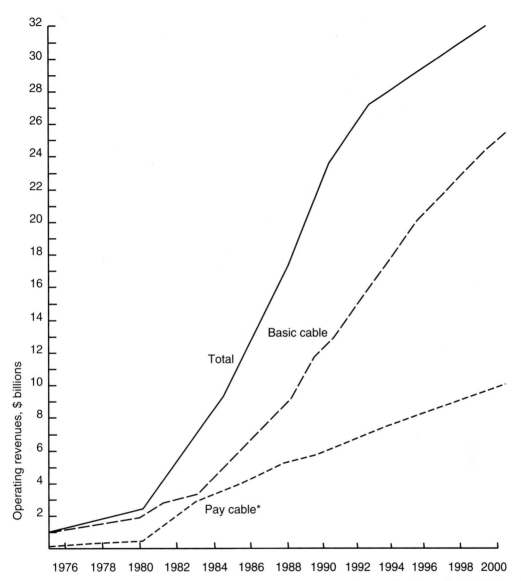

FIGURE 6-4 Cable television operating revenues, 1975–2000. *Source: The Kagan Cable TV Financial Databook*; Standard and Poor's industry surveys. Data are actual through 1990; 1991–2000 projections by the author. *Includes pay-per-view (PPV).

Advertising Revenues

The rise of cable advertising in recent years helped fuel cable's unprecedented revenue growth. While broadcast advertising, including radio and TV, hit a plateau, cable advertising moved to the forefront. A number of reasons lay behind this phenomenon.

- The number of cable households in the United States passed more than half the nation, enabling the medium to compete on more equal terms with over-the-air television and radio broadcasting.
- The sheer number of cable channels available to advertisers led to discounted pricing practices. Advertisers could buy a TV commercial or series of commercials on cable for a fraction of the cost of spots on a major network.
- Cable's unique demographic targeting, made possible by special-interest programming (such as CNN and ESPN), enabled advertisers to spend their dollars more efficiently on cable than on traditional television.
- New measurement techniques, such as Nielsen's "peoplemeter" (see Chapter 14), enhanced evaluation of cable audiences.
- Cable systems in major markets linked together to carry ads simultaneously, a process known as *interconnection*.
- Cable systems automated their operations to permit proper scheduling, management, and verification of advertisements on their channels, a process known as *insertion*.

These trends and their implications for cable management are documented fully in Chapter 13. However, we will examine here their effect on the bottom line: the enormous increase in the volume of cable television advertising. The relevant data are shown in Table 6-1.

Three overall trends emerge. The first is the validation of the growth of advertising as a revenue stream for the cable business. In 1980, cable advertising was a fledgling business, with about $58 million in total revenue (less than 5% of revenues in the industry). By 1985, cable ad volume had increased by nearly 15

TABLE 6-1 CABLE TV ADVERTISING REVENUES, 1980–2000
(Millions of Dollars)

Year		Network	Regional/local spot*	Total
1980	$	50	8	58
	%	86	14	
1985	$	634	181	815
	%	78	22	
1990	$	1809	737	2546
	%	71	29	
1995	$	3430	1803	5233
	%	65	35	
2000	$	5115	3135	8250
	%	62	38	

* Includes regional sports networks.
Source: Advertising Age, February 11, 1991, p. C-22; projections for 2000 by the author.

times. Advertisers spent over $800 million on cable, representing about 8% of all cable revenue. By 1990, cable advertising was a $2.5 billion enterprise, accounting for 15% of the cable revenue stream. By the close of the decade, cable advertising may eclipse $8 billion and account for as much as 1 in 4 cable revenue dollars.

The second trend provides tangible evidence that cable network programming has drawn advertising dollars from the broadcast networks and their leading stations. In 1985, for example, about $600 million was spent with cable's leading ad-supported networks (such as WTBS, CNN, ESPN, and MTV). Just 5 years later, advertising expenditures on cable networks had tripled. Madison Avenue spent nearly $2 billion on ad-supported cable networks, providing the cash flow for programming expansion into new channels (such as Turner's TNT) or into original programs (like original feature films on Arts & Entertainment, Lifetime, Discovery, and others). By mid-decade, the cable networks may bill over $3 billion in annual advertising.

The third trend shows that local systems have also profited from cable advertising. In 1980, few cable systems had the technology, sophistication, or staff necessary to generate advertising. Only $8 million was earned from this source, most of it by a few large suburban systems near New York, Los Angeles, and other large cities. Just 10 years later, regional/local cable advertising brought systems over $700 million in revenue, a 900 percent increase. (Remember: the number of cable systems increased by only a factor of 2 during this period!) Most cable systems have now joined the radio and television stations in their community in actively soliciting advertising and producing commercials. By the end of the decade, local cable may account for more than $3 billion in annual revenue. To build some perspective, this figure is roughly the same as the total volume of the cable industry in the early 1980s!

Pay-per-View Revenues

Like advertising, pay-per-view revenues have been targeted for explosive growth by cable management. The growth of PPV as a revenue stream in cable is traced in Figure 6-5.

In 1980, pay-per-view was a $9 million business, paced primarily by one-time special events, such as attractive boxing matches. By 1985, PPV had grown more than threefold, to $33 million in revenue. The lineup of PPV events was expanded to include feature films and high-profile musical events. Fueled by the arrival of more multichannel cable systems and addressable converter technologies, the PPV business grew to over $200 million in gross revenue by 1990. Today, PPV annual revenue is in the $350 million range.[9] By mid-decade, the anticipated multiplicity of cable channels promised by compression technology and the marketing of impulse PPV point to annual PPV revenue projections of $800 million or more. Should the number of addressable PPV homes reach half the TV market by mid-decade, PPV could exceed $2 billion in annual billings.[10]

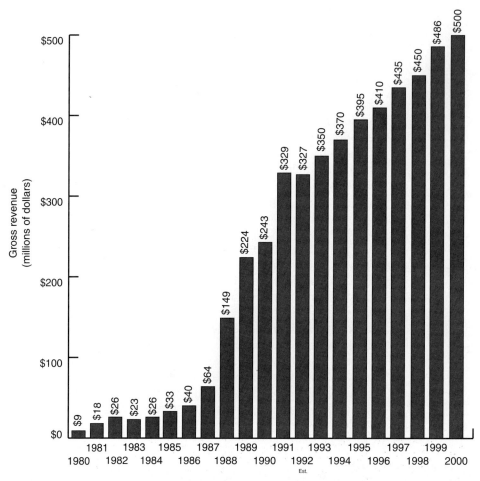

FIGURE 6-5 Gross pay-per-view revenues, 1980–2000. *Source: Cable World*, November 30, 1992, p. 31-A; projections to 2000 by the author.

Miscellaneous Revenues

Additional sources of cable revenue include installation charges, fees for additional cable outlets in the home, rental of such items as remote control devices and tuners for digital audio, and income from the lease of studio facilities and equipment. Together, these sources account for about $2 billion in annual revenue. However, unlike advertising and PPV, these sources of income do not change very much annually, at least as a proportion of gross revenue. In addition, these revenue sources do not have much gross profit potential. Installation requires costly crews. Converters cost money and can be damaged by the user. Leased access requires crews to train potential users of studios and editing suites.

As a result, the revenue from miscellaneous sources is roughly the same as a percentage of gross revenue today as it was in 1980. That year, about $150 million was earned from these services, representing about 6% of gross revenue. In 1990, roughly $1.2 billion came from installation and rentals: still only 7% of total gross industry revenue.

Cable Revenue Streams: 1980–2000

As you may have inferred by now, one reason cable is an attractive media business is its multiple revenue streams. Unlike radio and TV stations, which are totally reliant on advertising sales, cable makes money in many different ways. Leading industry analyst Paul Kagan has assessed this phenomenon and predicted how those streams will account for cable revenue in coming years.[11] His analysis is illustrated in Figure 6-6.

In 1980, nearly two-thirds of cable revenue came from basic subscriptions. As cable began to reach many American homes, most chose to subscribe. Pay cable accounted for 30% of revenue. Installation and rental fees were in their usual range of 5% to 8%; pay-per-view and cable advertising were largely a pipe dream.

By 1990, just over half of all cable revenue came from basic subscriptions. Pay-cable revenue had shrunk to about 28% of industry gross. The advertising slice approached 4%; pay-per-view revenue, just under 2%.

Projections for 2000 suggest that basic will remain at just over half of all industry revenue. However, the pay-cable business may shrink to 15% or less. Losses in pay TV will be offset by pay-per-view and advertising sales, each of which is expected to approach 10% of the total gross revenues in the cable TV industry.

CABLE EXPENSES

Despite its sizable revenue potential, cable is an expensive enterprise, requiring a significant up-front capital commitment. Capital expenditures include the costs of the head end, feeder, and drop network; satellite dishes and microwave horns; studios and other origination technology; and converters to lease to subscribers. Cable systems also pay pole and conduit rental charges to local utility companies, copyright fees for the importation of distant signals, royalties to program services, and franchise fees to local government units. Cable finance typically involves a comparatively large long-term debt which is gradually retired as the number of homes passed increases and subscriber rates escalate. High interest rates, escalating equipment costs, disappointing subscriber figures, especially the high incidence of churn (subscriber disconnects), and increases in franchise fees have contributed to inconsistency and fluctuation in the financial performance of cable stock.

In addition, cable's rather poor public image (and performance) has led to an increase in personnel and training costs and marketing expenses, not to mention professional fees for attorneys, accountants, and even a few physicians.

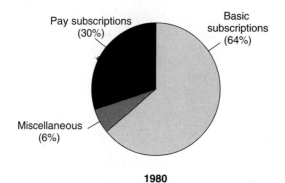
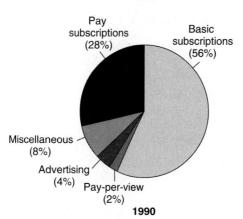
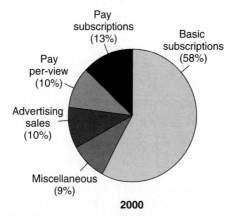

FIGURE 6-6 Cable revenue streams, 1980, 1990, 2000. *Source: Marketing New Media*, August 17, 1992, p. 2; projections for 2000 by the author based on estimates by Paul Kagan and Associates.

Capitalization and Construction

The major cable expense, capital investment in plant, varies widely according to the population density and the type of utility configurations in a given community. The industry rule of thumb is that capital costs of aerial construction—wiring a community by using existing telephone and utility poles—will range from $10,000 to $25,000 per mile. Underground plant costs can range from $50,000 to $500,000 per mile. In the process of wiring the larger cities, some entrepreneurs have found conduit space exhausted or existing wiring unacceptable for complex video transmissions, causing final costs to be as high as $1 million per mile.

Since cable has entered its maturity phase, one might expect new construction to be slowing. However, capital expenditures for physical plant remain high as existing systems rebuild for increased channel capacity, purchase equipment offering interactive capabilities for addressable converters, and replace decades-old coaxial cable with new fiber-optic lines.

One irony for the cable business is that just as it is completing its first major capital outlay—the wiring of America—the industry must embark on a completely new wave of construction to rebuild itself to compete with other businesses, most notably the telephone companies. Investment in fiber optics has been significant: as much as $15,000 per connection or "link," in the cable business. But replacing coaxial cable with fiber enables systems to offer such new technologies as additional pay-per-view (or "video on demand") channels, switched data (like computer interfaces), and personal communications services similar to cellular phone technology, which is expected to grow significantly in coming years.

Program Expenses

As in the broadcasting business, once construction is completed, the cost of programming tends to head the list of cable expenses. During the 1980s and into the 1990s, the costs of programming to the cable operator increased dramatically. At the same time, management found itself under increasing pressure to pass on the increases to its subscribers, who demanded such attractive channels as CNN, The Movie Channel, and USA.

The most costly channels are, of course, the pay-television services. Pay services are marketed to cable system managers on a negotiated split basis. That is, the cable operator and the pay service sign a contract to divide the revenues for each pay household. Typically, 60% goes to the operator and 40% to the pay-cable network.

Advertiser-supported networks are another growing expense for cable operators. In the early 1980s, like the three commercial TV networks, ESPN, WTBS, and other cable networks paid their affiliates to carry their shows. Today, the situation is reversed. Most ad-supported basic cable services cost between 5 and 25 cents per subscriber per month. Networks which own more than one cable service "bundle" their offerings. For example, CNN costs about 25 cents per month per sub. However, Turner Broadcasting System, the parent company of CNN, will bundle CNN with Headline News for an additional 5 cents per month.

Program expenses also include fees negotiated by local broadcasters and cable systems in their area as a consequence of the Cable Act of 1992. Finally, there are origination expenses for programming that is produced or bought to be aired by the cable system itself. While some systems feature full-bore news channels which emulate commercial TV stations, for most operators, origination expenses reflect less than 10% of the cable programming budget.

Surveys of cable managers (see Chapter 12) suggest that local cable programming is targeted for growth year after year. Also, program costs are likely to rise as a consequence of new federal regulations in the must-carry/retransmission content areas.

General and Administrative Expenses

General and administrative costs make up the next expense category for cable. Running a cable business involves rent, salaries and benefits for office personnel, utility costs, insurance coverage, travel and entertainment, dues, subscriptions and professional memberships, and supplies; and these expenses tend to be higher in cable than in radio and television broadcasting. First, cable operators need larger secretarial and clerical staffs to handle consumer inquiries, billing, and so forth. In the cable business, these staff people are known as customer service representatives (CSRs). Second, the cable billing process requires a greater commitment to office automation. Most systems have sophisticated computers for handling the complex tasks of subscriber billing, payment, connection and disconnection, and other functions.

Thus, cable systems usually have more phone lines, more computers, more clerical employees, more offices, even more pencils and paper than the typical radio or television station. All told, about 25 cents of the cable expense dollar goes to general and administrative costs.

Like programming, general and administrative costs in cable continue to rise. Pay-per-view requires more sophisticated billing procedures, plus additional phone lines and CSR personnel. Bringing systems into compliance with new regulations means more people and services. As cable matures, it is no longer simply a "turnkey" operation. The days of the community antenna are ancient history.

Technical Expenses

Technical expenses include technicians' salaries and the purchase of the vehicles, spare parts, and other equipment required for operating an engineering-based business. Unlike their broadcasting counterparts, many cable operators must subcontract a sizable part of their operations to outside services such as installation and repair crews and telephone answering and paging services. They must make payments to local utilities for telephone pole or conduit access to route their cables. They incur microwave and satellite costs when they import programming. Finally, most systems include franchise fees as part of their technical costs. In total, technical costs make up about 15% of annual cable expenditures.

You guessed it: technical expenses in cable are on the rise and will continue to increase through the 1990s. System rebuilds and upgrades require more and better-trained employees. And, as cable systems become more like broadcasters in local programming and advertising, they need more trained technical personnel: particularly engineers and studio technicians.

Marketing Expenses

The rough equivalent of radio and television sales expenses, marketing expenses include the salaries and benefits of sales personnel and the costs of advertising and promotion. Although marketing costs have traditionally been the smallest percentage of cable expenditures, the slowdown in pay revenues, the growing churn, dissatisfaction with cable services and prices, and other factors have created renewed interest in cable marketing. Growth of cable advertising has required larger and better-trained sales teams. More direct-marketing personnel (such as telephone and door-to-door canvassers) are needed to attract new subscribers and to convince pay subscribers to "lift" to multipay and pay-per-view. Cable systems are becoming more like broadcasters—sponsoring contests and giveaways for subscribers, planning community fund-raisers, and mounting promotional campaigns. Thus, although only about 15% of cable expenses are presently in the marketing area, it will not be surprising if by 2000 such costs account for 25 cents or more of the cable expense dollar. The commitment to cable sales and marketing is discussed in detail in Chapter 13.

By now, two trends in the cable business should be clear. First, the business has experienced phenomenal revenue growth in recent years, outpacing most American enterprise. Second, expenses have also escalated. In the view of the federal government, and of many local communities and cable competitors, expenses have not kept pace with profits, and cable managers have behaved like monopolistic profiteers. Industry leaders claim that their administrative expenses, debt load, and program costs have justified their rate increases, and that profit margins are consistent with those of other businesses in the entertainment sector of the economy. Who's telling the truth? Let's try to shine some light on the murky world of cable profits.

PROFITS AND PROFIT MARGINS

Getting a fix on profits is a bit more difficult in the cable industry than in the broadcast business. For one thing, cable's major trade association has never undertaken an annual financial survey. Most of the major cable systems also own networks and participate in other industry segments, making their profit from system operations difficult to decipher. In addition, cable's major players, the MSOs, have been extremely secretive about their financial situation in recent years, especially in light of inquiries from consumer groups and Congress about their alleged monopolistic practices.

However, some generalizations about cable profits can be made. Just as the number of cable systems lies somewhere between the number of television and radio stations in the United States, so cable profit margins are typically placed between the margins reported by those media. For example, as we traced in Chapter 4, in recent years the pretax profit margin for radio has ranged from 0% (with most stations losing money), to about 20% (for the higher-ranking FM stations in a major market). On the TV side, network-owned and -affiliated stations can return 50 cents on the dollar or more; smaller market affiliates and independents normally operate with profit margins of 15% to 25%.

The annual reports of publicly traded cable companies place their profit margins between those of radio and TV. For example, in 1988, Warner Cable (now folded into the Time Warner media conglomerate) reported profit of $541 million on revenues of $4.2 billion, for a profit margin about 13%.[12] Another Time Warner company, ATC, reported $145 million in profit on $812 million in revenue, an 18% return.

It would be unwise, however, to conclude that no cable system can match the profit margins of large, network-owned and -affiliated TV stations. Mature cable systems (with little carryover debt from their construction phase) in prime suburban areas (where pay, multipay, and pay-per-view households are plentiful) can return 50 to 100 cents on the dollar or more. By the same token, newer cable systems, those in need of a facilities rebuild, and those in less populated, lower-income areas often operate in the red in the hope of a better tomorrow!

Another point to keep in mind regarding cable profits is the nature of the business itself. As we will see, broadcasting businesses deal largely with intangible assets (good will, the value of a broadcast license, programming contracts, advertising spots, and so on). From a financial standpoint, this makes them comparatively cash-rich and liquid and often leads to the report of high margins on balance sheets. Cable, on the other hand, is rich in fixed assets (the cable itself, transmission equipment, trucks, field equipment, and so on). Since companies can "write down" or depreciate these assets, the result is often a smaller profit margin than one normally encounters in other parts of the entertainment sector (like the movie and record industries).

There is little doubt that cable can be and is enormously profitable. After all, millions of households write checks ranging from $20 to $200 or more each month "just for television." At a time when TV and radio advertising is flat, cable advertising expands. Add burgeoning pay-per-view revenues and the financial outlook for cable in the 1990s causes many executives in the industry to smile. However, the dual prospects of more vigorous regulation and enhanced competition temper the otherwise rosy financial outlook.

THE CABLE EMPLOYMENT PICTURE

With revenue and profit growth has come employment growth. While the job picture in radio and TV has been bleak at best, the cable work force has increased significantly in recent years. Don't apply immediately to the major MSO headquarters or to pay networks like HBO and Showtime, however: layoffs there were as commonplace as at the major broadcast networks in recent years. Let's examine the cable employment picture.

The total size of the cable work force is about 150,000.[13] Unlike broadcasting, however, cable is a decidedly blue-collar profession. Whereas in radio and TV, as many as 1 in 4 employees has managerial responsibilities, more than 80% of cable employees serve in nonsupervisory capacities.

The majority of cable employees are paid by the hour, rather than salaried. Industrywide, the typical hourly wage is about $11; the weekly wage, in the neighborhood of $450. These figures place cable workers near the low end of employees in communications industries, especially in comparison with higher-paid telephone workers, who often share the same utility poles and underground conduits. Not surprisingly, the Communications Workers of America (CWA) and

Cable customer-service representatives. CSRs are the front-line employees in cable systems. (courtesy of TCI)

TABLE 6-2 CABLE SALARIES

Job	Average salary	Cable System Large*	Cable System Medium†	Cable System Small‡
General manager	50,000	99,000	55,000	25,000
Chief engineer	45,000	85,000	40,000	22,000
Markets manager	43,000	90,000	45,000	18,000
Advertising sales manager	40,000	110,000	50,000	19,500
Consumer affairs director	39,000	75,000	35,000	—
Program director	35,000	75,000	37,000	17,500

* More than 50,000 subscribers.
† From 10,000 to 50,000 subscribers.
‡ Under 10,000 subscribers.
Source: Cablevision, June 7, 1992, pp. 114–120; *Cablevision*, August 26, 1991, pp. 33–38; projected to 1995 by the author.

other unions have targeted the cable work force for organizing activity. Today, about 1 in 10 cable employees is represented by either CWA or the International Brotherhood of Electrical Workers (IBEW).[14]

The employment picture is somewhat brighter for cable managers. Table 6-2 reports average salaries for six important management posts in cable television. Nationally, the typical general manager's salary is in the $50,000 range. Chief engineers rank second among managers, with an annual salary approaching $45,000. Marketing managers are next, followed by advertising sales managers, consumer affairs directors, and program directors.

Whereas in radio and television, salary is related to market size, the key determinant in cable is the number of subscriber households. Large systems (those with more than 50,000 subscriber homes) pay the best. For example, the general managers in large systems earn $100,000 per year on average, placing them at parity with radio and TV managers in large markets. Geography is also important to cable compensation. Systems in the Northeast and Far West pay best; those located in the Southeast and Midwest pay less.

One thing cable shares with broadcasting is its mixed record with respect to female and minority employment. Unlike broadcasting, cable management points proudly to work force percentages matching or exceeding national averages for women and ethnic minorities. In 1990, for example, the national work force was 45 percent female and 22 percent minority. The cable work force that year was 42 percent female and 24 percent ethnic minority.[15]

On the other hand, cable's lagging salaries temper these statistics. As we have seen, hourly wages for cable crews, many of which are composed of minority employees, lag behind those of telephone installers and similar wage earners. A cable industry report revealed that even in managerial positions, women earn 25% less than their male counterparts in comparable positions.[16]

There are some positive signs for readers contemplating careers in cable. While job growth has been flat in broadcasting, the cable work force continues to grow.

Also, though most of the growth is occurring on the operator side (that is, at local systems), after years of retrenchment, major programmers and MSOs have begun to add employees.[17] As we will see in Chapter 13, the major spur to that growth has been advertising and marketing. The future of cable employment opportunity would appear to be less "in the trenches" and more on the corporate and creative side.

COMPETITION AND COOPERATION IN THE LAST QUARTER MILE: THE TELEPHONE COMPANIES

As the new millennium approaches, it is clear that a major battle is looming. The prize is what some have called "the last quarter mile," the distance from the cable trunk line into the individual consumer household. As the cable companies raced to wire America, they paid little attention to the wire already in place: that of "Ma Bell," the telephone company. But in recent years, regulatory and technological developments have combined to create a new atmosphere of competition. In today's world of microelectronics and fiber optics, telephone lines can carry more than the low-level audio signals we are familiar with as routine telephone calls. They can transmit stereo audio signals, potentially thousands of television channels, computer graphics, and data. And, after nearly a half century of prohibition, the FCC and the federal courts are allowing telephone companies to try their hand at these expanded services. So, into the battleground of American consumer electronics marches a very large and strong army: the regional Bell operating companies, or RBOCs.

The Regional Bell Operating Companies

Long before there was concern about monopoly in the cable industry, consumer groups and the federal government were dismayed by the sheer size and power of American Telephone and Telegraph (AT&T). For 97 years the company, the largest in the world, held dominion over telephone service in the United States. Household choice was minimal: you could have a basic black phone or a white phone (the "Princess" model was a popular alternative made available in the 1960s). When it came to long-distance service, you had no choice at all: AT&T was the sole provider.

By 1984, following a series of antitrust actions initiated in the 1950s, the federal government succeeded in breaking up the "Ma Bell" monopoly. The breakup took the form of a modification of final judgment (MFJ) in the federal district court presided over by Judge Harold Greene.[18] As of January 1, 1984, AT&T's 22 local telephone services were spun off into 7 independent regional Bell operating companies (RBOCs). Though they are wholly independent of the parent company, the 7 RBOCs are also known as regional Bell holding companies (RHCs), or as the "Baby Bells."

Table 6-3 lists the RBOCs, their affiliated companies, total assets, and individual business and household connections, known in the trade as local access lines.

TABLE 6-3 THE RBOCs

RBOC	Subsidiaries	Access lines	Total assets*
Nynex	New York Tel, New England Tel	15,303,000	26.7
Bell Atlantic	Bell of Pennsylvania, Diamond State Tel, New Jersey Tel, the Chesapeake and Potomac companies	17,484,000	30.0
Ameritech	Illinois Bell, Indiana Bell, Ohio Bell, Michigan Bell, Wisconsin Bell	16,278,000	21.7
Bell South	South Central Bell, Southern Bell	17,500,000	30.2
Southwestern Bell	Southwestern Bell	12,105,000	22.2
U.S. West	Mountain Bell, Northwestern Bell, Pacific Northwest Bell	12,218,000	25.0
Pacific Telesis	Pacific Bell, Nevada Bell	13,800,000	21.6

* In billions of dollars.
Source: *International Directory of Telecommunications* (Essex, U.K.: Longman, 1990); *Telephone Industry Directory*, 6th ed. (Potomac, Md.: Phillips, 1992).

From the table, it is clear that these are huge companies with significant assets and access to vast numbers of American households. Each RBOC serves between 12 million and 20 million customers, and each reports assets in excess of $20 billion. By comparison, TCI (the largest cable company) serves about 10 million homes, Time Warner about 7 million, and the next largest systems fewer than 3 million each. Total annual gross telephone industry revenue for the RBOCs is approaching $200 billion: more than 10 times the gross revenues of the more than 10,000 cable systems!

Congress and the federal government are aware of this influence, and following the 1984 divestiture placed restrictions on the operations of RBOCs. For example, the Cable Television Act of 1984 prohibited telephone companies from also owning cable systems. The MFJ restricted the RBOCs to local telephone service and Yellow Pages directories.

But in recent times, some of the restrictions were modified; others were abolished. In 1991, the Federal Communications Commission cleared the way for the RBOCs to own domestic cable systems, though not in their immediate service areas. By this time, they were already major participants in the cable business overseas. In the United Kingdom, for example, the top-five cable operators include Nynex and Southwestern Bell.[19]

By 1993, the Baby Bells had been permitted to enter the information services business, including computer data, financial services, stock quotes, and news reports. In addition, the path was cleared for the RBOCs to offer video services on their lines, though, unlike the major cable MSOs, they could not own the program services themselves. These decisions, and others to follow, paved the way for the RBOCs to become either full competitors or full partners with cable systems in the immediate future. Following are some of the services on the immediate horizon for the RBOCs.

The National Information Superhighway Early in his term, President Bill Clinton advocated a new national system for the widescale distribution of audio, video, and data information. Dubbed "the national information superhighway," the proposal called for the development of a single-standard, open-architecture delivery system, which would be accessible to both business and public users, sort of a data equivalent of the interstate highway system.[20] The telephone companies were early entrants into this proposed system. In 1993, for example, Bell South announced plans to participate in a prototype of the new highway in North Carolina, linking Asheville, Charlotte, Greensboro, Raleigh, and Wilmington.[21]

Video Dial Tone The FCC's decision to permit telephone companies to offer video services to their customers is known as video dial tone (VDT). In theory, VDT will offer all the services common to the most modern cable system: hundreds of channels, basic and pay television, pay-per-view, and audio, data, and interactive services. By the mid-1990s most of the RBOCs were on record as being seriously interested in exploring video dial tone. In 1993, U.S. West sought permission from the FCC for a full-scale test of the proposed service in Omaha, Nebraska.[22] That U.S. West would be interested in probing the potential for telco-based entertainment offerings is not surprising. By the time of the announcement, the Baby Bell had invested $2.5 billion in Time Warner, the world's largest diversified entertainment company.

Personal Communications Services (PCS) Largely left out of the cellular revolution, cable companies have been at the forefront of the development of the next wave of mobile telephones: personal communications services, or PCS. PCS makes it possible for a person to send and receive telephone calls or computer messages via a single telephone number virtually anywhere in the world, using a combination of cellular, satellite, and traditional technologies (including telephone and cable lines). Early cable entrepreneurs in PCS included TCI and Cox Enterprises.

Not to be undone, the RBOCs, particularly Bell Atlantic, have vigorously pursued PCS development. Bell Atlantic serves a heavily populated area with much international business and expects PCS to be in heavy demand upon its rollout in the late 1990s.[23] Should PCS take hold, expect the other RBOCs to follow suit in the race to make the service widely available.

Megamergers: RBOCs and MSOs

The telecommunications industry was jolted in late 1993 by the announcement of a merger between Bell Atlantic and TCI.[24] Considered the largest corporate merger in American history, when completed the combination of the biggest cable company with a major RBOC would have created a $30 billion company.

For comparison, this is roughly three times the size of the entire radio industry and larger than the gross billings of broadcast and cable television.

This announcement was followed by the decision of major MSO Cox Cable to merge with Southwestern Bell. Other cable TV/RBOC alliances were expected. The new information superhighway appeared to be under construction.

However, by the time this book went to press, the Bell Atlantic/TCI merger had been called off and other alliances faced scrutiny before Congress, the courts, and among consumer organizations.

There is little doubt that the entry of RBOCs poses a danger to the system of broadcasting and cable which evolved in this country over the past half century. It is symptomatic of the trend traced in Part One: the merger of audio, video, computer, and data services into a single stream of information. Advocates for telco entry into video services point to the successful track record of the phone companies in delivering messages to consumers. There is little doubt that ours is the best telephone system in the world (as anyone who has tried to make a call overseas will attest). The telcos have the capital and resources to invest in costly research and development. They have a well-paid, well-trained work force.

On the other hand, there is nearly a century of evidence documenting classic monopolistic behavior by the telephone companies. Overcharging has been commonplace. There has been limited choice to consumers among both products and services. Only after the breakup, for example, did we get a broad range of telephone designs and such services as call-waiting, call-forwarding, and call-screening.

The fact of telco entry into television service is indisputable. Its long-term impact, however, is unknown.

SUMMARY

In 40 years, the cable television business has grown from its beginnings as an adjunct to broadcast television to an independent multimillion-dollar enterprise. With cable becoming available to most U.S. households, the focus of the business has shifted from construction to marketing and customer service.

Research suggests that audiences subscribe to cable to obtain more viewing choice and variety, particularly movies and sports. Cable households tend to watch more television overall than nonsubscribers. The additional choices offered by cable include distant television stations, advertiser-supported cable networks, pay-movie services, pay-per-view movies and events, local programs, and other programming services.

The backbone of cable revenues has been its basic subscribers. However, pay and PPV subscribers have become an increasingly important revenue source for cable management. Cable advertising revenues are growing, but still remain a fraction of the amount spent on broadcast television and radio.

Cable expenses have risen dramatically in recent years as construction costs have skyrocketed and subscriber rates have been disappointing. Consequently, cable profits have fluctuated wildly, falling sharply in the mid-1980s before recovering as a result of a slowdown in the need for new construction and falling interest rates.

Jobs in the cable industry, particularly at local systems, are increasing. Job categories on the rise include general management, marketing, and sales. While the number of women and minorities in the industry continues to increase, there is concern about a lack of female and minority participation in cable management.

Technological and policy developments have led to the participation of the seven regional Bell operating companies in video and related services. Services in development include a proposed national information superhighway, video dial tone, and personal communication services.

NOTES

1 Paul Kagan Associates, *Marketing New Media*, July 20, 1992, p. 6.
2 Paul Kagan Associates, *Marketing New Media*, April 20, 1992, p. 1.
3 See "Pay Per View Extra," *Cable World*, March 31, 1993, p. 1A.
4 A comprehensive review can be found in Dean M. Krugman, "Evaluating the Audiences of the New Media," *Journal of Advertising*, vol. 14, no. 4, 1985, pp. 21–27. See also Lee B. Becker, "Cable's Impact on the Use of Other News Media," *Journal of Broadcasting*, vol. 27, no. 2, Fall 1983, pp. 127–140; Richard Ducey, Dean Krugman, and Don Eckrich, "Predicting Market Segments in the Cable Industry: The Basic and Pay Subscribers," *Journal of Broadcasting*, vol. 27, no. 2, Spring 1983, pp. 151–161; Carrie Heeter and Bradley S. Greenberg, *Cableviewing* (Norwood, N.J.: Ablex, 1988).
5 See *Cable World*, November 30, 1992, p. 32-A.
6 "Changing Viewing Patterns Attributed to Cable," *Advertising Age*, February 11, 1991, p. Cable-23.
7 *Cablevision*, February 8, 1993, p. 42; 1995 projection by the author.
8 *Ibid*.
9 "Pay Per View Extra," *Cable World*, May 31, 1993, p. 1-A.
10 "PPV vs. Total Operator Revenues," *Cable World*, May 31, 1993, p. 1-A.
11 Paul Kagan Associates, "Marketing New Media," *Cable World*, August 17, 1992, p. 2.
12 See *The Kagan Cable TV Financial Data book* (Carmel, Calif.: Paul Kagan Associates, 1989), p. 89.
13 Projection based on *FCC 1990 Broadcast and Cable Employment Report*, CEE 009-01, p. 334; "Cable's Employment Picture," *Cablevision*, May 4, 1992, p. 84.
14 "The Union Press," *Cablevision*, May 4, 1992, p. 84.
15 *1990 Broadcast and Cable Employment Trend Report* (Washington, D.C.: U.S. Government Printing Office, June 11, 1991).
16 *Cablevision*, June 7, 1993, p. 118.
17 Richard Katz, "Jobs on the Rise," *Cablevision*, August 26, 1991, p. 39.
18 For a policy history, see "Telecommunications Services," *U.S. Industrial Outlook 1993* (Washington, D.C.: U.S. Government Printing Office, 1993), pp. 28-3, 28-4.
19 "Top U.K. Cable Operators," *Cable World*, May 10, 1993, p. 1.
20 "Another Telco Elbows onto Cable Bandwagon," *Cable World*, March 1, 1993, p. 6.
21 "Bells Weaving Alliances for Their Video Future," *Cable World*, May 17, 1993, p. 38.
22 "U.S. West Plans Video Test in Omaha," *Broadcasting and Cable*, June 28, 1993, pp. 40–41.

23 "PCS: Too Late, Too Little for Cable?" *Cable World*, May 24, 1993, p. 63.
24 "Face to Face with Ray Smith," *Broadcasting and Cable*, November 8, 1993, pp. 18–20.

FOR ADDITIONAL READING

Baldwin, Thomas F., and D. Stevens McVoy: *Cable Communications*, 2d ed. (Englewood Cliffs, N.J.: Prentice-Hall, 1988).

Garay, Ronald: *Cable Television*: A Reference Guide to Information (New York: Greenwood Press, 1988).

Johnson, Leland L.: *Telephone Company Entry into Cable Television* (Santa Monica, Calif.: Rand Corporation, 1992).

Jones, Glenn R.: *Jones Dictionary of Cable Television Terminology*, 3d ed. (Englewood, Colo.: Jones 21st Century, 1988).

Roman, James W.: *Cablemania: The Cable Television Sourcebook* (Englewood Cliffs, N.J.: Prentice-Hall, 1984).

National Telecommunication and Information Administration: *Video Program Distribution and Cable Television* (Washington, D.C.: NTIA, 1988).

Webb, G. Kent: *The Economics of Cable Television* (Lexington, Mass: Heath, 1983).

PART THREE

CORE PROCESSES

Part Three: Core Processes takes prospective media managers through the steps required to design, finance, build, or acquire a modern telecommunications venture. Chapter 7 details forms of media ownership, from small mom-and-pop operations to international communications conglomerates. Chapter 8 delves into capitalization, media construction, and acquisition strategies. Chapter 9 details the process of financial management in telecommunications. Chapter 10 documents the process of employee relations in media companies, from hiring and promotions to termination.

7

PATTERNS OF TELECOMMUNICATIONS OWNERSHIP

All applications shall set forth such facts as . . . the citizenship, character, and financial, technical, and other qualifications of the applicant to operate the station.

These words are excerpted from the application filed by prospective broadcasters for a new radio or television station. While recent deregulatory efforts have removed some specific requirements, the spirit of this statement remains: a telecommunications facility cannot be owned by "just anyone." This chapter traces the qualifications for and patterns of ownership in electronic mass media. The goals of this chapter are:

1 To trace the general and specific qualifications for media ownership
2 To list and define the major forms of media ownership, from independent mom-and-pop operations to large multinational corporations
3 To trace current issues and trends in media ownership, including increased media concentration, globalization, joint ventures, and the participation of minority groups and women in ownership

RESTRICTIONS ON OWNERSHIP

Business enterprise in the United States is based on classic models of the marketplace—from the Greek agora to New York's Orchard Street. In theory, our system provides each member of society the right to exchange goods and services in public, for a fair market price, free from government intervention. The common

165

denominator of U.S. commerce is competition; no single entity should control the supply or regulate the demand for any product or service.

Unfortunately, broadcasting, cable, and the other new media technologies are not like many other so-called free market industries. The reasons for this are various, but paramount among them are the related issues of scarcity, the public interest, and local monopoly.

Scarcity

In the classic model of the marketplace, anyone with goods to sell can set up shop, since space is "infinite." Think of the open-air markets of Rome or of your local drive-in flea market, stretching for blocks with a seemingly endless array of products. But television, radio, DBS, and other services which use the electronic spectrum, or airwaves, do not have the luxury of limitless space. Spectrum space is finite: only a certain number of channels are available for each use. Compounding this problem is the fact that some services require more spectrum than others: for example, the space occupied by a single television station could be used for 30 FM radio stations! Thus the concept of scarcity has dominated federal regulatory policy toward the electronic media since their inception.

Public Interest

Related to the issue of scarcity is the notion of public interest. In the flea market, the goods being sold are the property of the owner: a right guaranteed in the Constitution. But who owns the airwaves? In some countries, such as the United Kingdom and Canada, the government owns the spectrum. But in the United States, the FCC has traditionally maintained that the airwaves "belong to the people" and that spectrum users may occupy assigned frequencies only if they serve the interests of the general public.

While what constitutes "the public interest" has always been open to debate and interpretation, the doctrine has been consistently upheld as a rationale for FCC and congressional oversight of the operations and programming of broadcast and cable properties.[1]

Local Monopoly

In the case of cable television and telephony (including newer technologies, such as cellular telephones), there has always been concern over control of these services by a single company in a given community. Reluctantly, government authorities have recognized the impracticality in many cities of building competing telephone and cable systems. There simply would be too much duplication of effort, and too much disruption of community services for such processes as digging trenches and establishing aerial and ground rights-of-way.

In return for approving such local monopolies, federal and local government has maintained oversight organizations, such as the FCC's Common Carrier Bureau and local public utility commissions, to police the ownership and operations of media entrepreneurs. While new technologies permit new competition (as we will see), ownership restrictions remain a common component of the cable and telephone businesses.

Taken together, the scarcity of spectrum space, the requirements of public service, and the desire to preserve competition have governed requirements for ownership by prospective telecommunications entrepreneurs. From the beginning, applicants for radio and television stations have had to meet a variety of income, citizenship, and character tests. Following the lead of the FCC, cable franchising authorities have set forth similar selection criteria. The major criteria for ownership are discussed below.

CRITERIA FOR BROADCAST OWNERSHIP

The FCC has long utilized a set of standards to screen prospective owners of radio and television stations who file license applications. Like many standards, these are not definitive and have rarely been uniformly applied. They have been more strictly enforced when competing companies apply for the same frequency or when one group's license renewal is blocked by another group or individual seeking the channel. Despite inconsistencies in enforcement, these standards help identify the FCC's definition of "ownership in the public interest."

- Citizenship
- Character
- Local ownership
- Civic involvement
- Integration of ownership and management
- Diversification of management background
- Prior experience
- Operating plans

Citizenship

While what constitutes the "public interest" has been argued for years, there is no disagreement that it is at least "American." In other words, broadcast licensees must service U.S. citizens first and foremost. Translated into ownership policy, majority interests in broadcasting properties must be U.S. citizens. Thus, while Beverly Hills and downtown New York may have an increasingly foreign air of ownership, our radio and television stations are still "owned American."

Character

A broadcast applicant should have a "clean" background. Generally, convicted felons, forgers, perjurers, and other social transgressors have been blocked from entering the broadcasting industry. Once licensed, broadcast owners must strive to keep their slates clean; licenses have been forfeited and renewals denied to broadcasters known to have deliberately deceived clients, competitors, advertisers, or the FCC itself![2]

Local Ownership

All other things being equal, the FCC will generally look more favorably upon an application for a radio or television station that is filed by a prospective owner who resides in the city of license. This is known as the philosophy of localism. Underlying this proviso is the notion that a local owner will be more likely to have common interests with neighbors and friends, and thus be more responsive to their needs.

Like all ownership qualifications, this point is arguable. It is plausible that a local entrepreneur will be more responsive to the community. However, it is at least equally plausible that an outsider would make a significant commitment to local service in order to win local audience acceptance. In fact, some media scholars have noted that so-called absentee owners often provide more public service programming than do their local counterparts.[3]

Civic Involvement

Related to the localism criterion is the notion of civic involvement. The FCC has generally looked favorably upon potential broadcast owners with a record of civic interest and participation. This is particularly true at license renewal time. Faced with a competing applicant for a broadcast station, an "incumbent" may earn points at a hearing for having participated in telethons, scholarship and internship programs, bond drives, and other civic enterprises. Of course, in addition to being a plus in the eyes of the FCC, active civic involvement is a good business and public relations practice.

Integration of Ownership and Management

Since the licensee is ultimately responsible for what goes out over the air, the FCC maintains that if ownership and programming management are well integrated, licensees are more likely to be in control of their stations. While this criterion springs from the FCC's commitment to public interest, its merits are debatable. For one thing, as we have seen, it is quite easy to assign an office or title to a broadcast employee without a concomitant increase in salary or responsibility. In some cases, a policy of "window dressing" has been used to circumvent the integration dictum.[4]

In addition, many of today's radio stations are programmed by national services and feature tape machines and computers instead of disc jockeys. It is arguable whether local ownership in such cases would translate into local public service.

Diversification of Management Background

Diversification is based on the assumption that broadcast applicants are more likely to be civic-minded if they have a financial interest in businesses outside of broadcasting, newspapers, or other media. It is presumed that owners and stockholders with diverse backgrounds and a broader income base will be less likely to use the airwaves for personal aggrandizement or political manipulation.[5]

Although this theory may also be questionable, it has led to an industry which is owned by entrepreneurs with extremely varied backgrounds—including cowboy movie stars, politicians, insurance salespeople, and physicians.

Prior Experience

As noted earlier, it takes unique technical and entrepreneurial skills to operate a broadcast station successfully. Thus, the FCC has shown favor over the years to applicants with a record of past performance in broadcasting. Of course, this criterion must be balanced against the requirement for diversity. (Also, as noted earlier, minority organizations see the proviso as discriminatory—preventing new parties from participating in media ownership.) However, few can argue that the ability to turn a profit in an industry marked by high turnover, increasing competition, and continual innovation is a prime predictor of the success of a new broadcasting venture.

Operating Plans

Applicants often go to great extremes to show the FCC that theirs will be the best-equipped and best-staffed broadcast facility in the nation—or at least in the proposed city of license. Some applicants have gone overboard, proving that their employee bathrooms will be more sanitary than those of their competitors!

In practice, this dictum has translated into a policy of rewarding minimal financial and technical competence. In general, the FCC has used a vague yardstick for measuring potential operating ability: given an applicant's financial statement, background, and track record, how long could he or she operate the station without accepting a single dollar of advertising? Generally, the FCC has regarded an applicant with access to a cash escrow account sufficient to ensure operations for 3 months as having sufficient financial backing to launch a broadcast station.[6]

The above general criteria for ownership apply to those who would use the spectrum for local commercial broadcasting. Thus, the onus of proving satisfactory ownership backgrounds and interests falls mainly on radio and television entrepreneurs. However, many communities have applied the FCC guidelines in their local cable franchising agreements, and the National League of Cities has specified similar character and competence guidelines in its suggested "model" franchising plan for local communities.[7]

To foster localism, diversity, and competition, the FCC has also developed specific ownership requirements for broadcast and cable entrepreneurs. These regulations are spelled out in the following section.

SPECIFIC OWNERSHIP REGULATIONS

Multiple Ownership Rules

The FCC regulates the number of radio and television stations that can be owned by a single entity. In 1953, the maximum number was set at 7 AM stations, 7 FM stations, and 7 television stations, for a total of 21. In addition, no more than 5 of the television stations could be VHF. For over 30 years, the so-called rule of 7s guided broadcast ownership—the result being domination of the industry by the three national networks and their owned stations and the rise to prominence of leading broadcast groups such as Capital Cities, Cox, Metromedia, Gaylord, and Taft.

The explosion of the number of radio and television outlets in the 1980s and the move to deregulation led to modification of the multiple-ownership rules. In April 1985, the rule of 7s became the rule of 12s. Under the revised rules, an entity could own up to 12 AM stations, 12 FM stations, and 12 TV stations as long as the TV stations collectively served less than 25% of the nation's television homes. In the formula, UHF stations counted for only half of a market's TV homes. In addition, groups were permitted to own up to 14 stations in each service and to blanket 30% of the nation's TV households if two each of the AM, FM, and TV stations were controlled by minorities.

In the early 1990s, the FCC proposed even further expansion of the permissible number of co-owned radio and television properties.

Radio: The Rule of 18s In response to the rather bleak economic picture for the radio business traced in an earlier chapter, in 1992 the FCC proposed permitting a single company to own up to 30 AM and 30 FM stations. However, many small station owners vigorously protested, claiming they would be run out of business by large radio conglomerates.[8] Congress and minority groups were concerned about a loss of programming diversity.

The commission backed off, and instead increased ownership limits to 18 AM and 18 FM stations for a single owner. Under the new rules, a single owner can also hold up to 2 AM and 2 FM stations in the same market, if that market has more than 15 stations. In markets with fewer than 15 stations, a single owner may own up to 3 stations, but no more than 2 AM or 2 FM facilities. The rule of 18s may be short-lived, however. The FCC may expand the limit to 20 AM and 20 FM stations.

Television: From 12 to 24? In 1992, the commission proposed expanding the TV ownership cap to 24 stations serving no more than 35% of homes.[9] However, as this text went to press, the 1985 ownership cap remained in place—at 12 stations with 25% coverage. Alternative proposals were being introduced, including a cap of 18 stations and 30% coverage and ownership of any number of stations, so long as national coverage remained under 30%.

The reader is urged to consult the trade press for up-to-date information on specific ownership caps. The trend, however, is inescapable: multiple ownership of broadcast properties by a single entity is and will likely remain on the upswing.

The new rules have sparked a flurry of activity in sales of broadcast stations as existing groups seek to expand the number of stations they own, to move into bigger markets, or to spin off properties to other groups. The issues involved and the implications of this development are examined in more detail below, in the discussion of forms of media ownership.

The One-to-a-Market Rule Simply stated, the one-to-a-market (or one-to-a-customer) rule means that a single entity cannot own a radio and television station in the same market. The intent of this rule, of course, is to stimulate diversity and to avoid a local monopoly on information and entertainment.

Since first introduced in 1968 and modified in 1971, the one-to-a-market rule has reduced intramarket concentration and stimulated the number of broadcast group owners nationwide. At the time that the rule was adopted, it was not made retroactive, so companies that had title to two or more stations in their market were allowed to retain them. Such companies are referred to as grandfathered combinations.

Duopoly As with multiple ownership, there is evidence that the one-to-a-customer rule is an endangered species. The new FCC radio limits just discussed allow for a single company to operate two AM or FM stations in a market, a situation known as *duopoly*.

Part of the FCC's proposed rules on multiple ownership of television stations calls for allowing TV stations to buy a second operation in the same or an overlapping market. In addition, the commission is examining whether to permit new co-ownerships of radio and TV stations in the same market. However, as this edition goes to press, with the exception of grandfathered combinations, duopolies are still not permitted in television.

Since these rules were introduced, station owners in the nation's larger markets have scurried to gain control of a second station to create a formidable powerhouse.[10] The idea is to own the top two stations in a given format, in order to create an unbeatable ratings and advertising combination. On the AM side, owning two all-news or news-talk stations would be ideal; on FM, owning two leading country or rock stations would pave the way for long-term profitability and some immunity from competition.

As an example, consider the case of two leading radio groups, Viacom and Group W.[11] In 1993, Group W traded its two stations in Houston to Viacom, in exchange for the latter's combination in Washington, D.C. To sweeten the deal (and because Washington is a larger market), Group W tossed in $40 million.

The result was the creation of a "format fortress" in Houston for Viacom consisting of both KILT-AM/FM and KIKK-AM/FM, the market's leading country stations. In Washington, Group W added country leader WMZQ-AM/FM to WCPT-AM, an all-news station, and WCXR-FM, a classic rock station. Should a competitor plan to take on WMZQ, Group W could switch WCXR-FM to a country format, effectively negating the challenge.

Lease-Management Agreements Short of outright ownership of a second station in the same market is the possibility of a lease-management agreement (LMA).[12] In recent years, some struggling radio stations have leased portions of their airtime to a stronger station in the same market. In this way, the stronger station is able to add a bonus audience for its advertisers and to offer more than one kind of format for its sales force to sell. By paring its staff and reducing its overhead, the struggling station is able to stay in business.

The FCC has been keeping a close eye on LMAs, also known as local market agreements, time brokerage agreements, or radio joint ventures. Stations which lease their time do not sacrifice their public service obligation, and, as licensees, they maintain sole responsibility for the content which emanates from their towers. However, given the difficult economic climate for radio and the sheer number of stations now on the air, it is likely that LMAs will continue to operate in some fashion for the foreseeable future.

Caps on Multiple Cable Ownership: On the Horizon? It is ironic that, as radio and television ownership restrictions are being eased, cable ownership caps may be on the way. Until now, there has been no limit on the number of cable systems or subscriber households that could be controlled by a single company. The result has been the creation of huge multiple-system operators (MSOs) such as TCI, Time Warner, Continental, Comcast, and others discussed later in the chapter.

However, the re-regulatory fever which has gripped Congress and the FCC over charges of monopolistic behavior by cable companies has led to an attempt to curb the size and influence of the MSOs. In 1993, the FCC proposed cable ownership caps.[13] If passed in its present form, the rule would prohibit a single company from owning cable systems serving more than 25% of the nation's television households. Early reaction to the proposal was mixed, with one cable executive suggesting it was "no big deal" and no more than a "nuisance." Some have suggested that this attitude is behind much of cable's troubles, and may lead to early passage of the FCC proposal or perhaps to even more stringent curbs on cable ownership. Time will tell.

Cross-Ownership Rules

Cross-ownership regulations are designed to prevent the common ownership of more than one type of media outlet in the same market. As with multiple ownership, the rules are changing. Here's how things stood in recent years.

Broadcasting/Newspaper Cross-Ownerships Since 1978, common ownership of a broadcast station (radio or TV) and a newspaper in the same market has been illegal. As was the case with one-to-a-market regulations, existing cross-ownerships found to be operating within the public service guidelines were "grandfathered." However, the courts extended the FCC's mandate to allow the commission to break up existing cross-ownerships when sold and to review each grandfathered company's case when warranted. For example, in 1993 Rupert Murdoch was granted a waiver so that he could regain control of the failing *New York Post* and still retain ownership of television station WNYW.

Broadcast/Cable Cross-Ownerships Local television stations may not also own cable systems in their city of license. On this point, too, recent industry practice is different from regulatory theory: television/cable cross-ownerships can petition (and many have petitioned) the FCC for a waiver of this rule.

Network/Cable Cross-Ownerships A 1970 rule prohibited the "big three" commercial TV networks (CBS, NBC, and ABC) from owning cable systems. However, that rule was relaxed in 1992. Much to the dismay of some of their affiliates, networks can now own cable systems. However, CBS announced it had no immediate plans to enter cable.[14] NBC and CapCities/ABC were already in the programming side of the business. NBC owned CNBC and was planning to launch additional networks; ABC was the primary owner of ESPN, among other cable programming ventures. The "fourth network," Murdoch's Fox Broadcasting Company, is already a participant in cable, coventuring with TCI and other cable companies.

Cable/Telephone Company Cross-Ownerships Since telephone companies hold a local monopoly in their service areas, the FCC traditionally held that telephone companies could not also own area cable television systems. However, under provisions of the Cable Communications Act of 1984, Congress gave the FCC the power to grant waivers of this rule in areas where cable service might otherwise not be available to local residents. Thus, cable/telephone company cross-ownerships became possible in small, rural communities.

As pointed out above, the RBOCs have recently been permitted entry into the video services business, including ownership of cable systems. As of 1993, some telcos active in cable ownership included Southwestern Bell, which has purchased two cable systems in the Washington, D.C. area, and U.S. West, which now offers cable TV services in Denver (perhaps not coincidentally, the home of some of cable's most powerful companies, including TCI and Jones Intercable)![15]

The impact of the proposed mergers between cable MSOs and RBOCs with respect to this rule remains to be seen. However, most observers predict that RBOCs will be permitted to own cable systems so long as their phone companies and cable systems are not in the same service area. A second scenario would allow RBOCs to operate in the same area in which they provide telephone service so long as their cable operation is a new build, with residents having a choice between an RBOC's cable service or that offered by a preexisting cable provider. However, none of this had been codified as this book went to press.

A Cautionary Word

Clearly, the issues of multiple ownership and cross-ownership are among the most volatile in contemporary telecommunications. As the rules continue to be relaxed, a public outcry against monopolistic behavior may ensue (see below), followed by a regulatory backlash of new ownership curbs. So, it is likely that the rules will undergo another round of changes before the next edition of this book reaches print. Beware!

FORMS OF MEDIA OWNERSHIP

Assuming the applicant for a television or radio station or the franchisee of a cable system has met the criteria for ownership discussed above, the next step is to obtain the venture capital needed to build or buy the station. The capital-intensive nature of the telecommunications business has given rise to a wide variety of ownership patterns and types—from small, independent enterprises to global communications conglomerates. This section tracks these major forms of media ownership.[16]

Sole Proprietorships: Mom-and-Pop Stations

Sole proprietorships are small businesses which are owned and operated by a single person. Broadcasting has a long tradition of such ownership; in the industry, these stations are referred to as mom-and-pop operations. Many mom-and-pop stations are daytime radio stations, mostly in rural areas. Many also have interests in local newspapers and have obtained waivers from the FCC allowing them to operate local cable systems or newspapers.

There are many advantages of sole proprietorship in broadcasting, cable, or other telecommunications ventures. First and foremost, revenues and profits need not be shared with any other group or individual. No special corporate or local taxes are incurred, and the decision to expand, diversify, or liquidate can be made by a single person. On the employment side, working at small mom-and-pop stations permits great flexibility and freedom, away from craft and union regulations, albeit typically for low wages. Many present media managers fondly recall "paying their dues" at small stations in the hinterlands and credit such operations with nurturing and kindling their interest in a media career.

Of course, there are obvious limitations to a sole proprietorship arrangement. For one thing, the continuity of the company is dependent on the health and well-being of a single person. In addition, operating a broadcast or related facility is costly, but in mom-and-pop operations, station assets are limited to the venture capital and borrowing power of one owner. And it is that owner's money, personal assets (including house and car), and credit rating which are directly at risk. In addition, the profits of a sole proprietorship are taxed as personal income. This is a disadvantage, since the personal income tax rate is normally much higher than the corporate rate.

For these reasons, recent years have been marked by a sharp decrease in mom-and-pop television, radio, and cable companies. In their place, an amalgam of partnerships and corporations has evolved.

General and Limited Partnerships

A general partnership consists of the pooling of the capital and skills of two or more people as co-owners of a business. Like a sole proprietorship, a partnership is not a separate legal entity; thus liability extends to the personal assets of the partners, and each partner is taxed on his or her share of the partnership income at the personal income tax rate.

Partnerships have a limited life; they may dissolve at the death or withdrawal of a partner or upon the addition of a new partner. A limited partnership extends the general partnership to permit investors to become partners without assuming unlimited liability. Limited partners will usually risk only as much as their original investment. At least one member of a limited partnership must be a general partner, who assumes unlimited liability.

The major advantage of a partnership is profit and risk sharing. Partnerships allow media entrepreneurs to pool their assets and borrowing power as well as to share their losses. As with sole proprietorships, there are no special partnership taxes. The continuity of a partnership is determined by the health and longevity of the partners, and upon sale or transfer of ownership, the value of each partner's interest must be determined and settled.

Partnerships have their drawbacks. For one, they are at least partially based on the ability of all partners to get along with one another—no simple task in an ego-driven enterprise. In addition, partnerships are limited in their ability to generate capital. Raising new money means opening the partnership to new members who may or may not share the operating philosophy of the original investors. And partnerships can be difficult and messy to dissolve; determining the value of each principal's interest can lead to lawsuits, countersuits, and much ill feeling.

For these reasons, like sole proprietorships, limited and general partnerships are generally few and far between in contemporary telecommunications. The vast majority of our radio and television stations, cable companies, and new delivery and storage systems are owned by large corporations. The reasons are purely financial: filing articles of incorporation limits the liability of the principal owners of a business and widens opportunities for capital formation. Operating in the contemporary media milieu requires the protective armor afforded by large corporate structures.

Corporations

In the past 50 years, the media industry in the United States has been marked by an increasingly corporate structure. Unlike a sole proprietorship or general partnership, a corporation is a separate legal entity. It has a "life" of its own, apart from its owners; in the eyes of the law, a corporation is a person—it can make contracts and pay taxes and it is liable for bad debts. The strength of a corporate structure lies in the fact that creditors have claim only against the assets of the corporation; stockholders' liability is limited to the fixed amount of their investment. In addition, the corporation's unique identity allows for its continued life after the death of stockholders, the sale of stock, or the transfer of ownership.

In return for its legal status, corporations must comply with local, state, and federal regulations. They must be registered and approved by the secretary of state of the state in which they operate; if they do business in more than one state, they must comply with federal guidelines pertaining to interstate commerce.

There are two types of corporations: public and private. Private corporations generate capital from the assets of their investors; public corporations can sell stock to raise additional capital. Corporations can be mixed, with a percentage of control in public hands and a percentage in private hands.

THE TELECOMMUNICATIONS PYRAMID

There are three general classes of corporations in the contemporary telecommunications business environment: (1) stand-alone operations, (2) broadcast groups and cable MSOs, and (3) multinational media and entertainment conglomerates. These categories can be viewed as constituting an inverted pyramid of control and influence in telecommunications. Interestingly, many of those at the top began at the bottom, as small stations or groups. Keep in mind that these categories are merely convenient guidelines and are not mutually exclusive. For example, the point at which a group owner becomes a conglomerate is not always distinct, especially in this era of "merger mania."

Stand-Alone Operations

Stand-alone corporations are at the bottom of the media ownership pyramid. Essentially, they are the locally owned mom-and-pop media businesses, including small radio stations, cable systems, newspapers, and even a few television stations. These companies have filed articles of incorporation mainly for self-protection: to free their personal assets from direct liability. In many instances, such stations are wholly owned by a single person or a family and may incorporate under "pet" names or words formed by combining the names or initials of the major owners. *Broadcasting and Cable Yearbook* lists a number of such personal ownership touches, from Chermi Communications in Oak Creek, Colorado (owned by *Cher*yl and *Mi*chael Barry), to the Paul Bunyon Broadcasting Co. in Bemidji, Minnesota, to Cactus Broadcasting, owners of KXEW in South Tucson, Arizona.

Many of these stand-alone operations are group owners in a sense, operating AM-FM combos or maintaining a grandfathered television/radio or newspaper/radio/television cross-ownership. However, they are unlike group owners in that all holdings remain in the same city or area of license.

Broadcasting Groups and Cable MSO

Next in the hierarchy of media ownership are station and cable system clusters: groups of media facilities located in a variety of regions, but with a single home office. In broadcasting, these organizations are known as group owners; in cable, as multiple-system operators (MSOs).

Major Radio Groups As we have seen, relaxed multiple ownership rules have had a pronounced effect on the radio business. Today, most radio stations, especially those in the major markets, are owned by large groups. Table 7-1 lists the major radio groups in the mid-1990s.

TABLE 7-1 RADIO'S TOP-20 GROUPS

Rank	Company	Number of stations	Estimated audience*
1	CBS	21	14,720,800
2	Infinity	25	13,425,200
3	CapCities/ABC	18	11,838,400
4	Group W	18	11,809,500
5	Shamrock (Disney)	21	7,641,900
6	Viacom	14	6,228,700
7	Cox	14	6,090,400
8	Evergreen	11	5,680,300
9	Bonneville	15	5,564,100
10	Emmis	5	4,991,800
11	Greater Media	16	4,485,500
12	Gannett	11	4,314,100
13	Clear Channel	26	4,103,500
14	Susquehanna	16	3,820,300
15	Summit	7	3,593,000
16	EZ Communications	16	3,442,500
17	Jacor	13	3,349,500
18	Pyramid	11	3,043,800
19	Great American	13	3,040,900
20	Tribune	6	2,841,000

*Number of weekly listeners (12+), summer 1993.
Source: *Broadcasting and Cable*, December 6, 1993, pp. 84–86

Leading the pack are the stations owned and operated (O&O) by CBS. Many of the CBS O&Os are combinations in major cities, like WCBS-AM, an all-news station, and WCBS-FM, the king of oldies stations, in New York. CBS also has profitable combos in Chicago, Los Angeles, Detroit, Minneapolis, and St. Louis. Together, the CBS radio stations reach about 14 million people each week, and contribute over $200 million to the company's annual revenue (about $12 million per station).

Led by mercurial entrepreneur Mel Karmazin (and the company's most infamous radio personality, Howard Stern), Infinity has followed a path of increasing expansion in recent years to rank as the second largest radio group. By mid-decade, Infinity operated 25 stations and reached more than 13 million people on a weekly basis.

Group W, the Westinghouse radio group, is one of broadcasting's oldest, most respected, and most profitable. Group W operates powerhouse AM-FM combinations in Boston, New York, Philadelphia, Los Angeles, and Washington.

Other major radio groups include Capcities/ABC, Shamrock (the radio arm of the Disney empire), Evergreen, Bonneville, Susquehanna, and Jacor, to name a few.

Major Television Groups Some of the same names crop up when we look at the list of America's top TV groups in Table 7-2. Leading the way is CapCities/ABC, whose 8 stations reach nearly a quarter of all households with TV in the United States. ABC-owned stations include powerful and popular VHF facilities in New York (WABC-TV), Los Angeles (KABC-TV), Chicago (WLS-TV), Philadelphia (WPVI-TV), San Francisco (KGO-TV), and Houston (KTRK-TV), among others.

Next in rank is CBS, with 7 stations reaching 22% of U.S. homes. CBS owns stations in New York, Los Angeles, Chicago, Philadelphia, Minneapolis–St. Paul, Miami, and, for diversity, the comparatively small town of Green Bay, Wisconsin (WFRV-TV).

General Electric's NBC owns stations in the "big three" markets (New York, Los Angeles, and Chicago), plus Washington, Miami, and Denver. Together, NBC-owned stations reach 1 in 5 TV homes. A pioneer in independent television, Tribune (also a major newspaper publisher) controls 7 stations, reaching just under 20% of U.S. TV households. Among these are WGN in Chicago, WPIX in New York, KTLA in Los Angeles, WPHL in Philadelphia, WGNX in Atlanta, and KWGN in Denver.

Rounding out TV's top 10 are Rupert Murdoch's Fox Broadcasting Company (8 stations, reaching just under 20% of households), Silver King (formerly HSN Communications), Chris-Craft, Univision (whose stations target Hispanic-Americans), Gannett (also active in radio and newspapers), and Westinghouse's Group W, which we've previously encountered as a radio leader.

Leading Cable MSOs A glance at the bottom of Table 7-3 might suggest that cable is less concentrated than radio or television. The MSOs ranked from 11 to 20 in size serve between 600,000 and just over 1,000,000 cable homes each—

TABLE 7-2 TV'S TOP-20 GROUPS

Rank	Company	Number of stations	Percentage of U.S. TV households
1	Capital Cities/ABC	8	23.81
2	CBS	7	22.1
3	NBC	6	20.36
4	Tribune Broadcasting	7	19.57
5	Fox*	8	19.42
6	Silver King Communications	12	18.605
7	Chris-Craft/United Television	8	18.195
8	Univision	9	10.62
9	Gannett Broadcasting	10	10.335
10	Group W	5	9.85
11	Telemundo Group	6	9.355
12	SCI Television	7	8.815
13	Scripps Howard	10	8.645
14	Cox Enterprises	7	8.56
15	Hearst Broadcasting	6	6.77
16	A. H. Belo	5	5.83
17	Pulitzer Broadcasting	10	5.505
18	Disney	1	5.35
19	Providence Journal	9	5.28
20	Great American	6	5.26

*In 1994, Fox invested $500 million in New World Communications, a group of 12 additional stations.
Source: *Broadcasting and Cable*, March 22, 1993, pp. 29–31.

TABLE 7-3 CABLE'S TOP-20 MSOs

Rank	Company	Subscribers*	Percentage of U.S. cable households*
1	Tele-Communications Inc.	9,686,000	18.62
2	Time Warner Cable	6,807,330	13.09
3	Continental Cablevision	2,855,000	5.49
4	Comcast	2,852,000	5.49
5	Cablevision Systems	2,008,986	3.85
6	Cox Cable	1,722,007	3.31
7	Jones Intercable/Spacelink	1,586,233	3.05
8	Newhouse Broadcasting	1,321,806	2.54
9	Adelphia Communications	1,189,000	2.28
10	Cablevision Industries	1.175,422	2.26
11	Times Mirror Cable Television	1,126,775	2.16
12	Falcon Cable	1,122,000	2.15
13	Viacom Cable	1,116,000	2.14
14	Sammons Communications	940,400	1.80
15	Century Communications	917,000	1.76
16	Paragon Communications	838,341	1.61
17	Crown Media	813,720	1.56
18	Colony Communications	766,294	1.47
19	TeleCable	680,400	1.30
20	Scripps Howard	649,181	1.24

* 1993 estimated figures.
Source: *Cablevision*, April 5, 1993, p. 42.

John C. Malone, CEO of Tele-Communications Inc., cable's largest MSO. (courtesy of TCI)

only about 2% of the total number of subscribing households in the country. But the key to the list is at the top.

Cable is dominated by two huge multiple-system operators: Tele-Communications Inc. (TCI) and Time Warner. TCI boasts almost 10 million subscriber homes, accounting for 1 in 5 monthly cable TV bills. Time Warner owns systems serving about 7 million subscriber households, about 13% of total cable TV billing. Other cable leaders include Continental Cablevision, Comcast, Cablevision Systems, Cox Cable, Jones Intercable, Newhouse, Adelphia, and Cablevision Industries (which is not affiliated with Cablevision Systems).

The size disparity between larger and smaller MSOs is enhanced when the nation's largest (and most profitable) systems are taken into account. Among the top 10 individual systems are four owned by Time Warner (in New York, Orlando, Houston, and Honolulu). TCI owns three of the largest single systems, in the suburbs surrounding Chicago, Denver, and Philadelphia (Wayne, N.J.).

The power of the major MSOs also resides in the fact that, unlike the telephone and broadcast companies and TV networks, cable operators have historically been permitted to own interests in production companies, syndication firms, and other vertical ventures. This is why we'll encounter some of these same firms again, in the discussion of powerful media conglomerates.

Multinational Media/Entertainment Conglomerates

Communications conglomerates are those large corporations whose interests extend across all forms of media—from radio and television stations, cable and satellite interests, and production companies to magazines, newspapers, and book companies, and to electronics manufacturing and retail outlets. These companies are the major players in the telecommunications game.

The flurry of mergers and acquisitions which characterized domestic media in the 1980s also struck the international scene. In fact, recent years have seen a blurring of long-standing distinctions between types of media corporations and the creation of global giants, active on both the hardware (technology) and software (programming) sides of the business.

The End of American Dominance

One of the casualties of increasing globalization of the media may have been American dominance. The late Steven Ross, CEO of Time Warner, lamented this fact in a 1990 speech:

> Certainly, there is no doubt about the immense strength of our media and entertainment industry. Entertainment is . . . America's second-largest net export. But, unfortunately, as a nation, we seem to be missing the opportunity in front of us. Instead of creating partnerships, we're selling the store. Since 1985, foreign companies have spent over $35 billion to acquire a wide range of U.S. media and entertainment properties. I think this "fire sale" of American media and entertainment properties is a long-term disaster for the United States.[17]

While Ross did not live long enough to see the outcome of the "fire sale," there is little doubt that it took place.

Between 1985 and 1990, nearly 150 major American media corporations were purchased by foreign-based conglomerates.[18] Many became targets because of the debt load left over from an earlier period of takeovers and leveraged buyouts. Others were characterized by poor management. Still others simply sold out for a huge price, too good to be refused.

Figure 7-1 illustrates the trend of acquisitions of American firms by foreign-based corporations. Nearly 4 in 5 of the foreign takeovers occurred in the newspaper, publishing, and advertising businesses. In addition, 3 in 4 of the takeovers were by British or Canadian companies. But, from the standpoint of the television, cable, music, and entertainment businesses, the scenario was (to borrow from a recent popular song) "I think we're turning Japanese."

Rebuffed by American banks (many of which were having significant financial problems of their own), many American media companies increasingly sought capital from Japanese institutions. Thus, by the early 1990s, TCI alone was carrying over $3 billion in foreign debt: much of it to Japanese banks.

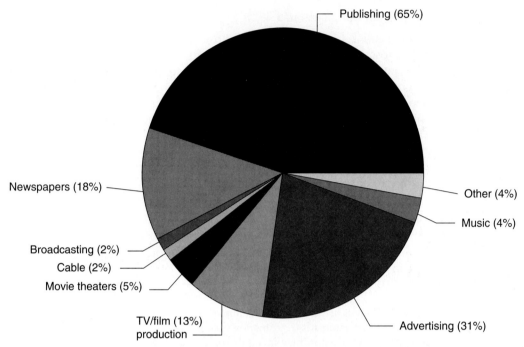

FIGURE 7-1 Foreign takeovers of American media corporations. 1985–1990. *Source*: Based on Table 1 in R. Carveth, "The Global Media Marketplace," *Communication Research*, vol. 19, no. 6, December 1992, p. 715. Total acquisitions: 144.

On a more public level, Japanese media companies made a concerted effort to acquire American-based programming to feed their VCRs, CD players, and walkabout stereos. In 1990, Matsushita (parent company of familiar brand names Panasonic and National, among others) acquired MCA/Universal, including its record companies, movie studios, TV production suites, and theme parks. Sony added the Columbia motion picture and television companies to its earlier acquisition of CBS Records (and the Columbia Record Club, among other recording and music properties).

The Japanese have not been alone in their pattern of media expansion. In the music business alone, the United Kingdom's Thorn EMI now owns Capitol; Germany's Bertelsmann A.G. owns the RCA and Arista labels; and Holland's Philips is the owner of Polygram, Island, and A&M. Perhaps one reason for Steven Ross' alarm was the fact that Time Warner is now the only major U.S.-owned record company.

INTERNATIONAL MEDIA CONGLOMERATES: THREE PERSPECTIVES

To provide some insight into the influence of the world's largest media corporations, let's examine three lists. The first is an alphabetical array of 16 multinational

Australian media magnate Rupert Murdoch (standing, right) at the headquarters of Sky Television in Europe. (Reuters/Bettman)

media companies, including their degree of vertical integration, home country, and gross revenues.

The second ranking is based on media activity in Europe, which has been a particularly volatile market in recent years owing to the end of the cold war, the increased privatization of media properties, and the long-delayed introduction of new technologies, such as cable and satellite television.

Finally, we return to the United States for one last look at the domestic media giants. This list ranks the firms active in American mass media on their gross revenues, and the proportion of those revenues which derives from their broadcast and cable activity.

The Global Giants

Table 7-4 presents a thumbnail sketch of the some of the world's largest, most vertically integrated media companies. Some of the names have already been mentioned; others may be less well known to the reader. However, each shares the characteristic of being a major force in global telecommunications. There are numerous ways to interpret this list. One is to examine total media sales, or gross revenues from operations.

On the basis of sheer volume, General Electric is the world's most powerful media corporation, with global revenue approaching $60 billion. However, the

TABLE 7-4 GLOBAL MEDIA GIANTS

Company Country	Bertelsmann Germany	CBS USA	Capital Cities/ABC USA	Disney USA	Fininvest Italy	General Electric USA	Hachette France	Matsushita Japan	News Corp. Australia	Paramount** USA	Sony Japan	TCI USA	Time Warner USA	Tribune USA	Turner Broadc. USA	Viacom** USA
Media segment																
Original films				•	•			•	•	•	•		•		•	•
Film library					•			•	•	•	•		•		•	
Theaters					•			•		•						
TV program(s)				•	•			•	•	•	•		•			•
TV station(s)		•	•		•	•			•	•	•		•			
TV network(s)		•	•		•	•			•				•			•
Basic cable network(s)	•		•						•			•	•		•	•
Pay-cable network(s)	•			•			•						•			•
Cable systems	•						•					•	•	•		•
Recorded music								•			•		•			
Theme parks				•				•					•			
Pro sports										•			•	•	•	
Publishing	•			•	•		•		•	•			•	•		
Audio/video equipment								•			•					
Gross revenues*	8.12	3.26	5.38	5.84	14.1	57.66	5.10	39.1	2.74	3.87	26.1	3.63	11.51	2.35	1.39	1.60

* 1990 figures; in billions of U.S. dollars.
** In 1994, Viacom acquired Paramount after a bitter fight with QVC Network.
Source: *Standard and Poor's Industry Surveys,* 1992, March 1992, p. L-20; *Hoover's International Business,* 1993.

influence of GE on media programming is comparatively limited. In fact, at the time of this writing, the company was reportedly investigating the sale of its TV network (NBC), while its forays into cable television (CNBC and the new Talk-America) were "small potatoes" compared with the activities of other companies.

Two other long-standing hardware giants are next in gross revenues. Both are from Japan and, as we have seen, both recently diversified their leadership in VCRs and audio equipment by buying American entertainment companies. Matsushita billed nearly $40 billion in 1990; Sony $26 billion.

The companies on the next rung derive most of their income from media programming. In 1990, America's Time Warner reported gross revenues of nearly $12 billion. A quarter of that income came from feature films (including the hugely successful *Batman*). The company's music and publishing divisions (from Warner Music to Time-Life Books) each contributed an additional 25% to earnings. The remaining Time Warner income consisted of revenue from cable television—as a major MSO, and as the owner of HBO, The Movie Channel, and newer ventures (such as Comedy Central and Court TV).

Italy's Fininvest, owned by Silvio Berlusconi, reported 1990 gross revenues in the $14 billion range. Berlusconi's properties include television networks, film production companies, newspapers, magazines, even the immensely popular soccer team AC Milan. Berlusconi is often compared with Ted Turner for his brash, outspoken style.[19] Like Turner, he has made many of his acquisitions with huge debt, which separates him from his European competitor, Bertelsmann. Unlike Turner, his politics are considered right-wing. In 1994, he was elected Italian Prime Minister.

Germany's Bertelsmann made its mark in the bookselling field. In recent years, the company has expanded its media portfolio to include record and tape clubs and cable television ventures. Its influence in America derives from its ownership of the RCA record label and Doubleday bookstores, among other holdings. Its Gruner+Jahr publishing subsidiary owns *Woman's Day* and *Parents*, plus many other popular titles.

A second way to interpret Table 7-4 is to examine the degree of vertical integration. On this basis, the world's most powerful media corporations include American-based Viacom-Paramount, Disney, and Time Warner, plus Japanese giants Sony and Matsushita, Berlusconi's Fininvest, and Australian Rupert Murdoch's News Corporation, the parent of Fox. Smaller but still highly integrated companies are Turner Broadcasting and TCI. By this criterion, the established TV network CBS seems comparatively small: it has sold off its recording, publishing, and movie holdings to concentrate on its core network and owned station divisions.

Any way you look at it, the media conglomerates listed in Table 7-4 are huge firms with vast international media activity and influence.

A Look at Europe

Table 7-5 lists Europe's leading media corporations. It should not be surprising that Bertelsmann and Fininvest lead the pack, with revenues from European media

TABLE 7-5 EUROPE'S TOP-20 MEDIA COMPANIES

Rank	Company	Media revenue*	Gross revenue*
1	Bertelsmann	6836.75	9610.84
2	Fininvest	6406.00	14102.00
3	Havas	3498.76	4696.38
4	Groupe Hachette	3320.99	5390.11
5	Reed	2435.63	2876.54
6	Axel Springer Verlag	2013.25	2217.65
7	RCS Editori	1958.90	2061.95
8	Pearson	1796.30	2822.57
9	Heinrich Bauer	1602.71	1710.54
10	CLT	1463.19	1463.19
11	VNU	1316.58	1462.48
12	News International	1180.25	1227.81
13	Canal Plus	1148.17	1240.34
14	Wolters Kluwer NV	1145.99	1273.26
15	RTL Deutschland	1144.58	1144.58
16	Daily Mail and General Trust	1132.45	1170.72
17	United Newspapers	1118.17	1433.16
18	Elsevier	1092.51	1213.90
19	TF1	1091.81	1158.63
20	Carlton Communications	961.73	1120.46

* 1991–1992 estimates; in millions of U.S. dollars.
Source: Media International, April 1993, p. 13.

operations exceeding $6 billion each. Havas and Hachette, two French publishing giants, are next, followed by Britain's Reed and Germany's Axel Springer Verlag. Recently, Reed (fifth in media revenue) merged with Elsevier (eighteenth). The pooled media revenues of this new combination place it on a par with the "two H's," Havas and Hachette, but still some distance away from the "big B's": Bertelsmann and Berlusconi (Fininvest).

Two other points can be made about this listing. The first is that government-owned operations, which are still large and influential in Europe (such as Britain's BBC, France's Antenna-2, and Germany's ARD and ZDF), are not included. Second, note that American and Japanese firms have been relatively unsuccessful in breaking into the European market. American investment has been hindered by foreign regulators who seem ever-mindful of the excessive influence of American habits and mores in films and TV shows. While their hardware sales in Europe have been strong, the Japanese set their sights in recent years on the larger and more lucrative U.S. media market.

As private media enterprise replaces government ownership, and as the newly democratized countries of eastern Europe develop their consumer economies, look for this list to become less provincial and more global: like that in the United States, for example.

American Electric: Media Ownership in the United States

Table 7-6 lists the largest media companies in the United States on the basis of gross revenues. Vertical giant Time Warner heads the pack, with gross sales above $5 billion, most from its owned cable networks and systems. Capital Cities/ABC is also a $5 billion company, but its powerhouses are its TV stations and networks.

Following the two giants are six companies which bill in the $3 billion range. Topping the list at $3.4 billion is newspaper conglomerate Gannett, which publishes *USA Today* as well as dozens of other daily and weekly newspapers (and also appears in the top ranks of radio and TV group owners). Cable giant TCI is next, with its $3.2 billion emanating entirely from its cable systems and networks. Global powerhouse General Electric earns more than $3 billion from domestic media (but only a small portion of its worldwide $60 billion); as does CBS and two newspaper giants, Advance Publications and Times Mirror. The latter two companies are represented in cable television through their ownership of a number of midsized systems.

Murdoch's News Corporation makes another appearance on this list. The Australian news tycoon's ownership of the various Fox Broadcasting properties places him among the top 10 U.S. media giants. Three other cable companies

TABLE 7-6   AMERICA'S TOP-20 MEDIA COMPANIES BY REVENUE*

Rank	Company	Total media revenue	Broadcast revenue	Cable TV revenue	Other media revenue
1	Time Warner	5229.0	—	3301.0	—
2	Capital Cities/ABC	5172.0	3712.0	489.4	128.3
3	Gannett Co.	3382.0	360.1	—	392.1
4	Tele-Communications Inc.	3206.0	—	3206.0	—
5	General Electric Co.	3153.5	3086.0	67.5	—
6	CBS Inc.	3035.0	3035.0	—	—
7	Advance Publications	3013.0	—	440.0	—
8	Times Mirror Co.	2763.3	94.2	403.8	—
9	News Corp.	2278.0	964.0	—	575.0
10	Knight-Ridder	1953.8	—	—	—
11	Hearst Corp.	1947.2	265.2	—	—
12	Cox Enterprises	1716.0	409.0	596.0	—
13	New York Times Co.	1703.1	66.5	—	9.5
14	Tribune Co.	1636.6	495.0	—	—
15	Viacom International	1459.3	159.1	1300.2	—
16	Thomason Corp.	1419.6	—	—	—
17	Washington Post Co.	1292.2	163.5	159.5	—
18	Turner Broadcasting System	1190.7	—	1190.7	—
19	E. W. Scripps	1161.6	245.4	225.2	—
20	Continental Cablevision	1127.0	—	1127.0	—

* 1991 estimates; in millions of U.S. dollars.
Source: Advertising Age, January 4, 1993, p. 18.

break the top 20, with annual domestic billings above $1 billion: Viacom, Turner, and Continental.

The bottom line of this quick trip around the world is simple. The world's media are increasingly dominated by large, vertically integrated, diversified multinational corporations. Is that a good thing? Concerns about media monopoly head our next section, which reviews current issues in telecommunications ownership.

CURRENT ISSUES IN MEDIA OWNERSHIP

Concentration

That media ownership has become increasingly concentrated, with fewer and fewer companies owning more and more media outlets, is not in dispute. However, whether such concentration limits the free expression of ideas, is contrary to the public interest, restricts entry of "new blood" into the media marketplace, and produces programming of poor quality is a source of major debate among media management, telecommunications lawyers, citizens' groups, academicians, and other interested parties.

Addressing these concerns, the Rand Corporation in 1974 published a comprehensive study of the effects of conglomerate ownership on rate controls, quality of programming, and corporate abuse and excess.[20] The researchers concluded that there was no proof that media concentration had negative effects and that, in fact, the form of media ownership generally has a small impact on economic and program performance.

Four years later, the Federal Trade Commission conducted an inquiry into media concentration. During 2 days of testimony, experts from the legal, academic, economic, and business professions traded accusations and offered evidence for and against the concentration of ownership in the print and electronic media. The results, published in two volumes, are contradictory and inconclusive.[21]

The subject has also received considerable attention in the academic press.[22] Some recent findings include the following:

• There has been a definite increase in the number of communications conglomerates.

• Through the 1980s, cable television, broadcast TV, radio, recordings, movies, and the print media became more concentrated. Of these, concentration in cable caused the most consternation among regulators.

• While the ownership of American media properties by foreign companies has increased, the pace of the increase has slowed. Thus far, foreign ownership has not seemed to materially affect the editorial independence of American media firms.

• While the United States may have lost its dominant world position in telecommunications equipment, its strategic advantage in content (popular movies, television shows, recording artists, and the like) enables it to continue to maintain a competitive edge in domestic and world markets.

The bottom line is that telecommunications is at the vanguard of the so-called new world order. No other business (except, perhaps, the fast-food industry) features the ownership, technology, infrastructure, and consumer demand to make it truly a global enterprise. The issues of concentration and foreign ownership will not go away, as the media moguls struggle to export their programs and technologies to eager audiences around the globe.

Joint Ventures and Coventures

The pace of media innovation and concentration, a shortage of venture capital (matched by burdensome debt loads), and the laxity in federal regulatory initiatives have stimulated a new ownership phenomenon: an unprecedented rise in the number of joint ventures among media companies.

Just a few years ago, telecommunications firms operated within a cloak of secrecy and according to a strict code of honor. Employees were fired for sharing rate or program information with their competitors; and companies moved quickly to acquire sole control of patents and copyrights to new programming or equipment proposals. The 1980s ushered in a new spirit of cooperation among many of these rivals.

The pooling of financial, technical, and creative expertise among media companies is known in the telecommunications trade as the *joint venture* or *strategic alliance*. Joint ventures have occurred among the large communications conglomerates as well as in the small media markets among local broadcasters, cable outlets, and newspapers.

The point of most joint ventures is to share risk, especially when new technologies are involved. For example, in 1993, the newsletter *Digital Media* compiled a list of strategic alliances in the field of interactive TV and multimedia. The list went on for over six pages of small type.[23]

In recent years, literally hundreds of joint ventures have matched media hardware companies, program suppliers, and distribution firms. Some examples:

• Hollywood's Carolco Films has teamed with cable conglomerate TCI to offer first-run movies on a pay-per-view basis (simultaneous with their theatrical release). Partners with Carolco in this venture and others include Japan's equipment firm Pioneer, Italy's Rizzoli, and France's Canal Plus.
• Time Warner, AT&T, and Apple Computer have met to form a strategic alliance for digital multimedia.
• Sega Enterprises, Time Warner, and TCI are teaming to offer video games over cable television.
• Sony, Matsushita, and Dutch giant Philips own equal shares in General Magic, which is working on a worldwide digital standard for computer and video software.
• Toshiba, Scientific Atlanta, and Itochu (another Japanese firm) are working on a cable TV box with computing capabilities.

Joint ventures are also on the rise at the local level. In Atlanta, Georgia, ABC affiliate WSB (a subsidiary of Cox Communications) joined forces with local cable companies to provide local news cut-ins throughout the day. Similar joint ventures between broadcasters and cablecasters are taking place in markets around the nation.

Such cooperation is quite a turnaround from the situation a few short years ago when broadcasters considered other groups, especially cable entrepreneurs, mortal enemies. As one media mogul observed as the rage to coventure engulfed him, "I'm in alliance shock." Expect that condition to continue through the rest of the decade.

Minority and Female Ownership

Another current ownership trend is a rise in the number of "nontraditional" entrepreneurs entering the telecommunications business. As a result of regulatory initiatives, some industry support, concentrated lobbying efforts, and the sheer multiplicity of channels, more women and members of minority groups are getting a share of the telecommunications marketplace.

In the early 1980s the FCC issued a report on the status of female ownership in broadcasting.[24] On the basis of a survey of over 900 AM stations, 700 FM stations, and 288 television stations (about 20% of the broadcasting station universe), the commission concluded that women participated in ownership in 68.8% of the AM stations, 69.4% of the FM stations, and 87.8% of the television stations. However, there is considerable debate about whether such figures indicate actual input by women into station management and operations. A follow-up report, the *Media Report to Women*, indicated that only about 50% of the women listed as owners in radio and television actually owned stock; the rest held positions on the board of directors, but had no stock and, therefore, no say in corporate actions.[25] Majority ownership by women was found in only 8.6% of AM stations, 9% of FM stations, and 2.8% of TV stations. In addition, where majority control existed, more than three-quarters of the majority owners were the relatives of other owners. The women holding stock in broadcast corporations were, more often than not, the wives, mothers, aunts, and sisters of those who controlled corporate affairs.

The major regulatory impetus for the increase in ownership by members of minority groups occurred at the FCC during the Carter administration. In 1978, the commission adopted a policy statement which proposed the granting of tax certificates and the allowance of "distress sales" of stations to members of minority groups in cases in which station licenses were designated for revocation. At the same time, the Small Business Administration began a policy of granting low-interest, long-term loans to minority media interests, and an industry group— the Minority Small Business Enterprise Committee (MSBEC)—was formed with the same goal. Similar initiatives were made in cable by the National Cable Television Association's Department of Human Resources. As a result, minority ownership in telecommunications increased such that by the mid-1980s there were at least 100 broadcast stations in which blacks held a controlling interest.[26] In addition, about 20 stations were owned by Hispanics. On the cable side, nearly 50 cable franchises were owned and operated by blacks, Hispanics, and Native Americans.

A Supreme Court ruling in 1990 affirmed the constitutionality of the "distress sale" rules, which allow a seller (to a minority buyer) to defer capital gains and attain other tax advantages. A 1991 review found that the policy had resulted in greater minority ownership of media properties.[27] Perhaps the most well known of the new minority owners in TV is Cook Inlet, a partnership of more than 6000 Alaskans, many native Eskimos. In 1987 Cook Inlet acquired an interest in WTNH in New Haven, Connecticut, and in 1988 the group became a major radio player by acquiring 11 radio stations from First Media Corporation. Black-owned Queen City Broadcasting (with partners Quincy Jones, O. J. Simpson, Julius Erving, and Patrick Ewing, among others) is the parent of WKGW-TV in Buffalo, plus other radio, TV, and cable properties. Granite Broadcasting, which boasts Oprah Winfrey among its investors, owns four TV stations: two ABC and two NBC affiliates.[28]

Despite some good news, there is a consensus among media executives—black and white, male and female—that the industry is still dominated by white males. While minority ownership of media properties has certainly increased, the increase is small: from 1% when the regulatory initiative was begun to about 3% today. By and large, minority-owned properties have not been among the most profitable or the most heavily involved in joint ventures.[29] Media properties owned and operated by blacks and women remain relatively rare. The huge barriers to entry faced by new competitors in this age of globalization and vertical integration are likely to keep minority and female ownership proportionately low for the foreseeable future.

SUMMARY

Ownership of telecommunications firms takes many forms and is subject to a variety of national and local restrictions, including financial, character, technical, and legal criteria.

General qualifications for broadcast ownership include demonstrated evidence of U.S. citizenship, good character, local participation, civic involvement, integration of ownership and management, diversification of ownership, prior media experience, and satisfactory financial and operations planning. Specific ownership regulations apply to both broadcast and cable entrepreneurs.

In radio, groups may not own more than 18 AM and 18 FM stations. In television, groups are limited to 12 stations reaching 25% of TV homes. There are no limits on the number of cable systems one company can own. Ownership regulations for radio, TV, and cable are expected to change in coming years.

Cross-ownership rules have generally restricted ownership in the same market of broadcast stations and newspapers, broadcast stations and cable systems, telephone companies and cable systems. However, these rules have been streamlined or relaxed in recent years.

Forms of media ownership range from small, individually owned mom-and-pop operations to large multinational media/entertainment conglomerates. In between are general and limited partnerships, broadcast groups and cable MSOs.

Important issues in media ownership include trends toward increased concentration of ownership, the rise of joint ventures between companies, and increased (though limited) participation by women and members of minority groups in media management.

NOTES

1 See, for example, Pat Aufderheide, "After the Fairness Doctrine: Controversial Broadcast Programming and the Public Interest," *Journal of Communication*, vol. 40, no. 3, Summer 1990, pp. 47–63; Lawrence W. Etling, *Listener Perception of Localism in Audio Programming: A Qualitative Study*, unpublished doctoral dissertation (Athens: University of Georgia, 1993).
2 Frederic A. Weiss, David Ostroff, and Charles Cliff III, "Station License Revocations and Denials of Renewal," *Journal of Broadcasting*, vol. 24, no. 1, Winter 1980, pp. 69–77.
3 See, for example, Michael O. Wirth and James A. Wollert, "Public Interest Program Performance of Multimedia-Owned TV Stations," *Journalism Quarterly*, vol. 53, no. 2, Summer 1976, pp. 223–230.
4 See *Window Dressing on the Set: Women and Minorities in Television* (Washington, D.C.: U.S. Commission on Civil Rights, August 1977).
5 See John C. Busterna, "Diversity of Ownership as a Criterion in FCC Licensing Since 1965," *Journal of Broadcasting*, vol. 20, no. 1, Winter 1976, pp. 101–110.
6 See Erwin G. Krasnow and J. Geoffrey Bentley, *Buying or Building a Broadcast Station* (Washington, D.C.: National Association of Broadcasters, 1982), p. 26.
7 See Nancy Jesuale and Ralph L. Smith (eds.), *CTIC Cablebooks* (Arlington, Va.: Cable Television Information, 1982); Robert E. Jacobson, *Municipal Control of Cable Television* (New York: Praeger, 1977).
8 "FCC/Radio Ownership (2); Scales Back March Rule Changes," *Dow Jones/News Retrieval Service*, August 6, 1992.
9 "Double the Number," *Broadcast Sales Executive Summary*, vol. 5, no. 21, May 25, 1992, p. 1.
10 "Duopoly: Cautious Approach," *Broadcasting and Cable*, April 19, 1993, pp. 17, 58.
11 "Viacom, Group W Swap Radio Stations," *Broadcasting Cable*, July 12, 1993, p. 20.
12 "FCC Sees Future for LMAs," *Broadcasting and Cable*, February 3, 1992, p. 36.
13 "FCC Would Cap MSOs at 25% of Homes," *Broadcasting and Cable*, June 28, 1993, p. 9.
14 "Approval Granted," *Broadcast Sales Executive Summary*, June 22, 1992, p. 1; *Business Week*, June 15, 1992, p. 41.
15 "The Telcos Are Coming," *Broadcasting and Cable*, March 1, 1993, p. 10.
16 The major forms of ownership are compared and contrasted in T. C. Carbone, "Option for Ownership," *Management World*, October 1982, pp. 36–37.

17 Steven J. Ross, "The Global Media and Entertainment Markets of the 1990's and Beyond," speech delivered before the Economic Club of Chicago, Illinois, December 18, 1990.
18 Rod Carveth, "The Reconstruction of the Global Media Marketplace," *Communication Research*, vol. 19, no. 6, December 1992, pp. 705–723.
19 See, for example, "I Came, I Saw, I Bid," *The Economist*, March 14, 1992, pp. 79–82.
20 Cited in *Proceedings of the Symposium on Media Concentration*, vols. I and II (Washington, D.C.: Bureau of Competition, Federal Trade Commission, 1979).
21 *Ibid.*
22 See, for example, Paul M. Hirsch, "Globalization of Mass Media Ownership: Implications and Effects," *Communication Research*, vol. 19, no. 6, December 1992, pp. 677–681; Joseph Turow, "The Organization Underpinnings of Contemporary Media Conglomerates," *Communication Research*, vol. 19, no. 6, December 1992, pp. 682–704; W. Kim and E. Lyn, "FDI Theories and the Performance of Foreign Multinationals Operating in the U.S.," *Journal of International Business Studies*, vol. 21, 1990, pp. 97–112; D. Waterman, "A New Look at Media Chains and Groups," *Journal of Broadcasting and Electronic Media*, vol. 35, no. 2, Spring 1991, pp. 167–178.
23 S. Yoder and G. Zachary, "Digital Media Business Takes Form as a Battle of Alliances," *Dow Jones/News Retrieval*, July 14, 1993.
24 Cited in "6 of 1634 Radio Stations in FCC Are Women-Owned—But Study Implies 69% Female 'Ownership' in Radio and 87.8% in TV," *Media Report to Women*, November–December 1982, pp. 1–10.
25 *Ibid.*
26 See "Minority Tax Certificates: Doing the Job," *Broadcasting*, April 8, 1991, pp. 68–70.
27 *Ibid.*
28 "Black Broadcasting '91: 30 Who Stand Out," *Broadcasting*, September 9, 1991, pp. 51–67.
29 Lawrence Soley and George Hough, III, "Black Ownership of Commercial Radio Stations: An Economic Evaluation," *Journal of Broadcasting*, vol. 22, no. 4, Fall 1978, pp. 455–467.

FOR ADDITIONAL READING

Alexander, Alison, James Owers, and Rod Carveth (eds.): *Media Economics: Theory and Practice* (Hillsdale, N.J.: Lawrence Erlbaum, 1993).
Bagdikian, Ben H.: *The Media Monopoly*, 4th ed. (Boston: Beacon Press, 1992).
Botein, Michael, and David M. Rice (eds.): *Network Television and the Public Interest* (Lexington, Mass.: Lexington Books, 1980).
Compaine, Benjamin M., Christopher H. Sterling, Thomas Guback, and J. Kendrick Noble, Jr.: *Who Owns the Media? Concentration of Ownership in the Mass Communications Industry*, 2d ed. (White Plains, N.Y.: Knowledge Industry Publications, 1982).
Dates, Jannette L., and William Barlow: *Split Image: African Americans in the Mass Media* (Washington, D.C.: Howard University Press, 1990).
Krasnow, Erwin G.: *Buying or Building a Broadcast Station in the 1990's* (Washington, D.C.: National Association of Broadcasters, 1991).
McGee, John S.: *In Defense of Industrial Concentration* (New York: Praeger, 1971).
Peterson, Richard A. (ed.): *The Production of Culture* (Beverly Hills, Calif.: Sage, 1976).
Rucker, Bryce: *The First Freedom* (Carbondale: University of Illinois Press, 1968).

8

ENTERING THE TELECOMMUNICATIONS MARKETPLACE

The larger the amount you are looking for and the more traditional the institution, the more they count on the calibre of management. . . . If they have experience and are well thought of in the trade, if they have done things in the past that have worked well, if they have a good management team around them, and if they seem to be able to stick with things and make them happen, it's much easier for us to feel confident about the loan they need.

Dana K. Cassell[1]

This statement was made by the vice president of an investment banking firm specializing in loans to cable entrepreneurs. In the preceding pages, we discussed the general business climate for telecommunications and the patterns of ownership which typify media firms. Now, we get to the specifics of media management—getting into the business. Specifically, we will trace how media ventures are started, acquired, and capitalized.

The goals of this chapter are:
1 To describe the various strategies utilized in establishing a new or acquiring an existing telecommunications facility
2 To identify common sources of financial backing (investment capital) for new telecommunications operations
3 To describe methods of station and system valuation and market analysis utilized to set the value of a telecommunications enterprise

ACQUISITION STRATEGIES

There are two ways a prospective entrepreneur can enter the media marketplace: (1) by establishing a new station or system or (2) by acquiring an existing facility. Each method has its merits and pitfalls. First we'll examine how a media enterprise is built from the ground floor up.

Starting at Square One—The New Station or System Start-Up

Unlike their colleagues in the consumer product industries, radio or television entrepreneurs cannot simply set up shop by erecting a transmitter and beginning to broadcast. A cable operator cannot simply select a desirable community and begin to construct the physical plant. As we have described throughout the text, all media managers face technical, legal, and financial barriers which inhibit their "free" enterprise. The process of starting a new broadcast or cable venture is mapped in Figure 8-1.

FIGURE 8-1 Steps in establishing a new telecommunications facility.

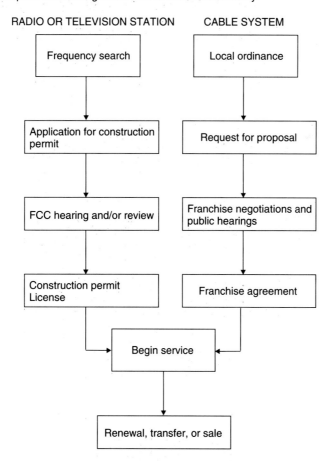

The New Broadcast Station

Step 1: Finding the Frequency The first hurdle facing new broadcast managers is that they must find available space in the electromagnetic spectrum to permit the construction and operation of their stations. The proposed signal must be sufficiently strong to cover a geographical area wide enough to ensure an audience (and thereby sustain profitability) but must not cause undue interference with an existing station or service.

One of the FCC's most difficult tasks is that of policing the use of the spectrum. Our constitutional obligation to free speech dictates a policy of plurality—as many voices as possible should fill the air. At the same time, technical limitations force the commission to limit the strength and direction of the many signals on the airwaves to prevent unnecessary interference. As a result, the FCC has adopted elaborate guidelines for managing the spectrum. As traced in Chapter 1, radio station licenses are granted in various classifications which govern the power, directionality, and operating hours of their facilities. In addition, for FM and TV stations, the FCC maintains a *table of assignments,* making only certain channels available in an area for these uses. Similar restrictions apply to the use of spectrum for new technologies. For example, in the case of LPTV, the commission will grant licenses only to applicants who prove that their signals will not interfere with any current or *future* signal in the proposed area.

The upshot of the FCC's regulatory policies is that the burden of proving that the construction of any new telecommunications service which uses the airwaves will meet all requirements rests squarely upon the shoulders of the new management team. It is their responsibility to find an available frequency for AM, or to consult the table of assignments for FM and TV. They must conduct tests which demonstrate that their new operations will not curtail any existing broadcast or other use of adjacent spectrum. In the commission's words: "[the] applicant must make his own search for a frequency on which he can operate without causing or receiving interference from existing stations and stations proposed in pending applications."[2] Normally, interference beyond 10% of the area already served will not be acceptable.

Not surprisingly, the initial step for the new broadcast owner is to engage the services of both a broadcast consulting engineers' firm and a communications law firm licensed to practice before the FCC. The engineering firm will tackle the problem of finding or selecting an available frequency. The law firm will expedite processing of the station's application.

Occasionally, a qualified engineer is already on board; that is, he or she is working for the media company, perhaps at a sister station or in another media division. More frequently, the new broadcast entrepreneur seeks an outside consulting firm. The task of the firm, once hired, is to find available spectrum space (the *frequency search*) and once found, to conduct tests relative to signal propagation and interference (*specs,* or *performance tests*).

Consulting engineers can be located in a variety of ways. A primary source is the trade press. Publications such as *Electronic Media* and *Broadcasting and Cable* maintain classified advertising sections touting engineering services. In addition, engineers specializing in the broadcasting industry are listed in *Broadcasting and Cable Yearbook* and *Television Factbook*. Engineering firms often rent booths at industry trade shows—in particular, the annual convention of the National Association of Broadcasters. And a good track record by an engineering firm is often communicated between media managers by word of mouth.

Selecting a good law firm is as critical as selecting a good engineer, since the firm will be needed for the long term—to represent the station in such matters as license renewals, personal-attack cases, and meeting affirmative action and EEO requirements. Law firms are also listed in the trades and directories mentioned above. Most reputable firms boast membership in the Federal Communications Bar Association (FCBA).

Step 2: The Application for a Construction Permit Once the prospective broadcaster, in consultation with his or her engineers, has selected an available frequency, the next step is to work with the lawyer to apply for a construction permit (CP). The application is a detailed form which inquires about the citizenship and character requirements for ownership discussed in previous chapters. In addition, the applicant must provide comprehensive details about the transmitting and operating equipment the station will use. New applicants must also prepare a detailed equal employment opportunity program documenting the recruiting, training, and advancement plans for staffing the station.

While programming and commercial requirements have been streamlined or eliminated entirely in recent years, prospective applicants must also provide documentation about how their proposed service will operate in the public interest. At the time the application is filed, prospective broadcasters must give legal notice of their intentions to operate by taking ads in local newspapers. In the case of license renewal, intention to apply for renewal must be aired.

Step 3: The FCC Hearing Since there are few unused radio and television frequencies remaining, it is rare that the FCC will guarantee a CP without review. The mechanism for official scrutiny into a CP application is the FCC *hearing,* which is presided over by an administrative law judge (ALJ). In the case of applications for a new station, hearings may be conducted in the city of proposed operations; in renewal cases, they are generally held in the FCC's main offices in Washington, D.C. A hearing may be scheduled for a variety of reasons. According to the commission:

> When it appears that an application does not conform to the Commission's rules and regulations, that serious interference would be caused, if there is protest of merit, or if there are other serious questions of a technical, legal, or financial character, a hearing is usually required.[3]

In scheduling the hearing, the FCC gives public notice of why and when it will be held. The hearing notice allows applicants, competitors, consumer groups, and other interested parties 60 days in which to prepare their best case. Once the ALJ has heard evidence in the hearing and has made a decision, all involved parties have 30 days in which to contest the decision and request further oral arguments. After new arguments are heard, the ALJ, the commission, or the review board of the commission may retain, modify, or reverse the original decision. "Final" decisions, like all FCC rulings, may then be appealed to the courts. The appeals process usually begins with the U.S. District Court of the District of Columbia. Once an application passes favorably through all hearing and review procedures, the FCC will issue the CP.

Step 4: Construction Permit and License If the application is granted, the new broadcaster may request specific call letters. Applicants have 60 days from the issuance of the CP to begin construction and must begin operations within a year. Applicants who cannot meet those deadlines may apply for an extension, spelling out the reasons for the delay and giving an estimated timetable for completion.

When construction is completed, the new licensee can begin engineering and equipment tests. If technical tests are successful, the applicant may file for an actual license and must demonstrate compliance with all the terms and conditions spelled out in the filing of the original application. Only when the license is formally granted may the new station manager begin program testing (that is, broadcasting words, music, and advertisements over the air).

Step 5: Broadcasting The display of a license in the control room or transmitter site and the beginning of actual broadcasting do not mean that "anything goes." Throughout the duration of the license period, broadcasters must continue to live up to the terms of their original applications, or face a tough renewal hearing. In addition, broadcasters must conform to all local and federal statutes relating to the conduct of business. And most will agree to adhere to self-regulatory decisions from industry groups, such as the National Association of Broadcasters (NAB), and respond to initiatives taken by consumer groups, or the complaints of newspaper critics. In theory, serious business, technical, or editorial violations can lead to the forfeiture of a license at any time. In practice, most failures to comply with regulations are examined at license renewal time.

Step 6: License Renewal When a broadcaster files for license renewal, the FCC will accept arguments from community residents about the operations of the station—its programming, employment, and public service performance—in the form of a *petition to deny*. Petitions to deny have become a valuable regulatory tool utilized by citizens' groups to lobby for such things as changes in broadcast employment and programming practices relative to women and members of minority groups. If the petition is deemed by the commission to have merit, hearings may be scheduled, at which time the licensee is back at step 3. Should the petition be examined and found not to merit reopening the application for exhaustive examination, a renewal for another term of 5 or 7 years is granted.

In recent years, presidential initiatives to reduce bureaucratic paperwork and to deregulate the broadcast industry have led to use of what is known as a "postcard" renewal form (see Figure 8-2).

The New Cable System In a way, cable entrepreneurs face a "double whammy" when entering the construction phase. On the one hand, its development as a public utility (along the lines of local power and telephone service) has led cable to be controlled largely by local governments. In addition, the Cable Act

Federal Communications Commission
Washington, D.C. 20554

APPLICATION FOR RENEWAL OF LICENSE FOR COMMERCIAL AND NONCOMMERCIAL AM, FM OR TV BROADCAST STATION

Approved by OMB
3060-0110
Expires 5/31/91

For Commission Fee Use Only	FEE NO.	For Applicant Fee Use Only
	FEE TYPE:	Is a fee submitted with this application? ☐ Yes ☐ No
	FEE AMT:	If No, indicate reason therefor (check one box): ☐ Nonfeeable application
	ID SEQ:	Fee Exempt (See 47 C.F.R. Section 1.1112) ☐ Noncommercial educational licensee
For Commission Use Only	File No.	☐ Governmental entity

1. Name of applicant

Mailing Address

City | State | Zip Code

2. This application is for: ☐ AM ☐ FM ☐ TV

(a) Call Letters: (b) Principal Community: City State

3. Attach as Exhibit No. _____ an identification of any FM booster or TV booster station for which renewal of license is also requested.

4. Have the following reports been filed with the Commission.

(a) The Broadcast Station Annual Employment Reports (FCC Form 395 - B) as required by 47 C.F.R. Section 73,3612? ☐ Yes ☐ No

If No, attach as Exhibit No. ______ an explanation.

(b) The applicant's Ownership Report (FCC Form 323 or 323-E) as required by 47 C.F.R. Section 73,3615? ☐ Yes ☐ No

If No, give the following information:
Date last ownership report was filed ________
Call letters of station for which it was filed ________

FCC 303-5
May 1988

FIGURE 8-2 Postcard renewal form (FCC).

of 1992 empowered the FCC to oversee rate structures and program offerings among the nation's 10,000 plus local systems.

Whereas in the case of broadcasting it is usually the broadcaster who needs the services of outside consultants (engineers, lawyers, and so forth) to expedite the licensing process, in the case of cable, it is usually the community which first seeks outside help. This is the result of a variety of factors. For one thing, the field of cable is changing so rapidly that it is unlikely that any community governing body could have the expertise required to put together a franchise agreement. Second, it is unlikely that a local government could have full knowledge of cable services in communities of similar size and makeup, but located far away. Thus most communities begin the process of cable franchising by retaining the services of *cable consultants.* Cable consulting firms are typically staffed by lawyers, technical experts, and financial analysts. They may be found, like broadcast consultants, in the trade press or through personal sources. Many communities engage the services of local residents who teach courses and administer curricula in mass communications at colleges and universities in the area. The services of an academic consultant are often cheaper than those provided by professional consulting firms.

Of course, cable operators have their own consultants, too, including legal experts, financial forecasters, and technical wizards. As you trace the steps in establishing a cable company in a community, note that often it is not as much a case of the entrepreneur meeting with the community as it is one of "my consultant" calling "your consultant."

Step 1: Local Ordinances The primary task of a municipality seeking new (or upgraded) cable service is to draft a set of local laws governing the proposed system. Such documents are known as local ordinances, and they serve to set the minimum level of cable service anticipated by the community.

Local ordinances for cable television generally include the following items:[4]

- *Definitions.* Legal definitions of key terms—for example, *cable system, franchiser, franchisee*
- *Application procedures.* Timetable for application, application forms, processing procedures
- *Selection criteria.* Financial and technical requirements, proposed program services, and other criteria the community will use in selecting among competing applicants
- *Monitoring procedures.* Methods by which the cable system will be supervised and reviewed during the length of the franchise agreement
- *Fee structures.* Franchise fees, rate schedules, rate negotiation, and increases
- *Ownership changes.* Procedures governing transfer or sale of system, forfeiture of operations, and the possibility of "buying back" the system by local government units
- *Term of the agreement.* Length of the agreement between the government unit and the cable operator
- *Program services.* Requirements for public access and for educational and institutional channels; provisions for free drops to police stations, fire houses, and other locations
- *Other provisions.* Discussion of ethical business practices, hiring requirements, and other areas specific to the community
- *Future services.* Opportunities for new telecommunications services (such as personal communications service, interactive TV, and high-definition TV) and expectations for such services from local cable companies

Step 2: The Request for Proposal (RFP) Once the local government unit has finalized its local cable ordinance, the next step is to draft a *request for proposal* (RFP). The RFP abstracts the requirements of the ordinance, itemizing all the community's needs relative to cable service, and provides the application form to be completed by all prospective cable operators. The RFP also usually contains a general description of the community. It generally calls for financial, character, and other background information about the cable applicant, as well as a comprehensive list of program and public service channels that will be provided by the applicant. Once the RFP has been developed and approved by the government unit, the community issues public notice that it is "offering itself" to cable entrepreneurs. This is known as being *open for proposals.* Public notice of an RFP is usually made in the form of classified ads taken in the trade press and/or local newspapers. Occasionally, a city will send a representative to an industry trade show or meeting. At this point, the action really starts, as prospective cable operators line up, each claiming that its cable performance will be clearly superior to that of the competition. The competitive drama of the courtship between a community and its rival cable suitors is played out in the form of *franchise negotiations* and *public hearings.*

Step 3: Franchise Negotiations and Public Hearings When a community has narrowed its prospective cable franchisers down to two or three leading candidates, it may enter private, formal negotiations to work out final details. Or the community may invite competing applicants to make their cases at a formal public hearing. In this way, all interested parties can meet the prospective cable operators face to face. Hospitals may make their needs for a channel known. Representatives of a local university may attend to request their own services, such as access studios and free drops. Individuals active in the arts may want to be assured that access will be provided for their events and activities. And area broadcasters may send representatives to make sure that their operations will not be unfairly curtailed by the granting of the cable franchise.

After public or private debate, the community governing board will settle on a first choice. The next step is to hammer out all the formal details of the binding contract between the company and the community—the *franchise agreement.*

Step 4: The Franchise Agreement The franchise agreement, which is a binding legal contract, specifies all technical, financial, engineering, and programming stipulations that have been hammered out in the ordinance and RFP phases of the franchising process. In addition, it may state the methods under which the performance of the cable company during the franchise term will be evaluated, and give a timetable for construction, improvement, renovation, and expansion of services. Failure to live up to the expressed terms of the agreement can lead to forfeiture or sale of a cable system by the operator.

Both the cable entrepreneur and the community must have some foresight in drafting the terms of the franchise agreement or they may find later that they have built in "planned obsolescence." (For example, franchise agreements signed before the advent of multichanneled, interactive cable services have in some instances hindered the availability of these new services to community residents.) This is particularly true in cases in which the franchise term is set at 25 or 30 years.

Step 5: Renewal, Transfer, or Sale At the completion of the term of the franchise agreement, the community must decide whether to renew the agreement with the cable operator or to reopen the negotiation process with a new company. If the community is satisfied with the existing cable operator and plans to continue the relationship with that firm, the process is known as *negotiated renewal.* If the community asks for a new cable entrepreneur to make a pitch, the process takes the form of a *competitive renewal.*

Generally, if cable performance has been satisfactory in the programming and service areas, most communities favor negotiated renewal, which spares them the time and expense of beginning at square one. However, the community cannot be capricious in seeking a negotiated renewal. It must address such concerns as the following:

- Were the terms of the initial agreement fully adhered to?
- Did the cable company maintain and upgrade its physical plant throughout the franchise term?
- Did the company provide the range of services and programs that were promised? And did it provide new services that became available during the term of the agreement?
- Did the company act in good faith in its business dealings with subscribers and community businesses?

If the members of the cable advisory committee, city council, or other community monitoring board cannot respond affirmatively to these questions, then the community might be better served by asking competing companies to bid. This course is not without its problems. Reopening the franchise procedure is generally costly and time-consuming. In addition, the community must ascertain whether the services promised by a new operator will actually be better than the services already provided by the incumbent. And there is the practical matter of taking steps to ensure that cable service is not interrupted during the renegotiation process.

In most cases, if the community enters negotiations with another cable operator, it will arrange for the transfer of assets from the original to the new operator. In this way, service and billing practices will continue and the new operator will be able to phase in its new hardware and program services.

The New Radio, TV, or Cable Operation: A Final Word

While cable and broadcasting are different types of telecommunications operations, entrepreneurs in both follow similar steps in beginning a new business. The role of the outside consultant is paramount. The provisions of national and local government are critical. The prospective company must become aware of the needs, interests, and desires of the community it intends to serve. And once the company is established, owners must remain in touch with the community and must keep abreast of new technologies pertinent to their operations and with new developments in government regulation.

A second point about new start-ups is that they may be an endangered species, especially in broadcasting and cable. As we have seen, space on the radio and TV bands is nearly exhausted, with more than 11,000 radio stations and 1500-plus TV facilities already on line. Cable TV, now available to over 98% of U.S. television households, is rapidly approaching its own saturation point.

Room for new broadcast facilities can be made through technical changes. For example, in 1993, the FCC proposed expanding the AM bandwidth to 1605 kHz (up from 1505 or "15" on the AM dial). The move could free an additional five channels and allow for hundreds of new stations.[5] Priority for the new channels would be given to existing stations which suffer from interference in the crowded lower portion of the band.

On the cable side, Congress and the FCC have used regulatory changes to open the market for new builds. It is now permissible for local municipalities to allow more than one cable company to construct a system in a given community, a situation known as an *overbuild*. And, as we have seen, Congress and the courts may soon allow telephone companies to compete with cable in the delivery of broadband communications services.

Acquiring an Existing Broadcast or Cable Facility

The road to starting a new broadcast or cable system is, to say the least, long and tortuous. By now, you may have reached the decision that it is easier to enter the media marketplace by taking over an existing concern than by trying to start a new one. If so, you are not alone. In recent years, the number of stations and cable systems changing hands has grown at a spectacular rate.

Broadcast Station Sales Figure 8-3 illustrates the pace of radio and television station sales over the years. Looking back (or to the right in the figure), at one time radio and television stations were rarely bought and sold. From the early 1950s to the late 1970s, between 200 and 400 stations were sold on average each year,

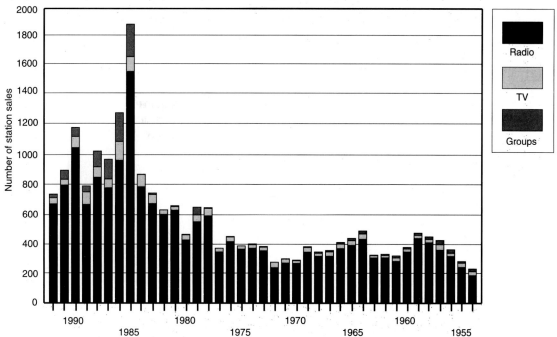

FIGURE 8-3 Station sales, 1950s to 1990s. *Source: Broadcasting and Cable*, February 8, 1993, p. 40.

Modern facilities like this TV station (outside and inside views) make attractive investment properties. (courtesy of Fox Broadcasting Company)

representing under 1 in 10 properties. The sales were mostly made between owners and their heirs, or between the major broadcast companies (such as the networks and the major cross-owners), as they reached or realigned their permissible number of holdings (7 each in AM, FM, and TV).

In the 1980s, radio and television stations were a central part of the "Reagan revolution." Freed from cumbersome regulations designed to inhibit station sales, many broadcast owners "cashed in" and sold out. A range of venture capitalist firms (some with questionable ethics, as it turned out) stepped in to acquire broadcast properties. They were attracted by the virtues of media companies traced earlier (low personnel costs, high cash flows, and lots of ego gratification, to name three). The result was that turnover in ownership of radio and TV stations skyrocketed.

During the decade, more than 10,000 radio stations changed hands. Put another way, virtually every station changed ownership at least once (some changed hands three or four times in just a few years). On the television side, over 700 commercial stations (about 3 in 4) acquired new owners. The volatility extended to group owners as well. For over a decade (from the mid-1960s to the late 1970s), group sales were almost unheard of (there were no such transactions in 1972, 1975, and 1977, for example). In 1985 alone, more than 200 station groups were sold.

By the early 1990s some stability had returned to the station marketplace. Yet today, still more stations are changing hands each year (between 700 and 1000) than at any other time prior to the 1980s.

Cable System Sales The pace of activity in cable sales has been nearly as volatile as the situation in radio and television. Like broadcasting, cable system sales were few and far between from the 1950s through the 1970s. Many of these "first generation" cable systems held long franchising terms, were burdened with debt from the costs of construction and programming, and were in private hands. The situation changed markedly in the 1980s, as investors were drawn to the new consumer status symbol that cable had become. Many cable companies went public; that is, they sold stock to finance their expansion. This attracted the investment community, which was drawn to cable for many of the same reasons that brought it to broadcasting.

The result was a decade of unprecedented volume in cable system sales. One good indicator is the dollar volume of cable sales, tracked each year by analyst Paul Kagan.[6] In the 1970s, the total dollar value of cable system sales was typically under $100 million a year. By the late 1980s, annual cable system sales topped $7 billion! Again, as with broadcasting, the pace of sales has slowed somewhat in the 1990s. However, the fact remains that the prospect of new ownership is commonplace in contemporary telecommunications.

Reasons for Sale Some reasons that stations and systems are bought and sold include the following:

- Financial mismanagement
- Failure to live up to the terms of a license or franchise agreement
- Availability of the "right" buyer offering the "right" price
- Retirement or death of principals or partners

Occasionally, a station or cable system will face immediate financial peril. Perhaps the bank has called its mortgage due. Creditors may have initiated lawsuits to recover unpaid debts. In such instances, a facility will be placed on the market to avoid bankruptcy.

The FCC may deny an application for renewal if a station has engaged in discriminatory hiring and promotion practices. A cable system may not have expanded to a promised level of service or may have failed to extend service to new suburban areas and housing tracts, and the community may have voided its franchise agreement and initiated a new RFP proceeding. Sometimes, turnover in broadcast ownership occurs for personal reasons. The retirement or death of principals forces corporate reorganization, which often leads to ownership changes. Or an owner may, after years in broadcasting, "burn out" and decide to try another line of work. Media ownership may also be affected by family feuds, including marital break-ups and sibling rivalries. Finally, the allure of a life in telecommunications has brought a steady flow of entrepreneurs. An owner may sell out simply because the offer made by an eager investor is too sweet to ignore.

While each sale of a media property follows a unique course, the pattern generally conforms to that provided by Figure 8-4.

Media Brokers and Analysts

In all cases of acquisition the central questions are: "Which stations or systems are for sale?" and "How much are the facilities worth?" It should not be surprising that, to get answers to these questions, prospective buyers use the services of specialty consulting firms. The key consultants are (1) media brokers and (2) financial analysts.

The business of *media brokers* is the sale or transfer of broadcast or cable properties. Like real estate brokers or stockbrokers, they maintain a list of clients whose properties are for sale. In return for a share of the selling price (usually a 5% commission), they try to put sellers in contact with potential buyers and generally supervise the entire sales process.

Media *financial analysts* specialize in establishing the value of a given station or cable system. They calculate the existing sales and profit base of the company and estimate its future worth. If employed by the system or station for sale (or the media broker), they calculate the asking price; if employed by the prospective buyer, they estimate the offering price; if employed by the buyer or seller's bank, they calculate the net worth, or assets, of the facility.

Often the services of the broker and financial analyst are combined. In fact, the leading media brokers (Blackburn and Company, Chapman Associates, and Burt Sherwood and Associates, to name a few) provide a full range of services to their

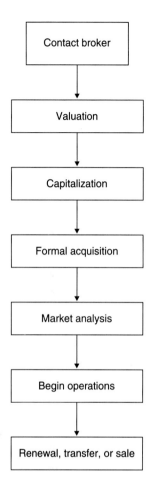

FIGURE 8-4 Steps in acquiring an existing telecommunications facility.

clients, including asset appraisals, financial forecasting, capital formation, tax planning, and special research studies. Not surprisingly, given the volume of station and cable system sales, media brokering has become a growth industry in recent years. Figure 8-5 charts the number of media brokers and financial services organizations from 1965 to 1993.

During the placid times from the 1960s to the dawn of the 1980s, between 60 and 80 firms provided assistance to buyers and sellers of radio, TV, and cable properties. At the height of the volatile 1980s, nearly 200 brokers had entered the business. Similarly, from the mid 1960s until the early 1980s, fewer than two dozen companies provided financial assistance and potential backing for media ventures. By 1990, that number had nearly quadrupled. Again, things seem to have cooled off thus far in the 1990s. Today, a prospective seller can choose from over 125 listed brokers; a prospective buyer can make appointments with nearly 80 financial consultants.

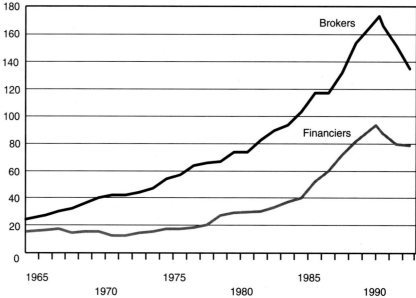

FIGURE 8-5 Media brokers and financiers. *Source: Broadcasting and Cable Yearbook,* for each year indicated. Data from 1965–1990 courtesy of Professor E. Funkhouser, North Carolina State University.

MEDIA VALUATION

Of all the services offered by media brokers, the most important is *valuation*, the calculation of the net worth of a broadcast station or cable system. A variety of benchmarks are used in making these vital calculations. This is the next step on the path to acquisition traced in Figure 8-4.

Approaches to Valuation

Suppose you decide to sell one of your most valuable possessions—your car, for example. Let's say that it's 5 years old. The mileage is approaching 80,000, but the engine is sound and there are no oil leaks. The body is not in showroom condition: there are a few visible flaws in the paint, and a rust spot or two is starting at the base of the passenger's side door. The question is "How much are you going to ask for it?"

One approach might be to check the want ads to see the asking prices for similar makes and model years. From those guidelines you add or subtract a few dollars here and there to adjust the "average" price to your asking price.

A more complicated method might involve taking the original price of the car and using a commercially available price list, noting the rate at which it has depreciated each year. Since you have made occasional improvements to the car along the way (a new set of tires, for example) you add the value of those improvements to your selling price. At the same time, you subtract a few dollars for each flaw the car currently has. Ultimately, you come up with a fair price.

A third method might simply be to put a value on the "best" attribute of the car. "A great road car," you say, and set your price. Or "starts every day." Or "great second car." Maybe even "hot wheels."

Believe it or not, the complex models and terminology used in setting the value of telecommunications facilities are actually extensions of these three basic approaches. In short, the value of a station or cable system is set using the following methods: (1) comparative evaluation, (2) econometric modeling (regression analysis), and (3) calculating multiples.

The Comparative Method Many brokers begin the price-setting process by examining the figures for recent sales of similar stations and systems in similar markets. This comparative process enables the broker to get ballpark figures for the assets of a station—most typically, an average high, average middle, and average low selling price.

Since the sale of broadcast properties requires FCC approval, the terms of each transaction are in the public domain. In fact, they are listed each week in the back of *Broadcasting and Cable* magazine in a section called "For the Record." Station sales histories are also documented in Duncan's *American Radio,* BIA's *Investing in Television,* Kagan's *Cable TV Financial Databook,* and other sources listed at the end of this chapter.

The creativity involved in the comparative method lies in the criteria utilized for comparison. On what basis might construction and sales prices be compared? A few common benchmarks used and some examples of each are as follows:

- *Population profiles.* Audience and subscriber base, TV households, homes passed
- *Type of station.* Quality and strength of signal, age of station, broadcast bandwidth
- *Programming.* Network and program affiliations, pay cable offerings, superstations, television program and film inventory, music library
- *Geographic profiles.* Sunbelt, Midwest, East Coast
- *Lifestyle profiles.* Young married, college, retirement community
- *Revenue base.* Station sales and revenue figures, retail and wholesale business in the area, average annual family income, individual buying power, location of shopping malls and convenience stores
- *Competition.* Competing broadcast and cable stations, area movies, amusement parks

Other factors being equal, the media broker might set a higher price than average for a television station in a larger market. A VHF station might be worth more than a UHF. A network affiliate may command a higher price than an independent. A station in the sunny Southwest could command a higher price than one in the "snowbelt" states of Wisconsin or Minnesota. A station located in a community with a stable downtown business district encircled by four or five suburban malls might be priced higher than one in a decaying central city with poor mall occupancy rates. And a station with few competitors would bring a higher price than one in a market flooded with TV stations, radio stations, and newspapers.

Since no two telecommunications businesses or communities are precisely alike, the comparative method is usually just a first step in valuation. Brokers and financial analysts follow up the data obtained in the comparative study with models or formulas designed to calculate a more accurate price. This process is known as econometric modeling.

Econometric Modeling (Regression Analysis) The econometric approach to media valuation utilizes a set of models based on a phenomenon known as *statistical regression.* In a very basic sense, the regression phenomenon is a mathematical "law" which states that repeated administrations of a test or measure lead to scores that get closer to the mean or average score, to the extent that each of the measures is related (correlated). You need not be a statistician to have experienced regression. Your high school counselor probably suggested that you take the SAT test more than once and that after two or three administrations, you would have a better sense of your "true" scholastic aptitude. Most people score better on their second or third SAT. This is the phenomenon of regression. Perhaps you did poorly on the first test in a college class. You said to yourself, "I'll do better next time, since I know what to expect." And you did score higher on the next test (so long as it was based on similar material and featured a similar format). This, too, was regression. Have you ever been interviewed for a job? Perhaps the interviewer put down your résumé and cover letter, which listed the dozens of attributes that made you qualified for the position, and asked simply, "What makes you think you can handle this job?" Without missing a beat, you gave precisely the answer that the interviewer was looking for—that snippet of prior experience or academic preparation perhaps. You were hired, in part, because of regression.

Regression analysis enables the business planner to inventory all the variables that might affect the sales price of a broadcast station and cable system and, from these, to ferret out the *best predictors* of the operation's value. Some regression equations are *proprietary*—they remain the trade secret of an individual broker or financial analyst. Occasionally, regression formulas are printed in scholarly journals by economists and business-oriented journalism and communications professors. Some models become industry standards and are well known to broadcast and cable managers and controllers. As media companies struggle for an edge against their competitors, new regressions are calculated frequently. As an example, let's examine the results of regressions based on the sale of television stations in the United States.

Regression Example: Broadcast Television

In the early 1970s, Paul Cherington, Leon Hirsch, and Robert Brandwein analyzed nearly 200 TV station sales between 1949 and 1965.[8] Their regression equations took into account the following variables: market size, network affiliation, age of station, type of ownership, and type of facility (UHF or VHF). They concluded: "Overall, the selling price of a TV station was primarily dependent on *market size;* age of station and network affiliation had substantially lower explanatory power; and type of seller exerted relatively little weight on the sales price."[9]

The *number of television homes* was found to be the best predictor of television sales price, accounting for nearly half of the explanatory power of each regression equation. The power of market size as predictor held for both the top-50 and the top-100 markets.

A follow-up study by Robert Blau, Rolland Johnson, and Kenneth Ksobiech examined data on the sale of 134 stations sold between 1968 and 1973.[10] They found that *net broadcast revenue* was the best predictor of a station's selling price. This makes sense: the station making the most money should be the one with the highest selling price. However, contrary to expectations, sales price was inversely related to station age. Newer stations commanded a higher price than older, more established facilities. And relatively speaking, stations in smaller markets commanded a higher price than those in more populous areas.

Two studies revisited the question of TV station value after the flurry of trading activity in the 1980s. In 1988, Benjamin Bates found that sales prices following TV deregulation were primarily determined by their audience size (average daily circulation), the size of the market (measured by the number of TV households), and the presence of cable in the market.[11] He found that TV stations were worth more if they were in the VHF band and were affiliated with a major network. Station value also increased if the buyer owned other TV properties.

In 1991, media analysts Dana Krug and Barry Sherman presented the results of an examination of station sales from 1980 through 1985.[12] Their findings were consistent with those of Bates. Television value was found to be a function of advertising revenue in the market, affiliation status, and ownership. In general, higher prices were associated with stations located in larger markets where high advertising revenues are possible. Again, network affiliates commanded higher prices. Interestingly, echoing and extending Bates's findings, station value increased if the seller owned other TV properties.

Although regression equations are mathematically elegant, they alone will not allow analysts to determine the exact selling price of a media facility. For one thing, different types of telecommunications facilities will require different models. For another, regression analysis is, by definition, "post hoc." It attempts to "explain" the selling price of previous transactions, but can offer only *projections* for future station value (always subject to error). For these reasons, station brokers, controllers, and financial analysts use the results of comparative analysis and regression equations only as ways to get "best guess" estimates of the present and future value of a media facility.

Multiples The multiple method is a simple extension of comparison and regression valuation. The broker will use the comparison method to obtain the "typical" or "average" sales price of a given station or cable company. He or she will then examine the results of a regression analysis to obtain the best predictor of station sales price. It is then a simple task to divide the sales price by the value of the predictor to obtain a multiple, or to divide the predictor by the sales price to obtain a ratio.

For example, in 1992, the average price paid for a TV station was about $3 million.[13] From Bates, we know that one key predictor of value is audience size. Thus, that year a station in a market of 250,000 people which sold for $3 million would have a multiple of 12.

Of all the possible multiples and ratios, two command the most attention in media valuation: estimated revenues and cash flow.

Revenue Estimates

Revenue estimates are made on the basis of the gross earnings of a media company in the fiscal year in which it is sold. The figure includes income from all sources: advertising sales, subscriber fees, royalties, rents, and so on. It is calculated before the payment of operating expenses, interest payments, taxes, and other expenses.

In recent years radio multiples of revenue have ranged from 1 to 4. Most AM sales have been in the 1.5 to 2 range; FM sales, from 2 to 4. For example, in 1992 Emmis Broadcasting sold all-sports WFAN in New York to Infinity for $70 million —considered a record price for an AM station.[14] The station's estimated revenue that year was $30 million. Thus, the multiple was 2.33. In 1993, Infinity bought KRTH-FM ("K-Earth 101") in Los Angeles for a whopping $110 million.[15] Its annual billings, in the range of $25 million, represented a revenue multiple of 4.4.

On the TV side, television multiples have ranged from 3 to 5 times revenue (the same range common among newspaper sales). For example, when Paramount purchased WKBD in Detroit, trade publications estimated that the price approached $125 million.[16] In 1993, WKBD's estimated revenues were $30 million. The multiple, therefore, was just above 4 (4.16, to be exact).

In cable, revenue multiples are usually reported in terms of the individual subscriber. That is, the sales price is divided by the number of total subscriber households in the community to produce "revenue per subscriber." Thus, a cable company with 10,000 subscribers which sold for $20 million would produce a revenue multiple of $2000 per subscriber. In recent years, most cable deals have ranged in the neighborhood of $1500 to $2500 per subscriber.[17]

While revenue estimates provide a good, quick estimate of media value, by far the most widely used indicator is cash flow.

Cash Flow

Cash flow is defined as the station or system's gross profit—that is, before the deduction of depreciation, amortization, interest expenses, and taxes. For a radio or television station, the critical ingredient of cash flow is, of course, its income from the sale of advertising time. In cable, subscriber fees are the main source of cash flow.

For a number of reasons, cash flow has become the most attractive means for valuing media properties. For one thing, as the saying goes, media companies "throw off a lot of cash." As with other businesses, collections are a problem in radio, TV, and cable. But each month, like clockwork, the majority of advertisers and subscribers will send checks to broadcasters and cablecasters. There is typically plenty of ready cash on hand.

Perhaps more important are the accounting principles used by media properties (detailed more fully in Chapter 9). In media companies, deductions for depreciation, amortization, and interest payments are often so large that they produce a bottom line which creates a "paper loss." But, from a cash flow viewpoint, the financial health of the same companies looks rosy and robust. Since media are popular with audiences, future cash flows can be calculated with optimistic growth rates. Many other businesses (automobiles and airlines, for example) are hard put to make such optimistic projections to their investors.

Radio In recent years, radio stations have traded at the rate of 6–10 times cash flow. As might be expected, FM stations have commanded the highest multiples—12 times or more in some cases. AM stations have considerably less, with 3–5 times cash flow commonplace in smaller markets. For example, in the early 1990s, Infinity acquired three stations from Cook Inlet (in Chicago, Boston, and Atlanta) for $100 million.[18] The combined cash flow of the stations was reported at $10.5 million. Thus, the sale produced a multiple just under 10 times cash flow.

The early 1990s saw a decline in the value of radio properties. Today, cash flow multiples of 7 to 8 are the norm.

Television In view of the regression research cited above, it should come as no shock that VHF stations, particularly network affiliates, command the highest cash flow multiples—in the range of 10–15. Independents associated with the Fox network are next most valuable; trading in the range of 9–12 times cash flow. Other independents, particularly those on UHF and without Fox, trade like radio stations, in the range of 5–7 times cash flow.

As an example, consider the recent sale of independent WGBS-TV in Philadelphia, purchased by Fox Broadcasting from Combined Communications.[19] The price was reported in the range of $50–$75 million. The station's current cash flow was projected at roughly $4 million. Using the high purchase cost estimate, the cash flow multiple was a huge 18.75. However, adding attractive Fox programming (from rival WTXF, owned by Paramount) was expected to double cash flow within a few years. Even with the high cost figure, this would bring the cash flow multiple down to under 10.

Cable Cable system cash flows generally follow trends in over-the-air television. Common figures today are in the range of 10–12 times cash flow. The value of cable systems reached their zenith in 1988 and 1989, when system sales brought an average of 13 times cash flow nationwide, and some individual sales topped 15 times cash flow.

With the reregulation of cable in the 1990s, there is potential for ever further devaluation of cable systems. One reason is the anticipated negative impact of rate rollbacks on cable system cash flow. For example, three cable giants—TCI, Time Warner, and Continental—expect rate reregulation to shrink their annual operating income from 8% to 15%.[20] This will cause cash flow multiples to decline accordingly.

Such concern may be real. In 1993, a group of 18 banks with large cable commitments wrote a letter to the FCC suggesting that many cable operators would soon go into default, since their loan agreements were based on expectations of increasing cash flow (and not the reverse).[21] Time will tell.

Of course, in the final analysis, the "value" of a station, cable system, or other media facility depends on how much the buyer is willing to pay for it—and how eager the seller is to unload—exactly the situation when someone decides to sell a car.

RAISING MONEY: CAPITALIZATION AND ACQUISITION FINANCING

After finding an available frequency, station, or system and determining its fair market value, the telecommunications entrepreneur must find the money needed to finance his or her media enterprise. This is known as the capitalization phase. Money for a new media facility is known as *venture capital*; money raised to take over an existing enterprise is referred to as *acquisition financing*.

As discussed previously, telecommunications is an innovative, high-tech industry. Thus, it requires significant amounts of venture capital. Since it is also a profitable industry, venture capital from a variety of investment sources has generally been available to the prospective entrepreneur. There are two types of capital needed to establish a new or to acquire an existing facility. *Internal capital* is the money raised through pooling the assets of a partnership, corporation, or joint venture. *External capital* is venture capital raised through outside sources.

External Capital

This section describes the most common sources of external capital for telecommunications enterprises.

Commercial Banks   As we have seen, commercial banks have significant interests in communications companies, including the major networks, radio and television stations, cable companies, and the whole gamut of new technologies.

Large "money center" banks are most likely to have experience and expertise in broadcast and cable investment and, thus, an interest in new telecommunications companies. Regional banks may also grant loans to newer firms, particularly minority enterprises. Small local banks tend to have less experience with minority ventures, and are also apt to have more stringent limits on the amounts and terms of their loans.

The influence of large multinational banks on telecommunications is immense. For example, one 1993 report revealed that cable companies carried outstanding debt in excess of $40 billion—two-thirds of which was due to major banks; including Bank of Boston, Mellon Bank, Morgan Guarantee Trust, and NationsBank.[22] On the television side, leading banks active in station finance included Chase Manhattan, Citibank, Chemical, First National of Chicago, and Security Pacific.[23]

In the end, the ownership trends traced in Chapter 7 are largely moot. Who owns the media? Banks, mostly.

Investment Companies   Investment firms specialize in short-term financing of mergers and stock transactions between large corporations. Thus, such firms are typically involved in capital formation during a transfer of broadcast station or cable ownership between group owners or multisystem operators (MSOs). In recent years, investment firms invested directly in new broadcast and cable ventures. They were drawn by the attractive cash flows in the media, as discussed above. And they were armed with interesting new means of acquisition finance, especially the notorious "junk bonds" of the 1980s.

Junk bonds were short-term, high-interest and high-yield bonds made available to aggressive entrepreneurs. As a mechanism for financing acquisitions, they were unmatched. The magic of junk bonds was that they were not issued on the basis of the assets of the media company (its tower, transmitter and tubes, for example) or even the company's ability to pay (based on earnings or profit margins). Rather, their value was measured by multiples of cash flow—that is, the projected future earnings power of the particular company. As a result, millions of dollars became available to media entrepreneurs with virtually no down payment or security. In the terms of the times, they were "highly leveraged transactions" (HLTs).

Few foresaw that advertising sales, the engine of media cash flow, would dip significantly as the recession of the early 1990s set in. Fewer still anticipated the public and congressional backlash against cable. Following the crash of the savings and loan industry, the jailing of Michael Milken, and new regulations against HLTs, many media companies woke up in the 1990s with a massive hangover: the large repayments due on their short-term high-interest debt.[24]

The junk bond market played a significant role in the takeover activity in mass media during the 1980s. While some companies survived the "morning after" of their junk binge (including Turner Broadcasting, McCaw Cellular, Viacom, and TCI), others faced financial ruin (including SCI Holdings, an investment firm created by Kohlberg-Kravis-Roberts and entrepreneur George Gillet to take over the six former Storer TV stations).

Today, investment firms retain their interest in financing media acquisitions, though they are naturally more conservative in their approach. Some major players from the investment community include Lazard Fréres, Lehman Brothers, Waller Capital, Hardesty-Puckett, and Paine Webber.[25]

Insurance Companies Insurance companies provide fixed-rate, long-term financing to media entrepreneurs. Their operations, governed by state laws, are generally very conservative. Insurance companies will not usually provide capital to partnerships and proprietorships, and they prefer large firms with a well-established track record in the communications field. Usually, their rates exceed the prime lending rate. The major advantage of insurance company financing is the long term generally available. Loans may be set for 10 years or more. Thus insurance companies have been an important source of investment capital for cable companies, whose commitment to construction and physical plant often delays anticipated profit for 5 years or more.

Pension Funds Like insurance companies, pension funds have strictly regulated and fiscally conservative lending practices. They are most likely to be a source of capital for small ventures. In recent years, the smaller asset base of pension firms, an increased demand placed upon their services by enrollees, and the growing costs of new media ventures have restricted these companies as a source of capital formation in telecommunications.

Commercial Credit Companies Commercial credit firms are fully secured, one-stop lenders. They are fully secured in that they will hold all the assets of the media facility as collateral during the term of the loan (rather than merely its stock). They are one-stop lenders in that they can provide equity capital (i.e., the minimum amount required by commercial banks and insurance companies and pension funds) or they can float the loan themselves.

Commercial credit companies are most often used in combination with other investment sources. Their money is comparatively expensive—interest rates can be as much as three or four points above the prime lending rate—and their terms are usually from 5 to 10 years. Some commercial credit companies active in media finance include General Electric Capital, AT&T Commercial Finance, Chrysler Capital, Heller Financial, and Westinghouse Credit Corporation.[26]

Seller Financing Occasionally, the seller of a media facility takes a note from the buyer; that is, the seller finances the loan. There are two main reasons for this practice. First, seller financing expedites the sales process. The sale is closed quickly; it involves less red tape and fewer complicated intermediate payoffs to financial consultants, brokers, accountants, and legal services. In addition, seller financing can help reduce the income tax obligation of the seller.

The advantages of seller financing to the media buyer are numerous. Sellers will typically provide a longer repayment period for the loan than would be available from a commercial lender, and will often float the loan at a lower interest rate. In addition, the seller will often reduce the purchase price of the station in return for an *assumption of existing indebtedness*. This can work out to the advantage of the buyer if the existing debt is for a long term, at a comparatively low rate. In such a case, assuming the debts of the seller may translate into a low up-front principal payment coupled with a relatively low interest payout.

Seller financing is most common in smaller media markets and when the seller is in a "loss" or "distress" situation. This should not be surprising, since commercial lending institutions are mainly interested in media facilities with good track records, generally located in major markets.

Internal Capital

There are two primary ways a media corporation can generate capital for expansion from within. The first method, the leveraged buyout (LBO), gained the kind of notoriety accorded junk bonds in the 1980s. The other method, going public (through sale of stock), is a time-honored tradition of American capitalism.

The Leveraged Buyout In a leveraged buyout, a public company is made private through the purchase of its outstanding stock, most typically by a group of its executives and managers.

The leveraged buyouts of the 1980s were financed almost entirely with borrowed money. In return for floating the funds, the venture capital firms providing the cash (Drexel Burnham and KKR, for example) were given an equity position in the media company. In addition, lenders were given warrants which permitted them to purchase additional stock in the company at a discounted price.

The biggest LBO deal of the decade occurred in 1984, when John Kluge purchased Metromedia, the company he had started in 1947, for $1.19 billion.[27] The next year, he sold the group of seven stations to Rupert Murdoch for over $2 billion. Another important LBO was that of Viacom in 1987 (parent of MTV and Nickelodeon) by Sumner Redstone, one of its minority investors.

While LBOs are still possible, the declining overall economic climate and new restrictive laws governing HLTs have made them less commonplace. Interestingly, private companies seeking capital have now gone in the other direction: selling their stock to the public.

Going Public: "The Reverse LBO" Facing imminent debt from junk financing, a number of newly privatized companies (including Viacom, mentioned above) moved quickly in the 1990s to go public and sell stock as a means of raising cash.[28] While going public has its benefits (a quick cash infusion, to be specific), it has its drawbacks as well.

On the plus side, sale of stock and securities permits unrestricted use of funds generated from the sale. Leading officers can be granted bonuses in the form of stock options, which have become a potent recruitment and retention incentive. Stock sales usually increase the company's net worth, making it more attractive when it seeks traditional funding from banks and commercial credit institutions.

On the minus side, selling stock subjects the company to the fluctuating whims of the stock market. Today's market is volatile and capricious, with memories of the "crash of 1987" still fresh in the public's mind. Public media companies are open to scrutiny by investors, financial analysts, journalists, and the general public. Their financial and personnel records are open to inspection. Finally, going public increases the likelihood of being taken over by a larger corporation—the case of the big fish swallowing up the smaller one.

The Financial Proposal

The days when new entrepreneurs could obtain capital on the basis of their prior success, their good name, or the general attractiveness of a new radio or television station are over. Today's tight money practices, an increase in construction and operating expenses, and the uncertainty caused by competition among media sources for audience attention have made the financial plan integral to the capitalization process. Generally, the plan should include nine elements.

- *Purchase price rationale.* Analysis of cash flow, using comparison or econometric models.
- *Five-year financial projections.* Regression models of anticipated station or system revenues, expenses, advertising sales, subscriber fees, equipment purchases, amortization and depreciation schedules, and other pro forma projections. Such projections, which should be realistic, well documented, and clearly stated, are often the key to obtaining financing.
- *Description of the proposed financing.* Documented sources of equity and financial capital, debt participation, and use of proceeds from the financing (hardware, purchases, reinvestment, etc.).
- *History and description of operations.* For new facilities, this will include the material on signal coverage, location, programming, and other data filed in a CP

application or RFP. In the case of a sale or transfer, a comprehensive description of the history of the station or system will be included.
• *Five-year financial history.* Documented evidence of the station's financial history: income, revenues, expenses, salaries, accounts payable and receivable, and other financial data. Of course, this material will be unavailable for new facilities and may be difficult to obtain in the case of mismanaged or previously unprofitable media ventures.
• *Biographical material.* Background and personal and financial history of major owners, managers, shareholders, consultants, and legal and acccounting personnel. All can shape the character of the company.
• *Operating strategy.* Proposed programming, publicity, sales and employment policies, management organization, and other operating plans.
• *Regulatory environment.* Discussion of applicable FCC and other local regulations.
• *Market information.* Audience surveys, analysis of the competition, advertising information, sales figures, and the like.

Of the nine elements, typically three receive the most attention by the lender. According to one media financial officer, "The cliché is, you look for *cash flow*, you look for *collateral*, and you look for *character*."[29]

FORMAL ACQUISITION: TRANSFER OF CONTROL

All transfers of broadcast licenses must be approved by the FCC. Thus, the acquisition process is not formalized until the commission has examined the transfer and made sure that the new ownership structure does not violate statutes contained in the Communications Act of 1934. Should the commission find that an unauthorized or unlawful transfer of control has taken place, a license revocation or denial of renewal could occur. In the past, transfers of control faced considerable scrutiny at the commission, but currently the FCC will grant approval of ownership changes so long as the criteria of citizenship, financial integrity, and local control of hiring and programming are met.[30]

In addition, the appearance of some unscrupulous owners during the junk bond era has led the commission to beef up its character standard.[31] While convicted felons were always shunned by the FCC as prospective owner-operators, language has been added to cover so-called white-collar criminals convicted of antitrust and anticompetitive statutes.

Formal transfer of control cannot take place until approval is officially received from the FCC. To ensure continuity of operations, many new owners sign short-term management agreements with existing operators.[32] This was particularly important in the days preceding broadcasting deregulation, when FCC approval could take as much as three years. Today, however, formal approval usually takes a matter of months.

In cable, the situation is slightly different. Depending upon the clauses in the applicable franchise agreement or the terms of the area's local cable ordinances, transfer of control may require approval of the local franchising authority, city council, or public utility commission. In some cases, transfer of ownership may not require approval of local government units at all. However, in most instances, the seller will be required to notify the appropriate governing authority and to demonstrate that the transfer of ownership will not interrupt or substantially diminish the area's cable service.

MARKET ANALYSIS

A key step in building or buying a potentially successful media enterprise is *market analysis.* Media managers must thoroughly research the community in which they will operate. After all, their success will be based on entertaining and informing the residents of that community. The cable entrepreneur hopes to sell and lease converters and to interest subscribers in pay and multipay service. The broadcaster wants the station to become an attractive advertising medium for local businesses. Both expect to become active participants in the economic and social life of their communities. This requires extensive market analysis, which is achievable in a variety of formal and informal ways.

Data on the population characteristics of a community are known as *demographics.* Such information includes household size and composition, age, sex, employment, religious affiliations, income, and other characteristics of the residents of a given community. Data on the leisure interests, lifestyles, opinions, attitudes, and other psychological aspects of community life are known as *psychographics.* Demographic and psychographic data are available from a variety of sources, both public and private. Common methods of assessing market characteristics include the following. (Full citations for texts and guides can be found at the end of the chapter.)

U.S. Bureau of the Census

Population data, compiled every 10 years and updated continually, are maintained by the U.S. Bureau of the Census. Complete records are available for the entire country and for most regions, states, and municipalities. Most public libraries keep copies of the most recent census data. In addition, computer tapes containing census information are generally archived in the computer centers of large universities. The standard starting point for utilizing census data is the annual *Statistical Abstract of the United States.* The prospective entrepreneur can use this volume to get summary demographics as well as to locate cross-listed sources of more in-depth reports.

U.S. Department of Labor, Bureau of Labor Statistics

Statistics from this government agency include current wage guidelines, regulations and developments, surveys of pay rates and schedules, and analyses of work stoppages (strikes).

Area Chambers of Commerce

Local chambers of commerce typically maintain a kit designed to interest new business or to acclimate new residents. The kit contains information about the population, educational system, employment patterns, and cultural facets of the community. Of course, since the goal of the chamber of commerce is to attract new businesses and residents, the view presented by such literature must be balanced by other sources.

Local Media

A preliminary feel for the economic and social life of a community, as well as a barometer of the activities of the media competition, can be obtained through regular reading of local newspapers and through watching and listening to area television and radio stations. At the beginning of a franchise or construction process, new owners, particularly if they are not long-time residents of the community, should subscribe to local press services.

Telecommunications Yearbooks

A seminal resource for media managers (and a frequent primary source for this text), the *Broadcasting and Cable Yearbook* lists all the major players in the telecommunications game—individual stations, networks, program sources, support and consultation services, even colleges and universities teaching broadcasting and cable courses. Summary data on audience and market characteristics are also included. The *Yearbook* is found in the library of virtually all telecommunications managers.

Another set of sourcebooks for broadcasters is the *Television Factbook*. Two editions are published each year. The first, called the *Stations Volume,* lists all U.S. television stations and gives their advertising rates as well as some audience information. The *Stations Volume* is also useful for obtaining general overview data on television sales, revenue, and income and for locating names, addresses, and other details about television executives and management personnel. The other *Factbook* annual, the *Services Volume,* lists and describes the variety of legal, financial, technical, and programming services available to television stations.

Cable financial adviser Paul Kagan publishes comprehensive current yearbooks and regular newsletters for that industry. Especially helpful for the new cable entrepreneur is the *Cable TV Financial Databook*. Data on cable's new competitor, the telephone companies, are compiled annually in Longman's *International Directory of Telecommunications* and Phillips Publishing's *Telephone Industry Directory*.

A useful compendium of TV industry statistics in attractive graphic form is *TV Dimensions,* published annually by Media Dynamics in New York. The growing international communications scene is surveyed in *World Guide to Television and Programming*.

Other Publications

American Radio   This is a useful compilation of ratings statistics, sales information, and financial data issued quarterly by radio analyst James H. Duncan, Jr. *American Radio* has become a widely quoted, seminal resource for information about the nation's radio stations and their listeners.

Investing in Radio/Investing in Television   These two annual publications are published by Broadcast Investment Analysts, a full-service media brokerage firm based in the Washington, D.C., area. Each includes a wide variety of useful data for more than 200 radio and TV markets, including ratings histories, market statistics, and station sales figures.

Sales and Marketing Management   This monthly publication provides information on a range of topics of interest to business. Of particular importance are its special reports, which provide the most recent data available on such things as manufacturing trends, consumer buying habits, population shifts, and employment statistics. Perhaps the most frequently cited issue is an annual *Survey of Buying Power,* which tracks consumer income and buying habits throughout the United States.

American Demographics   This monthly periodical tracks trends in consumer behavior, including audience use of and opinions about media. It is especially useful in keeping the media manager abreast of changing population shifts in the United States—including the increased participation of women and ethnic minorities in the marketplace.

Demographics USA   This new compilation from Market Statistics Inc., provides county-by-county population data on the top 250 metropolitan areas in the United States. The data are also made available on computer diskette.

Marketing and Consumer Research Firms

Of central interest to cable operators, television owners, and radio station managers are the media use and buying habits of their communities. Assessing these trends is the task of the major ratings services, such as Nielsen and Arbitron. (See Chapter 14.) In addition to the radio and television ratings services, some common sources of market data for the media manager are as follows.

Standard Rate and Data Service (SRDS) This company provides rate and audience information on U.S. communities (markets) to be used by advertising planners and buyers in developing their advertising strategies. Of interest to station owners and managers are SRDS indicators of market size and rank; demographic information; and listings of media competition and their affiliations, rates, formats, schedules, and even some common promotions. SRDS publishes separate editions for television, large-market radio, and small-market radio, along with volumes on newspapers, magazines, and outdoor advertising. Cable advertising rates and circulation data are available in a similar volume published by the NCTA, the annual *Cable Advertising Directory.*

Simmons Market Research Bureau (SMRB) One of the largest market research firms, SMRB publishes a number of volumes which document demographic and psychographic characteristics of consumers in the United States. Of particular interest to radio, television, and cable managers is the annual *Study of Media and Markets,* which includes data on such media habits as televison attentiveness and special events, multimedia audiences, and broadcast reach and frequency. In addition, Simmons provides analysis of consumer purchasing behaviors for over 20 product categories—from automobiles to women's beauty aids and cosmetics.

Advertising Bureaus

Each of the three main media forms studied in this text has its own advertising and promotion firm to aid in marketing, sales, and promotion. The television industry is represented by the Television Bureau of Advertising (TVB); radio by the Radio Advertising Bureau (RAB); and cable by the Cable Television Advertising Bureau (CAB). Each of these organizations maintains general information on the reach, popularity, credibility, and use of its medium. In addition, each gathers individual sales, audience, and other data for most markets and maintains a file of success stories for national, regional, and local advertisers.

SUMMARY

There are two basic ways for entrepreneurs to enter the telecommunications marketplace: (1) by establishing a new media operation (the start-up) and (2) by acquiring an existing media enterprise.

Steps in establishing a new broadcast station include a frequency search, filing an application for a construction permit with the FCC, attending public hearings, obtaining the formal construction permit, and making prebroadcast engineering and program tests. Steps in setting up new cable systems include establishing a local ordinance, filing requests for proposals (RFPs), bidding for a franchise, attending public hearings, and drafting a franchise agreement between the cable operator and local government units.

Ownership transfers and acquisitions occur for many reasons, including financial mismanagement and forfeitures. They may also occur for investment or profit-making purposes.

The critical step in successful acquisition is valuation—setting a fair market value for a media facility. Media financial analysts use various methods to analyze the value of stations and systems. The comparative method sets a price range based on recent sales of similar operations in comparable markets. Econometric models identify attributes of the facility that correlate highly with sales prices. Multiple methods use the results of comparison and econometric analyses to establish valuation. Common media multiples range from 5 to 15 times cash flow.

Investment capital for media enterprises can be obtained from a range of sources, including commercial banks and credit organizations, venture capital firms, and public offerings of stocks and securities. Obtaining capital requires a comprehensive financial proposal which details the personal and financial background of the entrepreneur and revenue and expense projections.

The final step in beginning a media venture is market analysis. Comprehensive market data are available from public and proprietary sources—from the U.S. Bureau of Labor Statistics to large market research companies.

We have now successfully mapped the territory for the prospective media entrepreneur, from initial idea to construction or acquisition. All that remains is to set up the books and hire the staff. These are the concerns of the next two chapters.

NOTES

1 Dana K. Cassell, "Opening the Vault: Raising Money in a Tight Market," *Cable Marketing*, September 1982, pp. 20–21.
2 *How to Apply for a Broadcast Station* (Washington, D.C.: Federal Communications Commission, March 1979), p. 2.
3 *Ibid.*, p. 4.
4 See Charles Spencer, "Franchising of Cable Television," *Louisiana Municipal Review*, November–December 1982, pp. 12–14.
5 Sean Sculley, "FCC Opens Up Expanded AM Band," *Broadcasting and Cable*, April 19, 1993, p. 14.
6 Data in this section are from *Cable TV Financial Databook* (Carmel, CA: Paul Kagan Associates), 1990, pp. 47–50.
7 The author is especially indebted to Professor Edward Funkhouser of North Carolina State University for access to his research on the rise of media brokerage and financial services.

8 Paul W. Cherington, Leon V. Hirsch, and Robert Brandwein, *Television Station Ownership: A Case Study of Federal Agency Regulation* (New York: Hastings House, 1971).
9 *Ibid.*, p. 66.
10 Robert T. Blau, Rolland C. Johnson, and Kenneth J. Ksobiech, "Determinates of TV Station Economic Value," *Journal of Broadcasting*, vol. 20, no. 2, Spring 1976, pp. 197–208.
11 Benjamin Bates, "The Impact of Deregulation on TV Station prices," *Journal of Media Economics*, (vol. 1, no. 3, 1990, pp. 5–22. Bates presents and extends these findings in "Valuation of Media Properties," A. Alexander, J. Owers, and R. Carveth, eds., *Media Economics: Theory and Practice* (Hillsdale, N.J.: Lawrence Erlbaum, 1993), pp. 91–113.
12 Dana A. Krug and Barry L. Sherman, "Bringing the Cash Cows to Market: Determinates of Television Station Market Value, 1980–1985." Unpublished manuscript, University of Georgia, January 1991.
13 See "Average Station Prices, 1986–92," *Broadcasting*, February 8, 1993, p. 39.
14 Data for this section are presented in "$70 million for WFAN," *Broadcasting*, February 8, 1993, p. 47.
15 Peter Viles, "Infinity Buys KRTH for Record $110 Million," *Broadcasting and Cable*, June 21, 1993, p. 12.
16 See Geoffrey Foisie, "Paramount Buys WKBD-TV," *Broadcasting and Cable*, July 21, 1993, p. 12.
17 "Big Deals," *Cable World*, August 30, 1993, p. 57.
18 "$100M for Cook-Inlet Stations," *Broadcasting*, February 8, 1993, p. 44.
19 See Geoffrey Foisie, "Fox Pulls Switch in Philly; ABRY Sells TV," *Broadcasting and Cable*, August 23, 1993, p. 12.
20 "Reregulation's Impact," *Cable World*, August 23, 1993, p. 83.
21 "Top Banks Warn FCC of Looming Cable Defaults," *Broadcasting and Cable*, June 28, 1993, p. 11.
22 "Public and Private Loans to the Cable Industry," *Cable World*, May 17, 1993, p. 40.
23 "TV's Top Banks," *Channels*, April 1989, p. 42.
24 Nancy Nichols, "Caught in the Junk Collapse," *Channels*, April 23, 1990, pp. 22–25.
25 See, for example, "A Look at Cable's Most Active Financiers," *Cable World*, May 3, 1993, p. 37.
26 "Busy Commerce for Commercial Credit," *Broadcasting*, June 4, 1990, pp. 50–52.
27 Paul Noglows, "With Other People's Money: The Best and Worst Deals of the 1980s," *Variety*, December 9, 1991, pp. 61–71.
28 "Reverse LBOs Still Captivate Investors," *Atlanta Journal*, August 29, 1991, p. C-4.
29 Cassell, *op. cit.*, p. 20.
30 See Robert Zuckerman, "Avoiding Unauthorized Transfers of Control," *Counsel from the Legal Department L-409* (Washington, D.C.: National Association of Broadcasters, April 1984), pp., 1–3.
31 Doug Halonen, "FCC Boosts Character Standards," *Electronic Media*, May 14, 1990, p. 3.
32 Erwin Krasnow, "Broadcast Stations Acquisitions: FCC Considerations," *Broadcast/Cable Financial Journal*, November–December 1991, pp. 21–30.

226 PART THREE: CORE PROCESSES

REFERENCE SOURCES FOR MARKET ANALYSIS

Books and Periodicals

American Radio
Duncan's American Radio, Inc.
PO Box 90284
Indianapolis, IN 46290

American Demographics
127 W. State Street
Ithaca, NY 14850

Broadcasting and Cable Yearbook
R.R. Bowker and Company
121 Chanlon Road
New Providence, NJ 07974

Cable TV Financial Databook
Paul Kagan Associates
126 Clock Tower Place
Carmel, CA 93923

Demographics USA
Market Statistics, Inc.
355 Park Avenue
New York, NY 10010

International Directory of Telecommunications
Longman, Inc.
1560 Broadway
New York, NY 10036

Investing in Radio
Investing in Television
BIA Consulting
PO Box 17307
Washington, DC 20041

Sales and Marketing Management
Bill Communications, Inc.
633 Third Avenue
New York, NY 10164

Telephone Industry Directory
Phillips Publishing
7811 Montrose Road
Potomac, MD 20854

TV Dimensions
Media Dynamics, Inc.
435 Fifth Ave.
New York, NY 10016

Televison Factbook
Warren Publishing
2115 Ward Court, NW
Washington, DC 20037

World Guide to Television and Programming
North American Publishing Company
401 North Broad Street
Philadelphia, PA 19108

Associations and Services

Cabletelevision Advertising Bureau
757 Third Avenue
New York, NY 10017

Radio Advertising Bureau
304 Park Avenue South
New York, NY 10010

Simmons Market Research Bureau
380 Madison Avenue
New York, NY 10017

Standard Rate and Data Service
3004 Glenview Road
Wilmette, IL 60091

Television Bureau of Advertising
477 Madison Avenue
New York, NY 10022

FOR ADDITIONAL READING

Alexander, Alison, James Owers, and Rod Carveth: *Media Economics: Theory and Practice.* Hillsdale, N.J.: Lawrence Erlbaum, 1993.

Bellman, Geoffrey: *The Consultant's Calling: Bringing Who You Are to What You Do.* San Francisco: Jossey-Bass, 1990.

Cherington, Paul W., Leon V. Hirsch, and Robert Brandwein: *Television Station Ownership.* New York: Hastings House, 1971.

Krasnow, Erwin G.: *Buying or Building a Broadcast Station* (2d ed.). Washington, D.C.: National Association of Broadcasters, 1988.

Litman, Barry R. "Economic Methods of Broadcast Research," in Joseph R. Dominick and James E. Fletcher, *Broadcasting Research Methods.* Newton, Mass.: Allyn and Bacon, 1985.

McGonagle, John J.: *Managing the Consultant.* Radnor, Pa.: Chilton, 1981.

Ryan, Bill: *Making Capital from Culture: the Corporate Form of Capitalist Culture Production.* New York: Walter de Gruyter, 1991.

Tangible Assets

In broadcast and cable accounting, assets are usually listed in order of their convertibility into cash. *Quick assets* can easily be converted into cash; *fixed assets,* which include things like land and equipment, are less easily converted.

Quick Assets

Cash Cash assets include the total cash in the bank plus petty cash used for miscellaneous purchases in the office. Advertising sales and monthly subscriber fees generate considerable cash assets in media businesses, making those companies very attractive to investors and entrepreneurs.

Notes and Accounts Receivable These represent the amounts owed to the company—by customers, advertisers, networks, or subscribers. Notes may also be due from owner-managers (who may borrow against their equity) or from stockholders in public companies.

Inventory This is the stock of finished goods ready for sale in a company. In broadcasting, unsold advertising availabilities are considered inventory. In contrast to other businesses, which have considerable material and labor investments tied up in their unsold inventory, a large part of broadcast inventory is merely "air"—namely, unsold advertising spots. This fact also makes broadcasting an attractive, cash-rich business to investors.

Cable television companies, on the other hand, maintain an inventory of unsold or unrented decoders and converter boxes, as well as equipment needed to install a drop at each subscribing household. Thus, unlike television and radio stations, cable systems typically have a large inventory of tangible assets.

In addition to finished goods, inventory includes supplies and materials necessary for day-to-day operations. In telecommunications, assets of this type include such items as tubes and transistors, wire, spare parts, paper goods, and maintenance supplies.

Fixed Assets

Property Buildings and grounds are the largest category of fixed assets in most media businesses. Unlike equipment and buildings, which lose their value over time, land generally increases in value. Thus, favorably located, established broadcasting stations often have considerable net worth because of their property. Stations with favorable transmitter locations can earn extra income by leasing part of their site to other broadcasters. Radio stations and many cable systems have an advantage compared with other businesses: they require very little land on which to operate. Television stations, on the other hand, need studio space, helicopter pads, and parking lots for remote trucks (and station employees' cars). Therefore, their land requirements are usually greater than those of radio and cable operations.

Telecommunications companies are rich in equipment assets, such as this mobile production truck. (courtesy of AVPC)

Equipment Telecommunications operations are rich in equipment assets. Included in this category are transmitters, antennas and related equipment, studio technical equipment, tape recorders and cameras, and mobile equipment, including remote vans, trucks, and automobiles.

Furniture and Fixtures These include office furniture, equipment, and fixtures. In broadcast and cable television, this category of assets also includes props, sets, and related production items.

Other Assets

Investments Investments directly concerned with the operation of the media business can be considered additional assets. As we have seen, many broadcast and cable companies have become investors in program production and syndication firms and similar joint ventures.

Prepaid Taxes, Insurance, Interest, Cash Advances Payments made before they are due can be considered station assets. Prepaid salaries and travel advances, even prepayment of news or programming services, are examples of this type of asset. Financially solvent media companies typically prepay taxes, insurance, and program costs to save on interest as well as to avoid anticipated increases in the cost of such services.

Intangible Assets

A second class of company assets are intangible items of somewhat difficult to determine but often substantial value. Traditionally, this category has included patents, trademarks, copyrights, and goodwill—the intangible value of a company's standing in the community. Over the years, the following items have been considered to add significantly to the net worth of a telecommunications facility.

Network Affiliation Contracts As discussed previously, in most cases an affiliated station is worth more than an independent station, and the affiliate of the network ranked number one in the ratings is worth more than the number two or three. Thus, the affiliation agreement between a station and networks adds to the net worth of a station.

The FCC License Since it is considerably more difficult to begin a new broadcasting station than it is to acquire an existing franchise, an FCC license itself is a primary station asset. In fact, many station sales over the years have amounted to auctions of the license, with little attention paid to the facilities, personnel, or programming of the station being acquired. With the number of new allocations by the FCC diminishing, the value of an FCC license will continue to appreciate.

Franchise Agreements In the same vein, a cable system which has a long-term agreement with the local municipality is worth more than one which faces lengthy and costly franchise negotiations.

Program and Advertising Contracts Also included in the assets of a broadcast station or cable system is the value of its contracts with program suppliers and advertisers. A local television station with a sizable collection of feature films, syndicated shows, and other programs can include the value of those shows in the calculation of its net worth. Similarly, stations holding lucrative long-term contracts with national and local advertising agencies can add the value of those contracts (over and above their face value) to their assets.

For cable systems, existing contracts with pay programming sources, microwave companies, and other suppliers can be considered intangible assets.

Goodwill The final class of intangible assets is goodwill—a telecommunications facility's public record. Goodwill includes a station's business policies (how well it pays its bills, treats its employees, etc.), as well as its image in the community (its involvement in civic events, fund raisers, charities, etc.). It is difficult to put a monetary value on goodwill, but the record of cable and broadcasting sales suggests that well-managed community-minded stations and systems are worth considerably more than stations of similar size but with a poor record in the areas of personnel management and public relations.

In an attempt to stem the rising budget deficit, the new tax law enacted by Congress in 1993 on the urging of President Clinton held particular benefit to media companies in the area of intangible assets.[2] In the past, the Internal Revenue Service regularly questioned the declared value of FCC licenses, network affiliation agreements, and program contracts; the new code permits such declarations. In addition, goodwill can now include a company's "know-how," trademarks, and trade names, as well as its work force and information base. Media companies are traditionally rich with such assets, from station IDs and slogans to skilled employees and considerable ratings and market analysis data. As a result, some media analysts have claimed that properties will rise at least 10% in value in coming years, owing to the contributions of intangible assets to their cash flow basis.[3]

Liabilities

Liabilities represent the various debts of a media company. The various classes of liabilities are as follows.

Notes Payable Usually the largest liabilities are outstanding loans to banks and other financial institutions. Such liabilities can be enormous, as has been the case in recent years with the unprecedented pace of mergers, acquisitions, and buyouts, much of them achieved with borrowed money.

Accounts Payable These are the total amounts owed to suppliers for merchandise, materials, supplies, and services. They include debts to equipment manufacturers, program suppliers, ratings companies, and outside consultants.

Payroll Payroll expenses include employee salaries, wages, and benefit plans.

Taxes Payable Taxes represent a major liability for telecommunications operations. Tax obligations of media companies include real estate and personal property taxes, federal income tax, state income tax, social security taxes, federal withholding taxes, state withholding taxes, and workers' compensation taxes.

In addition to these taxes, some municipalities have their own local sales taxes. A few municipalities have "head taxes" for companies that hire employees who live outside the city limits.[4]

Sales Commissions Payable Commissions earned by the sales staff are credited to this account pending actual payment.

License and Franchise Fees Two other classes of liabilities may appear on the ledgers of broadcast and cable firms. Many cable companies, particularly those in

FIGURE 9-1   Proposed FCC user fees. *Source: Broadcasting and Cable,* August 9, 1993, p. 6.

Cable		Broadcasting	
Cable Systems (per 1000 subscribers)..........$370		Class D Daytime.........................$250	
Cable Antenna Relay Service...................220	AM	Class A Fulltime...........................900	
		Class B Fulltime...........................500	
Telephone		Class C Fulltime...........................200	
		Construction permits.......................100	
Cellular (per 1000 subscribers)...................$60		Classes C, C1, C2, B.....................$900	
Personal Communications Service (per 1000 subscribers)......................................60	FM	Classes A, B1, C3,.........................600	
		Construction permits.......................500	
Local Telephone (per 1000 pre-subscribed access lines)......................................60		Markets 1 thru 10.....................$18,000	
Long Distance Telephone (per 1000 pre-subscribed access lines).....................60	VHF	Markets 11 thru 25.....................16,000	
		Markets 26 thru 50.....................12,000	
Competitive Access Provider (per 1000 subscribers)....................................60		Markets 51 thru 100.....................8,000	
		Remaining markets......................5,000	
		Construction permits....................4,000	
Satellite		Markets 1 thru 10.....................$14,400	
Geosynchronous (per satellite)..........$65,000		Markets 11 thru 25.....................12,800	
Low-Earth Orbit (per satellite)............90,000	UHF	Markets 26 thru 50.....................9,600	
Less than 9 meters (per 100 antennas)........$6		Markets 51 thru 100.....................6,400	
9 meters or more, transmit (per meter).......85		Remaining markets......................4,000	
9 meters or more, receive only (per meter)....55		Construction permits....................3,200	
VSAT (per 100 antennas)....................6			
Mobile satellite (per 100 antennas.............6		**Low Power TV, TV Translator, and TV Booster** $135	
		Broadcast Auxiliary............................$25	

Noncommercial broadcast stations are exempt from fees. Chart excludes fees for private and shortwave radio and several common carrier services.

(Earth Stations)

which there is no effective competition, pay franchise fees (up to 5% of their gross revenues) to the local governments in their service areas.

Whereas in the past, radio and television licensees have had the right to use their portion of the spectrum free of charge, this situation may change in the mid-1990s. Eager to find new revenues for the federal government, the same budget package which enhanced the value of intangible assets (above) added a new set of FCC user fees.[5]

The new user fees were proposed for all major media forms—including radio and television stations, cable systems, satellite services, and telephone companies. The fees were expected to add between $80 million and $100 million to the government's coffers. The full table of FCC user fees is included as Figure 9-1.

Depreciation and Amortization

Depreciation and amortization are the write-offs allowed media companies for their major capital expenditures for equipment and programming and for certain intangible assets. In general, fixed assets, such as equipment and buildings, are *depreciated;* intangible assets, such as program and affiliation contracts, are *amortized.*

Depreciation Depreciation is an expense deduction that allows the firm to recover its capital investment in equipment, buildings, automobiles, and other items throughout their usable life.

From 1961 to 1981, the broadcasting and cable businesses used the following depreciation schedules, which were set up by the Internal Revenue Service: electronic equipment, 6 years; land improvements (including transmitters and towers), 20 years; buildings, 45 years; office furniture and fixtures, 10 years; automobiles, 3 years; light-duty trucks, 4 years; heavy-duty trucks, 6 years. These guidelines were not favored by many financial managers in telecommunications. For one thing, the relatively long-term schedule of recovery allowed depreciation benefits to be eaten away by inflation. In addition, industry competition and innovation frequently forced stations and systems to reinvest in new equipment while they were still depreciating obsolete items.

The industry received relief in this area with the passage of the Economic Recovery Act of 1981. Designed to speed depreciation and thereby encourage capital investment, the act provided the following recommended depreciation schedule, known as the Accelerated Cost Recovery System (ACRS):

- *3-year property.* Includes cars and light trucks, as well as machinery and equipment used for research and development.
- *5-year property.* Includes all other outlays for machinery and equipment.
- *7-year property.* Includes most furniture and fixtures.
- *10-year property.* Includes most public utility property, including broadcast and cable installations.
- *15-year property.* Includes all other depreciable real estate and buildings, including studios and transmitter facilities.

In addition, the act allowed special first-year write-offs for new stations and systems and investment tax credits for firms investing heavily in new equipment.

Recent years have seen a number of new initiatives regarding depreciation of media properties. For example, the Tax Reform Act of 1986 created an Office of Depreciation Analysis, charged with the study of fixed-asset tax lives in various industries, including broadcasting and cable.[6] Depreciation rates, like other tax items, are subject to change. However, one source listed the most common rates at which most media properties depreciate their fixed assets.[7] The data appear in Table 9-1.

Amortization Amortization is the process by which media companies recover the capital costs for intangible assets, such as affiliation agreements, programming contracts, and goodwill expenses. Program expenses constitute the bulk of amortized expenses among broadcasting and cable firms. Over the years, the industry has utilized a variety of approaches in amortizing the cost of program contracts.[8]

TABLE 9-1 TAX LIVES OF FIXED ASSETS IN TELECOMMUNICATIONS COMPANIES

Fixed-asset category	Tax life (years)[1]
Land	–
Land improvements	15
Leasehold improvements[2]	
Buildings	31.5
Towers	15
Antenna system	5
Transmitter equipment	5
Studio technical equipment	5
Microwave equipment	5
Translator equipment	5
Mobile radio equipment	5
Satellite equipment	5
ENG equipment	5
Vehicles	5
Program production facilities	5
Program production materials	5
Furniture and fixtures	7
Office equipment	5
Test equipment	5
Tools	5
Spare parts[3]	1
Promotional materials[3]	1
Surplus property[3]	1

[1] Based on classes of recovery property from 1988 *U.S. Master Tax Guide*.
[2] Lives of leasehold improvements are shorter of 15 years or term of lease.
[3] Commonly expensed upon acquisition.
Source: *Broadcast Financial Journal*, May–June 1988, p. 5.

Straight-line amortization over the period of the run allows the company to deduct equal amounts for each year of the contract. For example, if a contract for a feature film costs $10,000 and the station is entitled to five showings over a 5-year period, the film can be amortized at a straight-line cost of $2000 per year. Generally, straight-line amortization is used for children's programs or for older, less costly programs, such as black-and-white shorts (*Little Rascals, Three Stooges,* etc.). The reason is that such programs will draw about the same audience whether rerun one or a hundred times. Each showing has roughly the same value to the station or network.

Another common method utilized by the industry is based on the fact that some programs are most valuable when first aired and become less and less valuable when rerun. Using *accelerated amortization,* the station can recover much of the cost of the program on its first run, with the remaining cost amortized in decreasing proportions. For example, given the same $10,000 contract for 5 years and five runs, the station might write off 50% of the cost in year 1 ($5000), 30% in year 2 ($3000), 10% in year 3 ($1000), and 5% each for years 4 and 5 ($500).

A variation of the accelerated method is the *sum-of-the-runs* approach. This method is based on how often the program runs during the contract period. A program shown frequently will be amortized more quickly than one left on the shelf.

Items other than programs which may be amortized include the network affiliation agreement and barter transactions. In addition, while cable firms are in the prematurity phase of construction and beginning operation, they may amortize the costs of their franchise application, pole, underground duct, antenna site, and microwave rental costs on the basis of costs projected for a fully operating system.[9]

As with depreciable assets, recent changes in the tax code have the potential to affect amortization of intangibles.[10] For one thing, to spur economic recovery among American companies, classes of intangible assets have been expanded. This is the good news. On the other hand, the Financial Accounting Standards Board (FASB), charged with overseeing financial reporting by businesses, has noted the tendency for broadcast stations and cable companies to take large write-offs on their program contracts. New rules prohibit media companies from taking greater amortization deductions than the "net realizable value" of their program contracts. In lay terms, companies cannot take a larger write-off than the shows could reasonably be expected to earn over the life of the contract. In essence, stations must calculate *now* what they think they will earn in advertising revenues from their shows over 5- or 6-year runs. This requires both a computer and a crystal ball!

METHODS OF FINANCIAL REPORTING

The assets, liabilities, revenues, and expenses of telecommunications firms can be reported in a variety of ways.

Journals and Diaries

The bookkeeping system at a broadcast facility will typically include a number of separate journals or diaries of business transactions. Of course, with the advent of modern computers, many of these journals do not exist in real physical space; they take the form of data on floppy disks and hard drives.

- *Cash receipts journal.* A chronological diary of all cash received by the business from sales, collections on account, and other sources
- *Sales journal.* A record of all sales showing credits to appropriate revenue accounts and debits to accounts receivable
- *Voucher register.* A record of all disbursements (salaries, commissions, etc.), showing credits to accounts payable and debits to each appropriate expense account
- *General journal.* A record of all transactions not included in the above journals—for example, depreciation and amortization, tax accruals, credit sales, and purchases

The books in which journal entries are transferred to accumulate information about the financial status of the company are known as *ledgers.* The process of making the transfer is known as *posting.* Media businesses typically maintain several different ledgers.

General Ledgers

The general ledger is the basic book of accounts, containing all individual accounts. Transactions taking place during the accounting period (month or quarter) are posted from the various journals to the appropriate general ledger account. The general ledger may also include the following additional ledgers:

- *Accounts receivable ledger.* Usually includes a separate listing for each sales customer (or agency)
- *Payroll ledger.* Reports records of hours worked, overtime, wage payments, and withholdings
- *Accounts payable ledger.* Contains a listing for posting all payments to creditors, usually segregated by major accounts
- *Plant ledger.* Lists tangible fixed assets such as land, buildings, and equipment

Maintaining the various journals and ledgers is largely a bookkeeping task. Extrapolating meaningful information from them about the financial health of the firm is an important management function. Thus the general manager is usually directly involved with the financial manager in preparing and interpreting the key status reports of the firm: balance sheets, profit and loss (P&L) statements, and financial ratios.

Balance Sheets and P&L Statements

The balance sheet reports the financial condition of the firm at the end of business on a particular day and compares assets and liabilities with those of a previous period. The balance sheet directly follows the accounting equation. Total assets balance with (are equal to) total liabilities plus net worth. Balance sheets are prepared on a regular basis—monthly, quarterly, semiannually, and yearly. Table 9-2 is a sample balance sheet.

A profit and loss statement is also known as an *operating statement,* or *income statement.* P&Ls compare revenues earned with expenses incurred over a given period. Table 9-3 is an example of a media P&L statement.

The balance sheet and P&L statement provide two different types of information to the financial manager. The balance sheet is much like a freeze frame of the business as it was on the last day of the accounting period. The P&L statement is like a videotaped record of the period, showing how the business got to the position occupied on the last day of the accounting cycle.

TABLE 9-2
Station KHYP
Balance Sheet
($000)

Assets

Cash	$ 67
Marketable securities	48
Accounts receivable	5,629
Broadcast program rights	3,800
Materials and supplies	128
Other current assets	277
Plant and equipment—net	6,291
Intangible assets	7,670
Other assets	221
Total	$24,131

Liabilities and Owners' Equity

Notes payable	$ 67
Accounts payable	1,877
Federal income taxes payable	2,329
Broadcast program rights	2,800
Interest payable	874
Deferred income and deposits	209
Long-term debt	3,998
Deferred income taxes	489
Other liabilities	21
Total current liabilities	12,664
Owners' equity	11,467
Total	$24,131
Debt-to-equity ratio	0.524802

Source: Format courtesy of Cox Enterprises, Inc.

From the balance sheet, the media financial manager can find out how much cash is on hand, how much is owed the firm, how much debt remains unpaid, and the present net worth of the operation. For example, the balance sheet for station KHYP at year's end (Table 9-2) reveals $67,000 in cash on hand, over $6 million in plant and equipment, and total assets above $24 million. However, on the debit side, over $12 million is owed a variety of creditors. Using the basic accounting equation, we can calculate that owner's equity (net worth) is equal to total assets minus liabilities: $24,131,000 − $12,664,000, or $11,467,000. Thus, the station is currently worth about $11.5 million.

Trends in revenue and expenses are revealed by the P&L statement. It shows whether program costs have exceeded expectations, whether advertising revenues are meeting expectations, and how much revenues may have outpaced expenses.

TABLE 9-3

Station KHYP
Income Statement
($000)

Revenue:	
Local	$ 9,250
National	10,145
Network	1,536
Political	126
Barter or trade	500
Other	211
Total	$21,768
Expenses:	
Technical	$ 1,499
Program production	1,730
Film amortization	2,785
News	1,875
Sales	2,880
Promotion	1,563
General and administrative	2,799
Total	$15,131
Gross profit	$ 6,637
Less: Depreciation	$ 885
Less: Amortization	225
Operating income	$ 5,527
Less: Interest expense	490
Pretax income	$ 5,037
Less: Income taxes	1,612
Net income	$ 3,425

Source: Format courtesy of Cox Enterprises, Inc.

For example, Table 9-3, the income statement for station KHYP, reveals total revenues approaching $22 million for the year, the bulk of which came from national spot advertisers. During the year, the station spent nearly $2 million to produce its programs, about $3 million to acquire and amortize syndicated product, and over $4 million on sales and promotional activities. It realized a gross profit of $6.6 million, and a healthy net income—after taxes, depreciation and interest—in excess of $3 million.

Margins and Ratios

In order to compare their business with a prior accounting period, with other companies, or with industry averages, media managers frequently make use of *financial ratios* or *margins*. Ratios, which are computed from data on the balance sheet and P&L statement, give a quick indication of the company's performance.

The two most important ratios used in media business are the *gross profit margin* and the *operating income margin*. The gross profit margin is derived in the following manner. First, the gross profit is calculated by deducting operating expenses from revenue reported on the P&L sheet. Then the gross profit margin is obtained by dividing gross profit by revenue.

For example, suppose a radio station generates $500,000 in revenue during a calendar year and its operating expenses during that period are $400,000. Its gross profit is $100,000 and its gross profit margin is 20%. Using Table 9-3, we see that station KHYP's gross profit margin is $6,637,000 divided by $21,768,000—a very healthy 30.4%!

The operating income margin takes depreciation and amortization expenses into account. First, operating income is calculated by subtracting depreciation and amortization expenses listed on the P&L statement from the gross profit. Operating income is then divided by revenue to determine the operating income margin.

A check of Table 9-3 indicates that at KHYP, the operating income margin is $5,527,000 divided by $21,768,000, or about 25%.

While it is perilous to generalize, well-run radio and television stations are known to operate at gross profit margins ranging from 30% to 50% with operating income margins in the 20–40% range.[11] Cable television and telecommunications systems operate at somewhat lower margins, primarily because of long-term debt and depreciation expenses arising from their extensive fixed assets. Thus, gross profit margins between 20% and 30% and operating income margins from 10% to 20% are considered standard industry performance numbers.

Liquidity and Leverage

Profit margins are most commonly used by financial managers in telecommunications to compare their stations and systems with others of different size or location. Additional ratios allow managers to assess the internal status of their own companies—specifically, liquidity and leverage. *Liquidity* refers to a firm's ability to convert its assets into cash—that is, its ability to pay its bills. Leverage refers to the proportion of debt to equity—that is, how much the company owes to banks and other creditors in relation to its assets or equity.

A common measure of the liquidity of a media corporation is its *current ratio*. The current ratio divides current assets (cash, accounts receivable, inventory) by current liabilities (accounts payable and taxes payable). The larger the ratio, the better.

Using KHYP as an example (Table 9-2), its current ratio is $24.13 million (total assets) divided by $12.66 million (total liabilities), or 1.90. If the station carried $3 million in additional debt (from rebuilding its transmitter, for example), the current ratio would dip to 1.54.

Another measure of liquidity is the *quick ratio*. The quick ratio removes inventory from the previous calculation. Thus, to determine the quick ratio, simply divide cash plus receivables by current liabilities.

A look at station KHYP (Table 9-2) indicates that cash on hand, securities, and accounts receivable total $5.74 million. The quick ratio is .45 ($5.74 divided by $12.66). Put another way, just under half of the value of the company is easily convertible to cash. A similar station with a greater debt burden (perhaps one recently purchased with borrowed money) would be less liquid (and worth less in the station sales market as a result).

Leverage is determined by dividing total debt by total assets. Suppose a television station has assets totaling $1.2 million and an outstanding debt of $500,000. Its *leverage* ratio is thus $500/1200$, or .42. Generally, the lower the leverage, the stronger the company's financial position. In this case, the station is somewhat strongly leveraged, suggesting it should not consider a major capital expansion until its debt is reduced.

A second measure of leverage, *debt-to-equity ratio,* is determined by dividing total debt by equity. Suppose a radio station has assets totaling $500,000 and liabilities of $400,000. Its owners' equity (net worth) is $100,000. If the station's total debt is $50,000, the station has a debt-to-equity ratio of .50, or 50%. A high debt-to-equity ratio (greater than 1.0 or 100%) indicates that a firm is strongly leveraged and has little cash for capital investments; a low ratio is usually a sign of fiscal conservatism (little long-term debt). In general, broadcasting companies have debt-to-equity ratios ranging from 25% to 60%; cable systems, from 40% to 75%. Since banks and other lending institutions recognize that broadcast and cable facilities have high construction costs, new ventures may be capitalized at debt-to-equity ratios between 1.0 and 1.5 (100% and 150%) Of course, lenders expect this sizable debt to be reduced quickly by the facility's cash flow and other quick assets.

How leveraged is station KHYP? A check of Table 9-2 indicates total liabilities of $12.66 million. Dividing that figure by total assets ($24.13 million) produces a debt-to-equity ratio of .52. This number puts the station in line with industry trends. It is comparatively liquid. However, with nearly $4 million in long-term debt on its balance sheet (perhaps it boasts new studios or a recently rebuilt transmitter), it would be ill-advised to undertake a major capital project in the immediate future.

Profit and Return

A final pair of figures important to media financial managers are a firm's net profit and return on investment. *Net profit* is calculated by subtracting taxes from operating income. A television station reporting $45,000 in operating profit but facing taxes totaling $15,000 will have a net profit of $30,000. Its net profit margin can be obtained by dividing net profit by sales. If the station billed $200,000 in advertising sales during the period, its net profit margin is .15, or 15%. This means the station is making 15 cents in profit for each dollar of sales. If a comparable station (perhaps a competitor in the same market) is making a higher profit on sales, say 25%, then the other station must be doing a better job of pricing, selling, and/or controlling expenses.

Return on investment (also known as return on equity) is obtained by dividing net profit by owners' equity (net worth). If the net worth of the station in the previous example is $90,000, its return on investment is .33, or 33%. Returns on investment of 50% or more have been common in broadcasting, which partly explains why so many of its disc jockeys, programmers, and producers are eager to get into ownership and management. Surprisingly, cable systems frequently offer similarly high returns on investment, since comparatively little of their working capital is owners' equity—the bulk is long-term debt. However, the slowdown in cable growth discussed in Chapter 6 has made banks and other financial institutions more conservative in their fiscal policies—requiring more investment capital from owners in order to secure a loan. This has had the effect of reducing return on investment, particularly in the early years of a system's operation.

Let's take a last look at station KHYP. Its income statement (Table 9-3) indicates net income after taxes of $3.43 million on total sales revenues of $21.77 million. Its net profit margin is .1575. In other words, the station returns about 16 cents in profit for every dollar of sales.

Return on investment is found by dividing KHYP's net profit ($3.43 million from the income statement) by its owners' equity ($11.47 million, from the balance sheet), for a ratio of .299, or roughly 30%. Thus, station KHYP is a solid, profitable enterprise which produces a satisfactory rate of return for its owners and investors.

There are many more indicators of the fiscal status of media organizations. Two important ones are cash flow and multiples, as discussed in Chapter 8. In addition, there are barometers known as *receivables turnover, inventory turnover,* and others. Each of these indicators, no matter how precise, merely describes the current status of a company. In order to predict future performance, managers must also be familiar with principles of economic planning and projection.

FINANCIAL PLANNING AND PROJECTIONS

The changing nature of media businesses in recent times has meant that financial planning has become increasingly important. The process of forecasting has become a central part of the job of broadcast and cable managers and financial analysts. Financial forecasting falls into three main classes:

- Internal forecasting
- Trend analysis
- Investment analysis

Internal forecasting is the financial planning that goes on within the firm itself on a department-by-department basis. Each area will normally set short-term and long-term financial goals and expectations. The sales department will prepare projections for spot costs, percentage of inventory sold, expected sales revenue, and anticipated ratings for existing and proposed programs. The programming department will forecast anticipated program expenses and costs for syndicated material. Forecasts will also be prepared by engineering, news, and other departments.

Trend analysis is the process by which media financial managers monitor the financial performance of their competitors to keep abreast of industry developments and changes. A small radio station is wise to track the financial and sales performance of its major competitors and to compare their spot costs, profit margins, expenses, and so forth with its own. Financial analysts at conglomerate operations also spend a good deal of time monitoring the performance of competing MSOs and station groups.

Figure 9-2 illustrates how one major media conglomerate, Cox Enterprises, tracked its competition through the 1980s. Pay particular attention to the gross profit margins near the bottom of the figure—from a "low" of 19% at the Tribune company in 1987, to the consistently high profit performance of Capital Cities/ABC (above 50% each year of the decade). It is also interesting to note that a number of the groups on the list have since changed hands. The former Metromedia stations are the core of the Fox Broadcasting Group, Taft has been broken up, and Storer is no more, its TV stations having been sold off and its cable properties purchased by TCI and Comcast.

Related to trend analysis is *investment analysis.* This is the process of monitoring industry trends with an eye for acquisition of media properties. Investment analysis is a common practice for networks, group owners, and MSOs. The trends in ownership traced in Chapter 7—the streamlining and/or elimination of group and cross-ownership restrictions, the rise of coventures and joint ventures, and so on—have led financial managers to virtually continuous planning for diversification. The bulk of their activities may be managing current properties, but many media executives keep a sharp eye on industry trends in search of investment opportunities in other stations, systems, and markets.

A number of business services firms provide investment analysis to the telecommunications community. Some were mentioned in the previous chapter, including Paul Kagan and Associates (which specializes in cable and related technologies), Broadcast Investment Analysts (BIA), and Duncan's American Radio. In addition, Standard and Poor's annual *Industry Surveys* include financial analyses of media companies, and the investment services firm Veronis, Suhler and Associates has garnered attention for its annual evaluation of communication companies, with projections for future industry performance.[12] Figure 9-3 illustrates the kind of media investment analysis used by managers and entrepreneurs to track their performance and plan for potential acquisition activity.

THE COMPUTER IN MEDIA MANAGEMENT

It is not surprising that the growing importance of financial planning in telecommunications has directly paralleled the rise of the computer in business, education, and government. In recent years, computers, particularly smaller microcomputers, have become a fixture in the operations of media businesses.

FIGURE 9-2 Broadcast industry margin analysis: broadcasting segments.

Sales	1983	1984	1985	1986	1987	1988	1989
CBS (TV and Radio Stations)	N/A	N/A	N/A	N/A	N/A	N/A	N/A
METROMEDIA	371558	N/A	N/A	N/A	N/A	N/A	N/A
TRIBUNE CO (Consolidated)	228739	282193	331664	466231	485276	505279	584326
CAPITAL CITIES/ABC	235754	271848	293717	N/A	N/A	N/A	N/A
TAFT (Results for FY ended 3/31/XX+1)	171464	189580	283353	281058	N/A	N/A	N/A
GANNETT	192874	232748	265480	351133	356815	390507	408363
STORER	167778	184714	N/A	N/A	N/A	N/A	N/A
LIN (Consolidated until 1986)	107333	148844	171671	154037	156196	158383	162378
BELO	79452	158554	163767	175635	165540	165343	177938
WASHINGTON POST	119807	136041	154513	167112	171396	180195	182545
TIMES MIRROR	116071	124714	128809	127301	110051	99045	102790
SCRIPPS-HOWARD	94402	106903	122892	191386	198203	214107	222627
MULTIMEDIA	125881	135319	107047	119277	133678	136947	136943
JOHN BLAIR	39867	89021	99372	N/A	N/A	N/A	N/A
VIACOM	119785	83598	83733	111340	132132	141199	146060
CHRIS-CRAFT	N/A	N/A	168264	199836	216656	230577	246741
OUTLET COMMUNICATIONS	N/A	N/A	50224	59852	63126	104223	97916
PRICE COMMUNICATIONS (TELEVISION)	N/A	N/A	9406	23933	35093	48431	32531
TOTAL	2170765	2144077	2433912	2428141	2224162	2374696	2501158

Income Before Dep & Amort (Gross Profit)

	1983	1984	1985	1986	1987	1988	1989
CBS (TV and Radio Stations)	N/A	N/A	N/A	N/A	N/A	N/A	N/A
METROMEDIA	128594	N/A	N/A	N/A	N/A	N/A	N/A
TRIBUNE CO	50716	58322	72774	93960	92249	107779	125971
CAPITAL CITIES/ABC	129540	149609	155161	N/A	N/A	N/A	N/A
TAFT (Results for FY ended 3/31/XX+1)	64183	69991	97590	102225	N/A	N/A	N/A
GANNETT	70826	92476	105129	137842	133002	145145	140666
STORER	59121	66835	N/A	N/A	N/A	N/A	N/A
LIN (Consolidated until 1986)	45308	64259	76323	79993	83130	82708	83910
BELO	35904	80605	76520	80794	73847	62146	78130
WASHINGTON POST	44693	55980	63659	76645	78202	72409	80335
TIMES MIRROR	66540	70407	69683	75683	62527	46654	45536
SCRIPPS-HOWARD	38919	47057	52983	71783	71849	75647	76560
MULTIMEDIA	44487	48837	34448	41680	50388	54724	55401
JOHN BLAIR	8004	23298	23130	N/A	N/A	N/A	N/A
VIACOM	22037	28467	29939	44607	57767	59487	59131
CHRIS-CRAFT	N/A	N/A	50464	47152	45774	39448	20714
OUTLET COMMUNICATIONS	N/A	N/A	15180	19308	22335	26850	25211
PRICE COMMUNICATIONS (TELEVISION)	N/A	N/A	2818	10940	14803	17253	11017
TOTAL	808872	856143	925801	882612	785873	790250	802582

Gross Profit Margin in %

	1983	1984	1985	1986	1987	1988	1989
CBS (TV and Radio Stations)	N/A	N/A	N/A	N/A	N/A	N/A	N/A
METROMEDIA	34.6%	N/A	N/A	N/A	N/A	N/A	N/A
TRIBUNE CO	22.2%	20.7%	21.9%	20.2%	19.0%	21.3%	21.6%
CAPITAL CITIES/ABC	54.9%	55.0%	52.8%	N/A	N/A	N/A	N/A
TAFT (Results for FY ended 3/31/XX+1)	37.4%	36.9%	34.4%	36.4%	N/A	N/A	N/A
GANNETT	36.7%	39.7%	39.6%	39.3%	37.3%	37.2%	34.4%
STORER	35.2%	36.2%	N/A	N/A	N/A	N/A	N/A
LIN (Consolidated until 1986)	42.2%	43.2%	44.5%	51.9%	53.2%	52.2%	51.7%
BELO	45.2%	50.8%	46.7%	46.0%	44.6%	37.6%	43.9%
WASHINGTON POST	37.3%	41.1%	41.2%	45.9%	45.6%	40.2%	44.0%
TIMES MIRROR	57.3%	56.5%	54.1%	59.5%	56.8%	47.1%	44.3%

FIGURE 9-3 Media investment analysis. *Source*: Standard and Poor's.

Company	Yr. End	Return on Revenues (%)					Return on Assets (%)					Return on Equity (%)				
		1986	1987	1988	1989	1990	1986	1987	1988	1989	1990	1986	1987	1988	1989	1990
ENTERTAINMENT																
AMC ENTERTAINMENT INC	†MAR	NM	NM	NM	NM	0.6	NM	NM	NM	NM	0.6	NM	NM	NM	NM	4.0
*BLOCKBUSTER ENMNT CORP	DEC	NM	9.5	11.3	11.0	10.9	NM	9.2	12.4	14.9	13.4	NM	20.0	21.6	28.5	26.3
*CBS INC	DEC	4.1	4.9	10.2	10.0	2.8	5.5	3.7	6.8	6.6	2.0	27.5	13.1	16.6	12.9	3.8
*CAPITAL CITIES/ABC INC	DEC	4.4	6.3	8.1	9.8	8.9	5.1	5.3	6.8	7.8	7.3	12.8	13.4	14.7	15.4	14.3
CINEPLEX ODEON	DEC	6.3	6.6	5.8	NM	NM	6.2	4.4	3.7	NM	NM	18.1	12.3	11.3	NM	NM
*DISNEY (WALT) COMPANY	SEP	10.0	13.6	15.2	15.3	14.1	8.2	11.3	11.7	12.0	11.2	19.0	24.0	24.8	26.0	25.2
*HANDLEMAN CO	†APR	5.5	6.2	6.4	5.1	3.3	11.3	12.7	13.5	9.9	5.6	20.2	22.3	23.2	17.7	10.3
*KING WORLD PRODUCTIONS INC	AUG	13.6	14.1	21.7	18.5	18.5	25.6	32.1	50.0	35.0	23.7	59.1	81.8	NM	NM	80.6
POLYGRAM N V	DEC	NA	NA	7.6	8.1	6.8	NA	NA	NA	10.3	9.4	NA	NA	NA	38.8	46.5
TURNER BROADCASTING -CL A	DEC	NM	NM	NM	2.6	NM	NM	NM	NM	1.4	NM	NM	NM	NM	NM	NM
VIACOM INC	DEC	NM	NM	NM	9.1	NM	NM	NM	NM	3.4	NM	NM	NM	NM	28.5	NM

Note: Data as originally reported. * Company included in the Standard & Poor's 500. † Of the following calendar year.

Functions of Computers

Computers are used to perform the following specific functions in media firms:

Traffic Automated program and operations logs are used to schedule programming and commercial announcements, as well as to verify that spots ran as scheduled (confirmation).

Sales Computers are used to list, plan, schedule, verify, and confirm availabilities, to prepare sales contracts, to calculate commissions, and to perform other sales-related tasks. In addition, most cable systems use automated billing services to handle the subscriber billing process.

Accounting The various ledgers, journals, and financial procedures discussed above may be fully computerized. Especially popular among small business managers are electronic spreadsheet programs such as Lotus 1-2-3, Quattro Pro, and Quicken.

Programming Computers are used to maintain program inventory, to track program ratings, even to replace the disc jockey or video switcher in placing program material on the air. The degree of program automation ranges from *live-assist,* where computers aid program personnel in their selections and decisions, to *fully automated* radio, TV, and cable operations requiring no flesh-and-blood program operators.

Types of Systems

Business automation systems have three general configurations:

On-Line, or Time-Sharing, Systems With time-sharing systems, the local station or system has its own video display terminals (VDTs) and printers, but the actual processing is done by a large mainframe (or host) computer at another location, linked by telephone, microwave, or satellite. The overnight ratings from Nielsen (see Chapter 14) are received in this way.

Distributed, or Batch-Processing, Systems With these arrangements, a minicomputer is in place at the station or system but is also linked to a mainframe host. The mini can perform some basic tasks, but once a day (usually late at night) it links up with the mainframe to perform more sophisticated tasks. Some automated billing and traffic systems began in this way. However, because of the rise of the microcomputer (see below), batch operations are becoming rarer.

Microcomputers Revolutions in microchips and related technologies have given rise to desktop and laptop personal computers which are as powerful as the mainframe computers of just a few years ago. As a consequence, virtually all media companies today use personal computers. Many are interconnected, or networked, enabling employees to communicate with one another (by electronic mail or "E-mail"), and allowing the various departments immediate access to the same information (such as a TV station's news scripts, programming, and commercial schedules). There are specialty bulletin board services (BBS, in computer parlance) for disc jockeys, newscasters, announcers, salespeople, and promotion executives. A morning-show disc jockey can even get topical jokes for tomorrow's program via personal computer (or fax machine) and telephone line (a service called "Telejoke," started by one underpaid TV professor).

Advantages and Disadvantages

Of course, computers by themselves are no guarantee of improved management and profitability. Like other technologies, their strength lies not in their capabilities but in how and why they are used. A computer will *not* remedy sloppy record keeping and accounting. A commonly used industry acronym is GIGO—garbage in, garbage out! Computers have the advantages of being fast, efficient, and increasingly inexpensive to own and operate. They have disadvantages as well. For one thing, despite advances in the user-friendliness of software and hardware, a good deal of specialized expertise is necessary to design and utilize programs. For another, there can be considerable employee resistance to computers, sometimes known as "technophobia." Employees may feel that their jobs will be rendered obsolete by the equipment. Or they may simply be in awe of the technology. In addition, there is concern that employees with too much of exposure to computer terminals may fall victim to boredom, stress, and such physical problems as migraines and backaches.

Ultimately, the decision to invest in automated technologies for accounting, sales, and programming applications must be based on the needs and particular management style of the media firm. It might be unwise for the owner of a radio station in a medium-sized market to terminate or retrain the traffic people, particularly if a manual accounting and traffic system has served the station well over the years. However, the larger stations, which face strong local competition and need an edge in marketing and sales, probably will have to be fully outfitted with the latest microcomputer technology.

A FINAL NOTE

Ronald Townsend, vice president of Gannett Broadcasting in Arlington, Va., has reflected upon the increasingly important role of financial management in telecommunications.

> The financial manager is more responsible than anyone else for minimizing surprises. He is in contact with more people at the station than anyone else, except the general manager, and consequently has to be communicating with those people, and know what is going on with talent contracts, sales strategies, the selection of new equipment, and other areas.[13]

As stated at the outset of the chapter, the days of the financial manager as an accountant with a green eyeshade are gone forever in media business.

SUMMARY

Financial management in telecommunications springs from the basic accounting equation: Assets = liabilities + owners' equity.

Quick assets are easily converted into cash; they include cash, accounts receivable, and inventory. Fixed assets include property, equipment, furniture, and fixtures. Intangible assets include an affiliation or franchise agreement, an FCC license, and community service, or goodwill.

Major liabilities in media firms are accounts receivable, payroll obligations, taxes, and franchise and user fees. Media firms may write off certain items by listing depreciation and amortization expenses. Tangible items such as equipment, buildings, and automobiles may be depreciated; intangible commodities such as program contracts, affiliation agreements, and goodwill are amortized.

Financial records may be kept in a variety of journals and ledgers, including a cash receipts journal, a sales journal, a voucher register, a general ledger, a payables and receivables ledger, and a payroll ledger.

Balance sheets and P&L statements are the financial reports most commonly used in telecommunications. They enable financial managers to evaluate the profit performance, equity, and debt position of an individual firm as well as to compare the financial performance of one company with others.

The computer has become an important part of media financial planning. Microcomputers use specialized programs to handle traffic, sales, bookkeeping, and programming tasks. There seems little doubt that the next generation of media managers will need to be familiar with the operation and application of microcomputers.

NOTES

1 Gene F. Jankowski, former president, CBS Broadcast Group, cited in "Failure Fallacies," *Broadcasting,* May 28, 1985, p. 72.
2 See "Tax Changes Could Heat Up Mergers," *Broadcasting and Cable,* August 16, 1993, p. 34.
3 *Ibid.*
4 Mark E. Battersby, "Those Other Taxes," *Broadcast Management/Engineering,* May 1983, pp. 87–88.
5 Kim McAvoy, "Congress Slaps Industry with FCC Tax," *Broadcasting and Cable,* August 9, 1993, pp. 6–10.
6 John S. Sanders, "Broadcasting Fixed Asset Tax Lives Under Reconsideration," *Broadcast Financial Journal,* May–June 1988, pp. 4–6.
7 *Ibid.,* p. 5.
8 See Donald M. Davis and Raymond L. Carroll, *Electronic Media Programming* (New York: McGraw-Hill, 1993), pp. 323–325.

9 See M. LaVoy Robinson and Kevin E. Roberts, "Cable Television Accounting and Financial Reporting," *Broadcast Financial Journal,* May 1984, pp. 20–22.
10 See Michael Sileck, "Valuation of Broadcast Rights: Science or Art?" *Broadcast Cable Financial Journal,* March–April 1992, pp. 34–36.
11 The author is indebted to William R. Killen, Vice President, Financial Analysis and Planning, Cox Enterprises, Inc., for his counsel and guidance in the area of financial yardsticks for media companies.
12 See "Media Revenue Up," *Broadcast Sales Training Executive Summary,* July 26, 1993, p. 1.
13 "Financial Managers Have Their Horizons Widened," *Broadcasting,* May 28, 1984, p. 74.

FOR ADDITIONAL READING

Broadcast Financial Management Association: *Operational Guidelines and Accounting Manual for Broadcasters.* Des Plaines, Ill.: BFMA, 1990.

Chorafas, Dimitris N.: *The New Technology of Financial Management.* New York: Wiley, 1992.

Marriott, Neil: *Management Accounting: A Spreadsheet Approach.* Englewood Cliffs, N.J.: Prentice-Hall, 1993.

National Association of Broadcasters: *Accounting Manual for Radio Stations.* Washington, D.C.: NAB Publications, 1990.

——————: *Accounting Manual for Television Stations.* Washington, D.C.: NAB Publications, 1990.

10

PERSONNEL MANAGEMENT AND EMPLOYEE RELATIONS

I try to get people who are better than I am at various elements of the business and then I do my damnedest to keep those people interested.

This is not an easy place to work. Everyone's kind of pushy. People want challenges and responsibility and we give it to them.

Quotes from media executives

In an earlier chapter it was stated that despite a fascination with electronic gadgetry, telecommunications is first and foremost a people business. The hardware of broadcast, cable, telephone company (telco), and satellite systems is utilized for *communication,* the transmission of messages between people with a minimum of static, interference, or noise. It is somewhat ironic that the principal agents of electronic communication have been characterized by a poor relationship between employees and management. As discussed in Chapters 4 through 6, the industry has been characterized by very high turnover, burnout, and attrition rates for employees. However, in recent years, telecommunications management has awakened to the importance of personnel organization, administration, training, compensation, and motivation. These are the focus of the present chapter.

The goals of this chapter are:

1. To describe the tasks and functions of personnel administration in media businesses
2. To describe models of station and cable system organization

3. To discuss policies of employee recruitment, training and evaluation, promotion, and termination
4. To describe typical employee compensation, benefits, and enrichment plans in broadcasting and cable
5. To discuss elements of employee communication, including employment handbooks, in-house newsletters, and other materials
6. To describe the status of labor-management relations in telecommunications, including labor unions and collective bargaining

MEDIA PERSONNEL: AN OVERVIEW

The dimensions and characteristics of the media work force were described in Part Two. As an introduction to the process of managing media personnel, it is useful to review some of these trends.

Overall, radio, television, cable, and related businesses employ about 250,000 people. The majority of these work at local television, radio, and cable facilities; the remainder, at networks and MSO headquarters. The telephone industry is much larger. Over 600,000 people are employed by the leading regional Bell operating companies (RBOCs), their suppliers, and related telco businesses.[1] Media firms require a broad range of employees, from specialized guild and craft workers (such as cable installers, broadcast engineers, and technicians) to academically trained managers and supervisory personnel (researchers and accountants, for example). Although salary levels are, overall, low or average compared with those in other industries, compensation can be excellent for a few highly skilled (or lucky) people with ability and a successful track record. Job turnover in telecommunications is alarmingly high when compared with turnover in industries of similar size. Finally, as has been the case in other industries, the 1990s have seen attempts to streamline or "downsize" the total media work force, with layoffs becoming increasingly commonplace at stations, networks, cable carriers, and telephone companies.

It is not surprising that in this highly volatile employment environment, telecommunications management has become increasingly concerned with methods of employee management, recruitment, training, and retention. Not too long ago, media employees were treated by management as little more than part of the equipment needed to sustain broadcast. Regulatory initiatives at the Federal Communications Commission (FCC) and the Equal Employment Opportunity Commission (EEOC), competition, and the realization that employee satisfaction is directly correlated with improved financial performance have led contemporary telecommunications management to adopt a much more enlightened view of the process of personnel management. While a few radio and television station managers still view their employees as an expendable budgetary line item, the majority of media firms have placed increased emphasis on the financial and personal well-being of the members of their work force.

FUNCTIONS OF PERSONNEL ADMINISTRATION IN TELECOMMUNICATIONS

Once a station or cable system has been authorized to begin operations, it might seem that staffing would be a simple task. In reality, personnel management is a complex process involving the following specific areas:

- *Company organization.* Developing station or system organizational structure and management flowcharts; delegating tasks to different departments and units.
- *Planning the work force.* Preparing payroll budgets; specifying job requirements; recording and analyzing information on employee turnover, absenteeism, and movement between different units of the organization.
- *Recruitment.* Preparing job descriptions, advertising, interviewing, making selections, describing terms and conditions of employment.
- *Training.* Developing procedures for on-the-job training, selecting personnel for out-of-house training programs and courses, following up and continuing training if necessary, evaluating training programs.
- *Performance review.* Developing appraisal methods, both formal and informal, and methods of getting employee feedback to management; determining type and frequency of employee-management meetings.
- *Wage and salary incentive systems.* Setting and altering salaries, premiums, and bonuses; setting schedules of salary increments; supervising overtime and other special payments.
- *Employee benefits.* Establishing plans for medical benefits, pensions, vacations, and other fringe benefits.
- *Industrial relations.* Overseeing labor agreements, formation of bargaining units, grievance procedures, and disciplinary procedures.
- *Health and safety.* Developing and monitoring adherence to safety rules and regulations, making arrangements for reporting incidents, setting up inspection procedures.
- *Communications.* Briefing employees on company policies and developments, publishing corporate newsletters, offering in-house telecommunications presentations (teleconferencing, A/V presentations, audiotapes, videotapes, films).
- *Personnel welfare.* Counseling; building and maintaining company morale. May include everything from employing a company psychologist to forming a station softball team.

As might be expected, personnel administration is a direct function of the size of the station or facility. Overall, roughly one-third of all broadcast and cable facilities maintain the services of a full-time personnel director.[2] The position of director of personnel is in place in about one-third of the radio stations and half of the nation's largest cable and television facilities employing 25 or more full-time people. In addition, virtually all networks, group owners, and MSOs employ full-time personnel directors.

The important task of personnel administration requires a good deal of planning. Thus, the first step in staffing a media facility is drafting a model of employee organization. This is known variously as a *table of organization,* a *management flowchart,* or a *table of reporting relationships.*

TABLES OF ORGANIZATION

Recall from Chapter 3 that media businesses tend to be *horizontally integrated;* that is, they require the skills of a variety of specially trained staffs (engineering, programming, sales) in order to function effectively. Also remember that media businesses have an *inverted-pyramid* ownership structure, with group and corporate ownership becoming increasingly common. Together, these two facts mean that individual broadcasting stations and cable systems must develop a detailed model of their internal organizational structure. All staff members must know their relationship to other departments, to whom they are directly responsible, and their place in the overall hierarchy of the firm.

In short, the purposes of the table of organization in media firms are:

1 To identify each of the units (departments) in the firm
2 To delineate the chain of command from top management to each unit
3 To describe the relationships among individual units

There are two major factors which directly affect tables of organization in telecommunications. These are *market size* and *type of ownership.*

It has already been pointed out that staff size, employee compensation, and job flexibility are directly related to the population of the community in which the station or system is located—its market size. In general, media firms in large markets (in the case of cable, systems with large numbers of subscribers) employ more people and pay better than do companies in small markets. However, large-market facilities are marked by departmentalization and specialization, allowing few opportunities for employees to switch departments and broaden their skills. In addition, there tend to be fewer ways to break into management for employees in larger-market facilities. Thus larger-market stations and cable systems are likely to have more complex and comprehensive tables of organization than do media facilities in smaller markets.

The type of ownership structure of a media firm also affects its organization. In general, the higher one goes on the inverted pyramid of media ownership, the more likely the station is to have a complex, formalized chain of command. Similarly, a cable system run by an MSO will be more likely to operate according to a carefully delineated model of employee relations than will a locally owned mom-and-pop facility.

Sample Organization Charts

Sample telecommunications organization charts appear in Figures 10-1 to 10-6. For each medium—radio, TV, and cable—we will examine the organization of a small-market, independently owned facility and the comparatively complex chain of command at large-market, group-owned operations.

Figure 10-1 shows how the staff is organized at a small locally owned and operated radio station. Four departments report to the station manager and general manager: traffic, programming, sales, and engineering. Notice how, for convenience, news falls under the responsibility of the program director.

Note the increasing complexity at a major-market radio station (Figure 10-2). Here, five department heads report to the program director alone: the public affairs director, the chief announcer, the news and sports directors, and the music

FIGURE 10–1 Small-market radio.

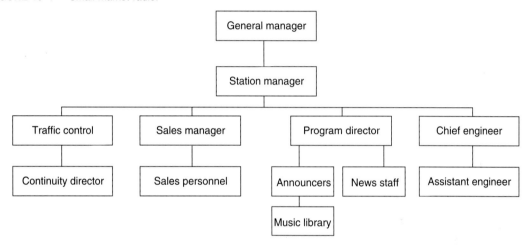

FIGURE 10–2 Large-market radio.

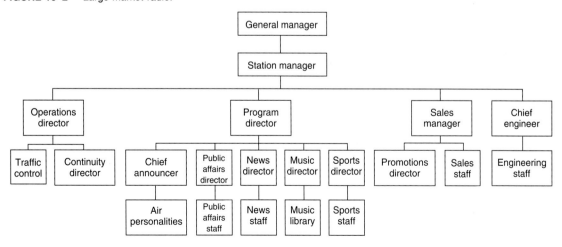

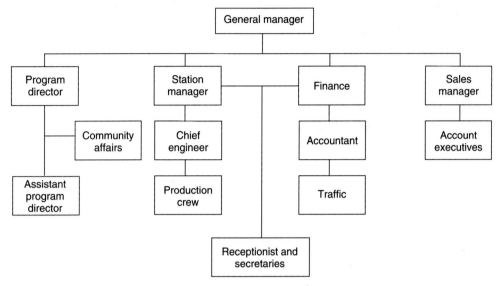

FIGURE 10-3 Small-market television.

director. Traffic and continuity functions are separated, as are sales and promotion.

Similar trends are seen in the charts of TV station management. Figure 10-3 describes the small-market television operation. Here, the program manager supervises all local programming, including community affairs. The controller supervises an accountant and the traffic staff. The sales manager handles both national and local sales accounts.

Things differentiate and specialize in the larger TV markets (Figure 10-4). The programming function may be divided between two department heads: one in charge of program acquisition (syndication), the other in charge of local programming (origination). Sales and promotion are commonly separated, as is the news department. In the sales department, separate staff people manage national and local spot sales. The financial function often comprises its own department, headed by the director of finance (or controller, in some large stations).

It should not be surprising that cable operations follow a similar pattern: as system size increases, so does organizational complexity. Figure 10-5 charts the table of organization at a small cable system (about 5000 subscriber households) Four staff heads report to the system manager: sales manager, program director, chief technician, and office manager. To keep the payroll down, the program director's staff includes student interns working in the local-origination (LO) studio who report to the LO producer.

Compare this "lean" management team with a major MSO (Figure 10-6). The home office includes nine specialized functional areas: finance (controller), employee relations, marketing, operations, public affairs, engineering, construc-

CHAPTER 10: PERSONNEL MANAGEMENT AND EMPLOYEE RELATIONS   257

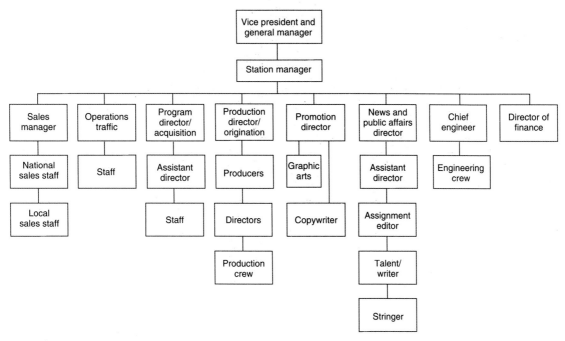

FIGURE 10–4   Large-market television.

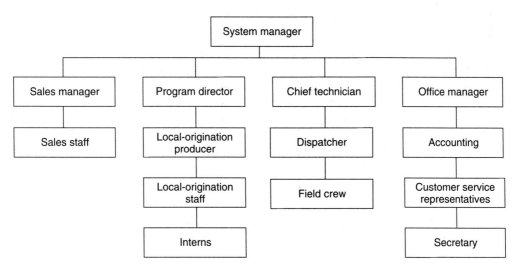

FIGURE 10–5   Small cable system.

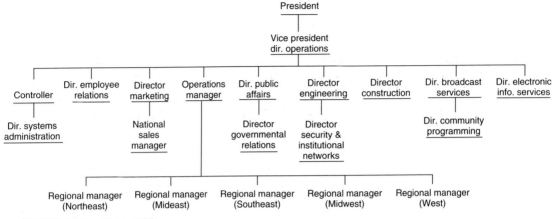

FIGURE 10–6 Cable MSO.

tion, origination (broadcast services), and electronic information services. In addition, individual system managers report to supervisors in their geographical region.

Reading the Charts

A number of points can be made about reading and interpreting organizational tables in telecommunications. The first is that no matter how carefully delineated and structured a facility is, its success is still dependent upon the *people* who bring the charts to life. As the cliché goes, "A chain is only as strong as its weakest link." Media organizations are only as strong as the weakest member of their management team.

Another point is that as complexity increases, stations and cable systems may too easily fall victim to a Theory X management structure (Chapter 2). Too much formality, departmentalization, and structuralization can weaken lines of communication among employees, lead to staff rivalries and competition for budgets and benefits, and create other problems.

Finally, note that as complexity increases, more and more layers are added between the chief officers and the employees, who make the company "happen" on a day-to-day basis. This can raise difficult problems of employee motivation and morale. Announcers, salespeople, customer service representatives, technicians, and other workers may justly feel that management is less interested in the work and the lives of the firm's employees than in profits.

CORE DEPARTMENTS IN TELECOMMUNICATIONS

As the tables of organization reveal, there are four core departments found at all media facilities: general management, sales, programming, and engineering. The basic functions of each department are as follows:

- *General and administrative.* Budgeting, secretarial and clerical functions, employee compensation, accounting.
- *Sales.* Sales of network, national, regional, and local availabilities; promotion, marketing, and research; advertising copywriting and production. In cable, the sales function includes subscriber recruitment, promotion, and customer service.
- *Programming.* Network and local program acquisition, production, and schedule. The function includes local news and public affairs (broadcasting) and local-origination programming (cable).
- *Engineering.* Station and system design, transmission, technical operations, and maintenance. In the cable business, engineering includes field crews for installation and repair of cable lines and equipment.

As facilities employ more people, the four basic units often branch and specialize into separate departments. The general and administrative function might be organized into the following separate staffs:

- *Controlling and accounting.* Direct control over the firm's budgeting, accounts payable, and accounts receivable functions
- *Traffic.* Responsibility for program and commercial scheduling (may also be under the supervision of the sales department)
- *Personnel.* Responsibility for employee recruitment, training, and evaluation
- *Office automation.* Responsibility for the operations of automated computer systems in use at the station

Members of the engineering staff, who report to the chief engineer, might be subdivided into these categories:

- *Studio engineers.* Charged with the maintenance and operations of studio equipment
- *Transmitter engineers.* Responsible for maintaining transmitter links, satellite links, and related equipment
- *Field engineers and technicians.* Common in cable, the crews responsible for the maintenance and installation of equipment away from the head end: distribution systems, cable drops, converters, and similar gear

Because of their increasing importance to telecommunications management, three of the core units are examined in detail in Part Four. These are radio and television program management (Chapters 11 and 12), sales and marketing management (Chapter 13), and audience research (Chapter 14).

JOB DESCRIPTIONS IN TELECOMMUNICATIONS

Once the firm has been organized into units with clear responsibility and reporting relationships, the next step in personnel management is to spell out the tasks of each employee. This entails the preparation of comprehensive job descriptions.

A good job description defines the overall purpose of the job in terms of the specific tasks to be performed. Too often, crises in management occur because an employee claims, "Sorry, it's not my job." Recent developments in technology have partially obscured the traditional boundaries between staffs in media organizations, making comprehensive job descriptions more essential.

While the preparation of job descriptions may seem to merely add bureaucratic red tape to the operations of a broadcast station or cable system, good job descriptions have at least three desirable results:

1 They give employees a clear understanding of their jobs.

2 They eliminate unnecessary overlapping of duties and responsibilities, enabling each person's work to complement the work of others.

3 They give management a better sense of what the organization does, thereby making it easier to determine the company's present course and to plan its future direction.

The main points that should be included in a job description are:

- The location of the job within the table of organization (division, department, branch, or section)
- The job title
- The job grade or level (such as apprentice, trainee, manager, director)
- The job titles of any individuals who *report* to the job holder, and the number of employees supervised
- A brief description of the overall purpose of the job
- The main tasks carried out by the job holder, listed in chronological order or in order of importance
- Details of the equipment needed to perform the job, including any special training or prior experience needed to use the equipment
- Special circumstances of the job, including shifts, night duty, weekend or fill-in requirements, travel requirements, and unpleasant or dangerous working conditions

The following sections describe typical telecommunications jobs.

Radio and Television

General Manager Charged with overall management and operation of the station, the general manager is responsible for all business and financial matters, including income, expenses, short- and long-range planning and goals, budgeting, forecasting, and profitability. The general manager must run the station in accordance with all federal, state, and local laws, including FCC regulations. In addition, the general manager is responsible for establishing and maintaining the station's image in the community it serves. When the station is a network affiliate, the general manager has an obligation to uphold the policies of the affiliation agreement and to maintain contact with network officials. He or she supports relationships with community leaders, advertisers, advertising agencies, program suppliers, and other outside station contacts. The general manager is in charge of producing advertising revenues and must constantly evaluate the effectiveness of the sales department. General managers also hire the major department heads, establish their goals, monitor their performance, and approve their budgets. General managers oversee each department's activities. They also approve all local programming and, with the program director, decide which syndicated programs are bought; approve newscast formats and editorial policies; and approve investments in new equipment and facilities.

Sales Manager   The sales manager, who reports directly to the general manager, is in charge of all sales activities of the station. Sales managers supervise the daily activities of all account executives and assign them to particular areas or accounts. With the general manager's approval, the sales manager sets the rates for the sale of all local and regional time.

The sales manager must be familiar with all aspects of marketing and advertising, including understanding and using market data, retail statistics, and audience research, and must maintain liaison with the national sales organizations, including the national sales representative, the Radio Bureau of Advertising (RAB), or the Television Bureau of Advertising (TVB).

Program Director   The program director, who reports directly to the general manager, is responsible for all network, local, and syndicated programming broadcast by the station. The program director must ensure that programming complies with all FCC regulations. Along with the general manager, the program director selects all syndicated programming and schedules, as well as all local programming and schedules. The program director administers the program budget allocated by the general manager.

Promotions Director   The promotions director reports to the sales manager and is responsible for the promotion and sales development activity of the station. The promotions director's activities include planning, execution, and evaluation of the station's two main classes of promotional and marketing activity: *on-air* (promotional announcements, contests, fund-raisers, station identifications, logo and graphic displays, etc.) and *off-air* (community events, personal appearances, tours and exhibitions, etc.). The promotions manager administers the annual budget of the department, including in-kind and cross-promotional activity with other organizations and sponsors.

Research Director   The research director, who reports directly to the general manager or sales manager, acquires and analyzes the information on which marketing and advertising decisions are based. The research director interprets national and local ratings data, conducts personal or telephone interviews, sends out mail-in questionnaires, conducts library study, and analyzes data obtained from government agencies. The research director must be familiar with the operation and functions of the station's mini- and microcomputer systems and applications software used in performing audience research and statistical analysis.

News Director The news director is responsible for all local news programming. He or she reports directly to the general manager or program director. In coordination with the assignment editor, the news director supervises daily assignments, ensures that scripts are ready for broadcast, and coordinates anchors and reporters, as well as sports, weather, floor, and camera personnel. The news director conducts regular meetings at which the station's news programs are screened, discussed, and evaluated.

Account Executives Account executives are responsible for the actual selling of available spots. They report directly to the sales manager. An account executive may be assigned to one or more areas (or one or more accounts) and is responsible for selling time to the advertisers so designated. Account executives are expected to meet or exceed sales quotas set on a weekly, monthly, and/or quarterly basis by the local sales manager and the general manager.

Community Affairs Director The community affairs director is responsible for developing a continuing program of station services in response to the problems, needs, and interests of the local community. The community affairs director represents the station at public meetings and, with the general manager and program director, coordinates the station's public affairs programming schedule.

Chief Engineer The chief engineer's task is to ensure that the station continues uninterrupted broadcast. He or she is responsible for ordering new equipment (with the general manager's approval) and maintaining it, directing all the activities of the junior or apprentice engineers, and conducting all regular and special tests as required by FCC regulations. The chief engineer must be available on an on-call basis in case weather or other emergencies threaten the continuance of broadcast.

Cable

System Manager The system manager is responsible for the overall operation and supervision of the cable system, including revenues and expenses, programming, sales, marketing, and engineering. The system manager is responsible for maintaining good relationships with local franchising authorities, utility commissions, and other government units and for operating the system in compliance with all federal, state, and local regulations. In addition, the system manager negotiates contracts with program suppliers and determines subscriber rates and available offerings and tiers. System managers are responsible for the evaluation and approval of all marketing and sales efforts, system hiring, and the supervision of customer service and field crews.

Office Manager The office manager, who reports to the system manager, is in direct charge of the staffing and operations of the system's retail sales office. The main duty of the office manager is to supervise the financial transactions of the company—in particular, the collection of installation, subscriber, and converter rental fees. The office manager supervises the customer service representatives directly, ensuring prompt and courteous handling of all accounts, complaints, and service requests; maintains employee personnel records; and, in conjunction with the system manager and the chief technician, schedules the work of installers and technicians.

Marketing Director The marketing director is responsible for the marketing, promotional, publicity, and advertising activities of the cable system. In conjunction with the system manager, the marketing director is involved with the selection of basic and premium program services and may negotiate with program suppliers. In addition, the marketing manager coordinates all national marketing campaigns and designs and implements local marketing and advertising efforts. The marketing director is also responsible for the production and interpretation of research on subscribers, including cable usage, program preferences, discount rates, uplift, and churn. Familiarity with marketing research techniques is considered essential for this position.

Local-Origination Coordinator The local-origination coordinator, who reports directly to the system manager, is responsible for the operations, maintenance, and scheduling of the system's facilities for local program origination. The local-origination coordinator supervises the studio production crew, and provides the training necessary to assist local groups in the preparation of local-access programming.

Chief Technician The chief technician is responsible for the day-to-day supervision and operation of the technical facilities and staff of the cable system. Chief technicians directly supervise the work of the system engineers, technicians, and field installers. They are responsible for ensuring the system's compliance with federal, state, and local utility and safety regulations, for carrying out all required equipment tests and preparing related documentation, and, in conjunction with the system manager, for selecting new equipment for purchase or lease.

Installers Installers are responsible for the installation and disconnection of cable in entities in the system's service area, including private homes, multiple-dwelling units, hotels, office buildings, and educational institutions. Their main duty is to run the cable from underground or aerial feeder cables to the subscriber drop point. Installers may also be responsible for the initial construction work for new plants, including digging ditches and stringing cable along telephone poles. In addition, they may be required to explain and demonstrate the operation of cable hardware to subscribers and to make minor adjustments to equipment in subscribers' homes or businesses.

STAFFING: RECRUITMENT, TRAINING, AND EVALUATION POLICIES

After determining the firm's organizational structure and preparing job descriptions, the next step in employee management consists of filling each position. This is the most critical task in employee management. Choosing the right person for the right job, and at the same time meeting federal employment regulations, can often mean the difference between the success or failure of a media enterprise.

Finding Prospective Media Employees

While it is occasionally said that getting a job in the media requires "knowing somebody," having a personal contact in a media facility is only one of many ways in which jobs are filled by telecommunications companies. The main ways of soliciting candidates for jobs in the media are as follows:

- *Internal.* Internal recruitment is done by means of a company search or through internal advertisements (job postings).
- *External advertisements.* These are usually of two types: (1) general advertisements in newspaper classified sections and (2) advertisements printed in telecommunications trade magazines, such as *Broadcasting and Cable, Electronic Media, Cablevision,* and *Cable World.*
- *Employment agencies.* These span a broad range, from agencies offering temporary office, clerical, and janitorial services to executive-search (headhunting) firms which represent upper-level managers, including sales managers, station managers, and general managers.
- *Training and educational establishments.* Like employment agencies, educational institutions span a range from specialized trade schools (such as the Columbia School of Broadcasting) to 4-year university programs and graduate degree programs in journalism, telecommunications, management, engineering, and business administration.
- *Internships.* Internships at media facilities are increasingly being used to train and develop prospective employees. Formal internship programs are in place in more than 50% of radio stations and over 90% of television stations in the United States. While most stations granting internships clearly specify that no permanent employment is guaranteed, internships occasionally blossom into careers for enterprising and competent students.
- *Other external sources.* There are a variety of other employment sources, including unsolicited résumés and letters, casual callers, recommendations from employees, professional or personal contacts, and even situations-wanted ads in trade periodicals.

The method chosen to select job applicants depends on a number of factors, including the type of job, the prior training and education required, and the reputation of the source in delivering competent employees. In practice, most telecommunications companies use a combination of the methods described above. Typical recruiting methods for different categories of media jobs are as follows:

- *Clerical and secretarial staff.* Local want ads
- *Technical and engineering workers.* Electrical engineering programs, technical schools and colleges, trade publications
- *Sales personnel.* Business, marketing, business administration, and advertising departments of colleges and universities; advertising agencies and representatives; word of mouth
- *Crew and production personnel.* Internal posting, college and university programs and internships, disc jockey (announcing) schools, trade publications
- *Upper-management positions* (general manager, system manager, station manager, etc.). National want ads, trade publications, executive-search firms, word of mouth

Equal Employment Opportunity and Affirmative Action Guidelines

A major factor in the job recruitment process is the consideration of regulatory guidelines. As stated throughout the text, the telecommunications industry has had a tainted record in the area of hiring and promotion of qualified women and members of minority groups.

It is instructive to note that the tide of deregulation of telecommunications has not included equal employment opportunity. In fact, this is one area where regulations have actually been augmented in recent years. The reason is simple: the industry remains one dominated by white males; the gender and cultural diversity of the country has not yet been reflected in the media work force.[4] For this reason, recruitment policies in the electronic media must follow specific FCC guidelines regarding equal employment opportunity (EEO) and affirmative action (AA). In addition, like all businesses, media companies must adhere to the regulations imposed by the Equal Employment Opportunity Commission (EEOC).

Federal law requires that employers ensure equal opportunity to all qualified job applicants without discrimination on the basis of race, color, sex, religion, or national origin. The responsibility for the implementation and enforcement of those rules resides with the EEO branch of the FCC.[5]

The commission's EEO rules involve two separate concepts: (1) assuring nondiscrimination and (2) affirmative action. The latter means that it is incumbent upon stations and cable systems to develop a "positive and continuing" plan to seek out and retain qualified women and minority applicants. EEO and AA guidelines span 10 areas. Telecommunications managers must develop nondiscriminatory policies in recruitment, selection, training, placement, promotion, pay, working conditions, demotion, layoffs, and termination.

The FCC has prepared a model EEO program that delimits the process of assuring compliance with EEO statues. The model calls for information in the following areas:

1 *General policy.* FCC regulations require a statement by the station assuring nondiscrimination in all personnel activity.

2 *Responsibility.* This guideline calls for the name and title of the individual in station management responsible for meeting EEO guidelines. Normally, large stations will designate an EEO officer. Smaller stations usually put EEO control in the hands of the station manager or general manager.

3 *Policy dissemination.* This rule requires posting of the station's EEO policy to current and prospective employees, in all job advertisements and recruitment correspondence.

4 *Recruitment.* This stipulation requires licensees to develop and document recruitment techniques utilized to find qualified women and members of minority groups.

5 *Training.* While training programs are not mandatory, the model EEO policy suggests that such programs will help a station to meet EEO requirements. Thus, in practice, many large stations develop extensive training programs and others maintain contacts with local colleges and universities for the purpose of generating scholarships and internships.

6 *Availability survey.* EEO guidelines require that stations become familiar with the percentages of women and members of minority groups in their markets, either in the work force or in the general population.

7 *Employment survey.* Stations must compile and report annual statistics on the station's numbers and percentages of women and minorities.

8 *Job hires.* Licensees must document the total number of new employees hired in the prior 12 months and provide breakdowns on the number of women and minority applications and the number of new hires of women and minorities.

9 *Promotion.* Stations must describe any promotional policies or practices which have benefited women and minorities.

10 *Effectiveness.* Stations should regularly review their progress in the EEO area and provide for revisions and alternative policies if EEO guidelines have not been met.

While the use of strict quotas is prohibited, the FCC has adopted a system of numerical checks to test compliance with its EEO requirements. Under the present guidelines, broadcast stations with 5 to 10 full-time employees are required to have 50% parity overall with the available numbers of minorities and women in the work force and 25% parity in the top-four job categories (officers and managers, professionals, technicians, and sales personnel). Stations with more than 10 full-time employees must reach 50% parity overall with work-force numbers as well as in the top-four categories.

The Cable Communications Acts of 1984 and 1992 includes similar guidelines for cable. Cable systems serving 50 or more subscribers with 6 to 10 full-time employees must have 50% parity with the labor force and 25% parity in the top-four job categories. Systems with more than 11 full-time employees must achieve 50% parity overall and in the top-four job categories.

The following example illustrates how these standards are applied. A station employing 50 people located in a market in which women comprised 40% of the total work force and minorities 20% would have its EEO policies reviewed if it failed to meet the following minimums:

- At least 10 female employees, or 20%, overall
- At least 5 minority employees, or 10%, overall
- At least 5 women, or 10%, in the top-four categories
- At least 3 minority employees, or 5%, in the top-four categories

The responsibility for the enforcement of EEO guidelines rests with the Equal Employment Opportunity Commission, but the FCC may impose several sanctions upon telecommunications firms which fail to adhere to its guidelines:

- Letters of admonishment
- Requirement of submission of progress reports
- Establishment of EEO goals and timetables for achieving them
- For broadcasters, conditional license renewal or renewal for less than the normal license term
- Fines (or "Forfeitures") up to $36,000

In 1993, the FCC beefed up its EEO reporting requirements.[6] A midterm review was added to broadcast stations, and cable operators were mandated to file more lengthy annual employment reports indicating their success in hiring and retaining women and minorities. In addition, EEO reporting standards were extended to some of the newer technologies we've discussed, including video dial tone (VDT) operations and multiple program distribution networks (such as wireless cable operators).

Broadcast stations and cable systems have occasionally viewed EEO requirements as burdensome, interfering with their rights to establish their own employment practices. However, the FCC reports that the program has been successful in increasing opportunities for women and minorities in the field.[7]

Selection: Interviewing, Audition Tapes, and Screening Procedures

Few jobs in telecommunications are offered without an employment interview. Increasingly, employment is also contingent on the preparation and submission of audition tapes, particularly for performance jobs, such as news reporter, anchor or radio announcer, or technical jobs, such as editor, technical director, or graphic artist. In addition, media firms may also use tests for evaluating and screening potential employees.

The Interview The purpose of the employment interview is to obtain and assess information about the prospective employee that will enable a valid prediction to be made about his or her future performance on the job. Other aims of the employment interview are to provide the candidate with information about the job and the company that will give the candidate a favorable impression of the company and its employees. Prospective managers and employees should recognize the key components of a good employment interview:

- It should be planned by both parties; that is, it should *not* be spontaneous, so that the job responsibilities, skills, personal characteristics, speaking and listening ability, and other relevant criteria can be adequately presented and evaluated.
- It should be conducted in an informal atmosphere. Stiff, structured formal interviews rarely reveal the character and qualifications of the potential employee and reflect negatively upon the character of the hiring firm.
- It should focus on the candidate, *not* the interviewer. Unfortunately, on occasion, media executives use the employment interview as a forum for showcasing their expertise and opinion. Such interviews do little to ferret out the skills and personal characteristics of the candidate.

The typical employment interview in telecommunications lasts from 30 minutes to 1 hour. Many media organizations require several interviews, first with personnel managers, then with the supervisor who would be directly in charge of the new hiree. In most cases, prospective employees are also required to meet with the station manager, general manager, or cable system operator, regardless of the department in which the new employee would work.

Audition Tapes Most on-air positions (such as announcer and newscaster) require the preparation of a résumé or audition tape. Audition tapes may also be required of editors, news photographers, producers, and directors. While criteria for evaluating an audition tape are subjective, some general points can be made:

- They should be submitted in a standardized format: 3/4-in U-matic or 1/2-in VHS videocassette for television jobs, 1/4-in full-track, open-reel or cassette audiotape for radio work.
- They should include the candidate's name, address, and telephone number on the tape box, and, if possible, on the tape or cassette itself.
- They should provide a cue sheet or log identifying the contents of the tape for the prospective employer.
- They should include the candidate's best work *first,* and the submitted material should match the programming format or news reporting style of the prospective employer.
- They should be expendable, since they will *not* generally be returned.

There are other factors to be taken into account when preparing and evaluating audition tapes. Tapes for production and technical jobs (news photographer, editor, graphics, and the like) should indicate the types of equipment used by the candidate in performing the work shown. The rapid changes and new developments in telecommunications hardware are making it increasingly difficult to assess the level of employee competence with various technical apparatus.

Finally, while tapes should be carefully produced, they should accurately represent the skills of the candidate. Excessively polished, overproduced, "jazzy" video tapes often signal to the program manager, chief engineer, or news director that there is less to the candidate than meets the eye. As one news director in the southeast put it, "Those kind of tapes are more glitz than grits."

Audition tapes and performance tests are often required for telecommunications jobs since they frequently demand expertise with technical equipment, such as this news teleprompter. (courtesy of Gannett Broadcasting)

Tests and Assessment More and more, telecommunications firms are utilizing specially designed assessment measures to judge the potential of a prospective employee. There are four types of test in use: intelligence tests, aptitude or attainment tests, personality tests, and performance tests.

Intelligence tests, including IQ tests, Stanford-Binet, and others, purportedly gauge the intellectual aptitude of the prospective employee. For a number of reasons they have fallen out of favor with most businesses, including telecommunications. First, there is much disagreement among educational psychologists about what the construct of "intelligence" actually is and whether IQ tests, even if administered accurately and consistently, measure it. In addition, intelligence tests have been criticized for being biased against members of minority groups (in their selection of words and phrases and in the examples provided in test questions). Critics contend that intelligence itself isn't measured; rather, the test is selecting on the basis of social class or family background. Social scientists are concerned about the validity and the reliability of intelligence tests. Ideally, an intelligence test would accurately reflect a candidate's intelligence (validity) and would provide consistent results (reliability). Unfortunately, such tests are hard to find.

Aptitude and *attainment* tests are designed to match a prospective employee with the type of position most suited to his or her abilities. For example, an aptitude test can ascertain whether an individual is best suited to hands-on jobs in production or engineering or desk jobs such as bookkeeping or word processing.

Since job opportunities are generally scarce and much sought-after in telecommunications, aptitude tests are rarely used in this business.

Personality, or *attitudinal,* tests attempt to assess the type of personality possessed by the applicant. It was stated earlier that success in media firms is typically identified with extroversion, aggressiveness and drive, enthusiasm, and creativity. There are many commercially available tests which can give an indication of an employee's personality traits, and some of these are in use in the personnel offices of major media firms. Like intelligence tests, personality tests vary widely, and their validity and reliability are questionable. However, they can be of use in making discriminations between a number of prospects with similar qualifications. This is increasingly the case in telecommunications, given the vast number of college-educated applicants seeking the relatively few available entry-level positions.

Performance tests are direct indices of the competence of a prospective employee. Perhaps the most familiar of these is the typing test administered to prospective clerical and secretarial workers. As indicated above, telecommunications equipment changes rapidly, with changes frequently occurring faster than educational and training institutions can alter their curricula to accommodate them. Thus it is becoming increasingly common for telecommunications firms to incorporate performance tests in their hiring practices. Performing tests are found in the following areas:

Word Processing Typing (or "keystroking") ability (at least 40 words per minute) is generally considered a level of minimum competency in most media firms. With the spread of office automation, familiarity with word processing and small computers may be expected of applicants for secretarial, accounting, sales, copywriting, programming, and news positions.

Communication Skills Telecommunications managers are constantly bemoaning the poor reading and writing skills of applicants for media positions. Thus, most large corporations, including many group owners, MSOs, and networks, have included reading comprehension, grammar, spelling, and punctuation tests as part of their recruitment regimen. Cable News Network in Atlanta has even gone so far as to administer a news and public-affairs test to ascertain whether its prospective employees are well informed on national and world events.

Technical Skills Finalists for jobs in the performance or technical areas of broadcasting and cable are often drilled in the studios of the hiring firm. Interviewees for announcing jobs may be given copy to read. Prospective disc jockeys may be given board drills in the production studios of the hiring radio station. A potential news anchor may be asked to deliver a mock broadcast with the regular coanchor to see if their styles, delivery, and personalities are compatible. A candidate for a news photography or editing position may be given raw footage and asked to assemble it and prepare voice-over copy within a specified time period.

The inclusion of performance tests in the recruitment and selection process is likely to increase in coming years. They have the advantage of being relatively easy to design and conduct and they allow direct comparisons between candidates to be made. In addition, unlike résumés and audition tapes, they are designed to test the specific skills that will be expected of the new employee. However, the performance tests for managerial positions (program director, general manager, system manager, etc.) have yet to be invented; upper-level management must still rely to a large degree on the financial performance and reputation of the candidate, as well as on their own intuition, in making these critical selections.

Training and Evaluation Procedures

Training Once hired, new employees may be trained in three ways: within the company—on the job, within the company—off the job, and externally, in professional and educational training settings.

Virtually all media companies provide on-the-job training, in which a supervisor directly assists trainees in mastering the skills of their new positions. On-the-job training has the advantage of immediacy, and it gives the new employee the opportunity for personal contact with the supervisor. However, unless such training is handled properly, it can be a liability. And, unfortunately, the frenetic pace at which many media companies operate means that new employees can become mere observers of, rather than participants in, the process. Such "training" is rushed and slipshod, and affords little opportunity for the employee to learn in an atmosphere conducive to inquiry and suggestions or to learn from occasional mistakes.

Off-the-job training within the company is increasing among media companies. With this kind of training, new employees learn to operate equipment during downtime periods. Such training may be informal, without direct supervision, or may be formalized under the direct tutelage and supervision of a senior staff member of professional trainer.

Television stations train new studio personnel during the period when network programs are being carried (such as afternoon soaps). Radio stations may train air personalities in their production studios when their air studio is in operation. If the station is automated, the air studio is used. More than one humorous episode has occurred when trainees accidentally "keyed" their efforts over the air, interrupting a live network feed!

External training, which is common in many large industries, has not been utilized to a great extent in telecommunications. Colleges and universities have argued for years that they have or could easily develop the facilities necessary to train professional broadcasters, much as their colleagues in the management and business areas have been doing for years in jointly administered programs. The growth of the computer industry has led leading corporations such as IBM and Apple to make large donations of free equipment to universities in exchange for an organized recruitment and training plan. Unfortunately, there are few instances in which broadcasters or cablecasters have made such donations and established a direct means for hiring and training new employees.

One leading media manager, Harry Mansfield, has commented on the overall lack of effective training in the telecommunications field. He notes that most parents of teenagers recognize that adequate training is needed before turning the keys to the family car over to the eager teen. Yet (media) managers often feel that a person can be given the keys to the station or system with only the barest of basic training. Mansfield maintains that "training is important. Everyone in the organization must be convinced of its value so that a true commitment to an effective training program can be made."[8]

Evaluation Once hiring and training have been completed comes the difficult task of evaluating employee performance. Performance review procedures in telecommunications span a continuum from purely subjective and informal evaluations ("She's really a super worker; he stinks") to attempts at highly objective, formalized evaluation methods which utilize elaborate paper-and-pencil measures requiring regular employee-supervisor interviews.

Both employees and managers should recognize the twofold purpose of performance appraisals. While they are unfortunately often viewed as a means of noting an employee's weak points, they also serve to identify individuals deserving rewards in the form of raises, merit pay, and promotions.

To increase the effectiveness of performance appraisals and to protect against lawsuits following dismissals, the following guidelines should be observed by media managers.[9]

1 Allow enough uninterrupted time.
2 Remain calm and objective.
3 Be helpful. Try to identify causes of poor performance and suggest ways to improve.
4 Encourage venting by the subordinate. Conduct a give-and-take discussion to resolve disagreements. If resolution cannot be reached, at least make sure each side understands the other.
5 Be honest. Do not gloss over performance shortcomings. Do not overrate accomplishments. Avoid superlatives.
6 State opinions as opinions, not facts. You are the supervisor and your judgment and opinion are important.
7 Do not use other employees as examples. The issue is this employee's performance in comparison with your communicated expectations.
8 Avoid emphasizing deficiencies that may be difficult or impossible to overcome.
9 Be sure the behavior discussed is job related.
10 Be specific. Use examples whenever possible. Avoid euphemisms.

A final note about evaluations:

- They should be *written,* and copies should be given to the employee and placed in the employee's personnel file.

- They should be based on objective, *results-oriented criteria*—that is, matching job performance against the expectancies spelled out in the job description.
- They should be *followed by interviews* between the employee and his or her supervisor to allow for feedback and criticism by both employee and evaluator.
- They should be *reviewed* by the evaluator's own supervisor.
- They should provide for an *appeal* or review procedure should the employee feel he or she was improperly or unfairly evaluated.

Promotions

The large degree of turnover in the telecommunications industry (charted in Chapters 4, 5, and 6) is in part the result of inconsistent practices in granting employee promotions. The highly specialized and structured nature of media firms enables qualified employees frequent and steady opportunities for promotion at lower levels of the organization (for example, from weekend fill-in to weekday announcer). However, promotion to higher-ranking management positions is frequently blocked because management typically brings in department heads from other markets. Indeed, it has become industry practice in broadcasting for major-market stations to staff openings in supervisory positions (news director, sales manager, program director, etc.) from other markets rather than to promote from within.

Thus, as we have pointed out, employees in telecommunications seeking career longevity with a single firm are most apt to find it in noncommercial media, in craft or guild positions, or in management positions at commercial operations in medium to small markets. Those aspiring to high-level management positions in the media centers (New York, Los Angeles, Chicago, etc.) should prepare themselves for considerable lateral movement from job to job, market to market, and firm to firm.

Recently, telecommunications managers have become more enlightened regarding the need for the kind of company loyalty which can come about only with increased opportunities for employees through training and promotion from within.

For example, in the cable industry, some MSOs have begun a process of cross-training their employees.[10] Participating employees rotate from customer service, to programming, sales, marketing, and other positions. This allows employees to learn about the other jobs in the company. Not only are they thus more informed about the overall operations of the firm, they also increase their own knowledge base and promotability.

Another large cable company has a new training program in which new mid-level managers are groomed to take on senior management positions. The middle managers attend the same meetings as senior management, and are invited to observe, analyze, and critique the decisions made by their superiors (under the careful supervision of a third-party management consulting firm).

Such innovative training and promotional opportunities are taking hold in broadcasting as well. Both radio stations and TV operations are taking notice of the need to keep and reward their best employees. But, as the next section reveals, separation from the station or system remains alarmingly high in the media businesses.

Disciplinary Action and Termination

Media employees most often leave on their own to seek better jobs elsewhere, but some are asked to leave by their employers.

A somewhat dated (1962) survey of termination policies in broadcasting is illustrative of the most common reasons named by managers for firing their employees.[11] The reasons given for termination of employees are diverse, but the one most frequently cited was the competence of the employee. Incompetence was cited as the reason for about 33% of all radio firings and 40% of television terminations. Again, the more specialized nature of the television business probably accounts for the higher premium paid by management for employee ability. The next most frequently cited reasons for termination were so-called management factors, which included cutting back staff size; ownership, management, or format changes; and station automation. These factors accounted for 36% of the terminations at both radio and television stations.

After management factors came reasons related to employee industriousness or application to the job. Terms used by managers included "inefficient," "unreliable," "lazy," "lack of effort," and "failure to improve," among others. This criterion was more prevalent in television, accounting for 3 in 10 firings, as compared to 1 in 10 in radio.

Personal characteristics were next on the list of reasons for employee firings. Included here were such factors as dishonesty, indebtedness, emotional immaturity, dope addiction, gambling, and alcoholism. Interestingly, discharge for personal reasons was more common at radio than television stations (18% compared to only 6%). Perhaps this reflects the closer contact between staff at many radio stations. It may also be that the glamour and prestige associated with television allows more liberal attitudes toward undesirable personal habits, including drinking and drug use.

"Interpersonal conflicts" were cited last by broadcast managers as reasons for employee termination. Included here were such adjectives as "troublemaker," "quarrelsome," "insubordinate," "belligerent," and "incompatible [with the rest of the staff]." As might be expected, this was more of a problem in the tight-knit radio environment, accounting for 6% of reasons given for termination. Only 2% of those discharged in television were fired for reasons involving interstaff relationships.

Of course, the workplace of the 1990s is vastly different from that of the 1960s. Social and personal pressures on the work force have increased. Thus, some new grounds for employee disciplinary action and termination have made it on the list. And, as we will see, problems such as alcoholism and drug abuse place an increasing burden on management to intervene and help, before the employee's personal life and job performance are irreparably harmed.

Below are common grounds today for employee disciplinary action and termination in telecommunications companies.

Grounds for Disciplinary Action

1 Reporting work time inaccurately
2 Tardiness or early quitting without authorization or notification of supervisor
3 Failure to perform work in accordance with recognized standards, either due to inability or lack of effort
4 Creating or contributing to unsanitary, unsafe or unclean conditions
5 Employment with another company without written consent
6 Contributing to a climate of racial or gender bias in the workplace

Grounds for Termination

1 Possession of a dangerous or lethal weapon
2 Deliberate tampering with or misuse of employment or company records (forgery, altering time sheets, altering employment information about another employee, etc.)
3 Theft of company property or funds
4 Possession, use, or sale of narcotics, barbiturates, hallucinogens, or other illicit drugs
5 Unauthorized consumption of alcoholic beverages during work hours, or reporting to work under the influence of alcohol
6 Excessive (usually three or more) unexplained absences from work
7 Persistent instances (usually five or more) of tardiness for work
8 Misuse or misappropriation of company property
9 Insubordination or disobedience
10 Indecent or immoral behavior
11 Failure to observe safety rules, or in any way deliberately creating a hazard
12 Commission of any crime resulting in a felony conviction
13 Willful and knowing disregard of FCC regulations
14 Willful, knowing, and repeated demonstration of racial bias or sexual harassment while on the job

Both employees and managers in telecommunications agree that turnover, including voluntary terminations and firings, is too high. In television alone, turnover has been estimated at between 10% and 20% of all employees in past years, and as high as 33% at some stations.[12]

The cost of such high turnover at telecommunications is obvious. For one thing, an industry facing constant innovation and change is slowed when burdened with the continual task of hiring and training. For another, high turnover inhibits the development of company loyalty, a factor that is frequently associated with high productivity. A related problem is that of poor employee morale. It is difficult to develop the sense of comradeship and teamwork needed to make any business, but particularly one involved in communication, successful when workers are continually competing for better jobs—either within the organization or outside.

Employee separation is one of the most frequent areas of litigation in modern media management. For telecommunications managers, discharging an employee demands careful planning, documentation, and execution. One leading personnel manager has provided two keys to govern employee disciplinary action and termination:[13]

- Supervisors should document all remedial efforts, including counseling sessions, probationary notices, and warnings
- Supervisors should document both favorable performance and remedial efforts on an equal and consistent basis.

As managers have learned too often from hard experience, improper, undocumented termination can lead to costly lawsuits, perhaps even six-figure settlements for former employees.

EMPLOYEE COMPENSATION, BENEFITS, AND ASSISTANCE PROGRAMS IN TELECOMMUNICATIONS

While a few employees violate the various rules and regulations specified above, most members of the broadcast and cable work force remain on their jobs long enough to earn a salary and to obtain the various benefits provided by their company. This section describes the different methods of employee compensation and the types of benefit and enrichment programs offered by media companies.

Methods of Employee Compensation

There are three types of employee compensation offered to employees of media businesses: hourly wages, salaries, and payments based on commission.

There are exceptions, but technical, craft, secretarial, and clerical employees are generally paid hourly wages. Professionals and managers (including most general managers, program directors, system managers, news directors, and similar individuals with supervisory or executive-level positions) are paid monthly or annual salaries. Sales personnel, including account executives and, on occasion, sales managers and general managers, are paid at least partly on commission, on the basis of advertising sales, or, in the case of cable, subscriber rates.

Wage payments are subject to the guidelines spelled out in the Fair Labor Standards Act and enforced by regional and local offices of the Wages and Hours Division of the U.S. Department of Labor. These regulations call for the payment of at least the federal minimum wage and overtime compensation of at least $1\frac{1}{2}$ times the employee's regular rate of pay for all hours worked during one workweek in excess of 40. Other regulations pertain to the use of child labor; equal pay without consideration of sex, race, national origin, or age; and requirements for record keeping on all major employment matters.

Telecommunications businesses are covered by the Fair Labor Standards Act, but the unique nature of some media jobs has led to certain exemptions. For example, in broadcasting, trainees and student interns are exempt, as are salespeople, whose primary work is outside the station. In addition, radio or television announcers, news editors, and chief engineers in small markets are exempt from overtime requirements. Also exempt are "professionals," defined as those whose primary activities are "intellectual and varied, rather than routine mental, manual, mechanical, or physical." This proviso generally exempts writers, narrators, interviewers, and even meteorologists from minimum wage and overtime requirements. Freelance personnel and consultants are likewise exempt from these regulations.

Interestingly, a recent court decision held that newspeople and news producers are not "learned professionals" as defined by law.[14] The judge noted that a specific academic degree was not always necessary to do broadcast journalism; that journalists generally learn their skills on the job and by experience. As a result, the defendant in the case—NBC—was ordered to pay additional overtime to several news anchors and producers.

To guarantee that media firms are following the letter of the laws covering wage payments and working conditions, the U.S. Department of Labor maintains field offices in each state and in over 100 major metropolitan areas. Unannounced spot-checks are common, particularly when complaints have been received from employees claiming unfair wage practices.

Upon inspection, managers must make employment documents available and must allow the employees to be interviewed without the presence of management. Should such an investigation conclude that, indeed, money is owed to the complainant, the manager has three avenues for recourse:

1 Pay the amount immediately.
2 Refuse to pay, and await a court proceeding.
3 Negotiate a settlement with the party or parties involved.

Salaries for administrators and managers, who are exempt from overtime requirements, are generally established as follows:

1 A base salary is set using comparisons with (a) other salaried positions and (b) similar positions in other companies. As aids in the computation of base salaries, personnel managers can use the results of nationwide salary surveys (conducted by the NAB and the Radio-Television News Directors Association) or FCC data.[15]

2 Base salaries can be scaled upward for individuals with (a) superior training and qualifications (education, years of experience, special skills, etc.) and (b) a good track record (sales quotas met or exceeded, good company profit figures, ratings, etc.).

Once hired, staff members may receive raises on the basis of seniority (years of service) and/or merit (for example, sales performance and high departmental morale).

There are five methods by which sales personnel are compensated, four of which are a direct function of their ability to sell time or subscriptions:[16]

1. Straight salary (like other supervisory and administrative personnel)
2. Salary plus bonus
3. Salary plus commission
4. Draw against commission
5. Straight commission

Very few salespeople are paid on a salaried basis (less than 2% in broadcasting). The few salespeople on salary are typically newly hired trainees or sales managers no longer servicing clients in the field.

Salaries plus bonuses based on sales are a slightly more common way to reward account executives and, again, are more common among sales supervisors and managers.

Salary plus commission is quite common among broadcast account executives and some managers still making sales calls. In this arrangement, a base salary is established and a percentage of the net advertising or subscription sales is added. For example, the base pay of a radio salesperson might be set at $20,000. The salesperson might receive an additional 3% of all advertising revenue for which he or she is responsible. Thus, $50,000 of net sales would add $1500 to the base salary, for a total annual compensation of $21,500. At present, about one in four media salespeople is compensated in this way.

Another common method of sales compensation is the draw against commission. Here the salesperson establishes his or her own minimum weekly or monthly payment (the *draw*) which is adjusted to reflect actual sales during the period. For example, a radio salesperson might establish a draw of $500 per week. This amount will be guaranteed to the salesperson as long as advertising sales exceed that floor value. Any sales brought in above that amount will generate a commission for the account executive. Continued billings below the draw will soon lead to unemployment. At present, about 30% of broadcast salespeople are paid on a guaranteed-draw basis.

Finally, there is the straight-commission method of sales compensation. This method is occasionally used with new salespeople to test their zeal and industry. Frequently, experienced, confident salespeople prefer this method because it can lead to salaries in the six-figure range. About 1 in 10 media account executives works on a commission-only basis.

Employee Benefit and Enrichment Plans

Like most businesses, telecommunications companies provide a range of benefits to their employees. As might be expected, the types and amounts of employee benefits provided are related to staff and market size. In general, media companies in larger markets and with larger staffs provide more comprehensive employee benefits than do small-market stations. And group owners and conglomerates are more likely to have a comprehensive benefits package than are independently owned stations or systems. Of course, there are exceptions. In some small markets, station owner-operators have adopted a paternalistic approach to their employees (most likely in the hope that this will keep them from moving elsewhere), and provide a wide range of benefits and bonuses.

The types of benefits available to media employees include the following:

• *Medical and surgical benefits.* Include hospitalization, and, in some cases, dental coverage and eyeglasses.
• *Life, accident, and disability insurance.* Includes short- and long-term disability plans, group life insurance, and accidental death or dismemberment policies
• *Savings and investment plans.* Include pension, profit-sharing, and thrift plans
• *Educational services.* Include tuition reimbursement plans, workshops, and training programs

Reports issued by the National Association of Broadcasters and the Broadcast/ Cable Financial Management Association indicate that broadcasting companies contribute in varying degrees to each of these benefits plans.

In general, television and cable coverage is superior to the employee benefits available in radio. Also, it can be argued that employee benefits programs in media companies, especially radio and television, are less generous than those available in other industries.[17] But things have been improving in recent years. Here are some examples:

Medical Coverage Nationwide, medical coverage is provided to nearly all full-time salaried employees in the United States. Good health care coverage is available to virtually all full-time employees in cable, television, and telephone companies. Yet, in radio, nearly 1 in 5 full-time employees is not covered either in part or in full by the station. Nationwide, 9 in 10 companies provide some sort of dental coverage. In TV and cable, 8 in 10 companies offer such care. But less than 4 in 10 radio stations help their employees to fix their teeth!

Insurance Some form of employee life and disability insurance is available to virtually all U.S. workers outside of telecommunications. Yet, only 9 in 10 TV stations make group life available and only 7 in 10 radio stations offer this benefit. Extending insurance benefits to dependents is possible in 20% of U.S. firms, but only 1 in 10 TV stations and 1 in 20 radio stations offer such protection.

Savings and Investment This is another area where the telecommunications industry has lagged behind other industries. Nationally, about half the nation's private companies allow some form of employee participation in savings and investment programs. About 1 in 4 employees can elect to share in the company's profits. In TV, 1 in 5 employees can participate in profit sharing, but in radio, the figure is less than 1 in 10. This may be a good thing, however, since we have seen that among many radio stations, there have been fewer and fewer profits to share.

Employee Enrichment If there is good news to be found in the area of employee benefits in telecommunications, it relates to training, education, and overall employee enrichment.

It is becoming more common for media enterprises to offer continuing educational benefits to employees. More than half of all television stations and a third of the nation's radio stations provide tuition reimbursements for employees completing university degree work. Employee workshops and clinics are also common in broadcasting and cable, with the majority paid in full by the company.

Also on the increase are employee assistance programs focusing on drug and alcohol dependency; psychological problems, including stress, anxiety, and depression; and financial problems, including excessive gambling. An NAB monograph on the subject of employee assistance programs pointed out that such plans have been slow in developing because of the persistent stereotype which acknowledges and sanctions heavy drinking and drug use among media professionals.[18] Ironically, there has been a perception, even among management, that people in the media are "different" from others and should be able to handle drugs and alcohol or ride through personal problems that others need help solving. In recent years, however, assistance programs have been established by each of the major networks as well as by many large media groups. The benefits of such plans clearly outweigh the costs, as former ABC Chairman Elton Rule observed:

> *If* there are outside-the-job problems, nothing you can change in the job situation will address them. Yet, if you do not do *something,* the problem can only worsen, until you must choose between demotion, termination, or acceptance of poor performance from people who at one time could do their jobs well. . . .[19]

Holidays, Vacations, Sick Leave, and Miscellaneous Benefits

The final class of employee benefits are those providing such things as days off for vacation, sick leave, and holidays.

In general, most media firms provide 1 day of sick leave for each month on the job. Some companies allow sick leave to accumulate, up to 30 days or more.

Since broadcasting stations and cable companies are in service every day of the year, media personnel are not given as many recognized holidays as are employees in civil service and other businesses. In 1990 the median number of paid holidays in radio, television, and cable was 7, compared to 9 for most civil service workers.

Holidays normally observed by media firms include Christmas and New Year's Day, Thanksgiving, and Independence Day. Other recognized holidays include Washington's Birthday, Memorial Day, Labor Day, and Veterans' Day. If an hourly employee is expected to work on a holiday, he or she is normally paid the overtime rate.

Vacation time is usually computed at the rate of 2 weeks' paid vacation per year for each year of service up to the fifth year; after that it moves to 3 weeks per year until 7 to 10 or more years of service are accumulated. Few broadcast and cable companies allow vacation time to accumulate from year to year if not used. Virtually all media firms allow employees to borrow against vacation time for personal reasons.

Thus, overall, employee benefits in telecommunications are not as comprehensive as those in other industries. This fact is particularly disturbing considering the income levels and profit margins of many media operations. However, recent years have seen an increase in employee benefits and assistance programs, profit sharing, and vacation and bonus plans.

INTERNAL COMMUNICATIONS

The term *internal communications* refers to the methods of in-house communication used by media managers to communicate corporate and personal information to their employees. The basic elements of internal communications are employee handbooks and house organs such as periodicals, newsletters, or closed-circuit television programs. Publicly held companies also issue annual reports to their stockholders.

The Employee Handbook

Employee handbooks are a useful way for firms to spell out their history and purpose, as well as their specific personnel policies. Virtually all large stations and cable systems, MSOs, and group owners and communications conglomerates make use of an employee handbook. The following is an outline of the contents of a typical employee handbook in telecommunications.

PROCEDURES MANUAL
Table of Contents

    I Welcome
   II The basics of the media
  III The mission and goals of the company
  IV Areas of responsibility
       A General organizational structure
          1 Breakdown of specific units
          2 Interrelationships between units
       B Departments and their functions
          1 Sales and promotion
          2 Programming
          3 News

        4 Accounting
        5 Engineering
        6 Production
        7 Senior management
V Employee relations
  A Reasons for promotion
      1 Creativity
      2 Initiative and productivity
      3 Community involvement
  B Reasons for demotion or dismissal
      1 Personal habits not conducive to personal growth (drinking problems, etc.)
      2 Dishonesty
        a Theft
        a Intentional misstatements pertaining to or affecting operations
        c Illegal gratuities
  C Grievance procedures
      1 Union employees
      2 Nonunion employees
  D Fringe benefits
      1 Insurance
        a Health, dental
        b Life, disability
      2 Pension
        a Contribution formula
        b Payouts at maturity
      3 Vacations and holidays
      4 Education (tuition reimbursement)
  E Retirement policy
  F Salary and wages—general guidelines for determination
VI The Federal Communications Commission
  A The function of the FCC
  B Communications Act of 1934
  C The station employee
      1 What is expected
      2 How the employee is affected
VII Summary
  A Personal objectives—how we can help
      1 The company as a stepping-stone
      2 The company as a career
  B Our corporate objectives—how you can help
      1 Profitability
      2 Community improvement
      3 Growth

House Organs and Other Internal Media

Large media firms with many employees and multiple holdings frequently publish periodicals or newsletters to keep employees aware of corporate policy; job postings; personal items, such as wedding and birth announcements; and the company's performance in sports and other social events.

In addition to print resources, a growing number of larger firms make use of closed-circuit television and computer networks for disseminating information and for training purposes. For example, the major commercial networks feed a separate, off-air channel to their affiliated stations in addition to the regular network service. This channel is used to feed alternative versions of news stories or packages (some with voice-over, others without it, for example) as well as to distribute promotional information and internal news.

Another means by which managers can communicate with employees without incurring travel expenses is teleconferencing. New satellite and fiber-optic technology has made video conferencing more affordable. It is becoming more and more fashionable for media executives to confab in boardrooms linked by satellite or special telco lines. Teleconferencing is used by such corporate giants in the media as the commercial networks, Cox, Gannett, Time-Warner, and TCI, just to name a few.

The power of microcomputers and the technology of local area networking (LAN) has led to an explosion of electronic mail as a tool of employee communication in telecommunications. E-mail systems allow employees throughout telecommunications firms to schedule meetings and exchange ideas, as well as to air their grievances. Employees must be careful when using E-mail, since "private" messages can sometimes be intercepted by others using the system.

Annual Reports

Publicly held communications companies (as well as many private firms) must report to their stockholders each year on their programming and profit performance. This typically takes the form of an annual report. The purposes of the annual report are threefold: to provide disclosure of financial and other pertinent data, to report on the company's past year events, and to provide reference material on the company. While the primary users of the annual report are the company's shareholders, annual reports are also a primary means by which public corporations provide the financial data required by the Securities and Exchange Commission and the Financial Accounting Standards Board. Annual reports are also distributed to potential investors, as well as to most public and university libraries.

A well-drafted annual report is a useful public relations device. It represents the best efforts of the corporation in the preceding year. For employees, annual reports can provide the "big picture" of the firm's goals, aspirations, and achievements. When well-written and informative, they can also help foster the kind of teamwork and goodwill necessary to fuel corporate growth.

Informal Communication

Despite management's best efforts to maintain open lines of communication between employees, there often exist rivalry and competition which affects the performance of a corporation. The clash of creative personalities in telecommunications firms frequently leads to conflict between different staff members or departments. Sales managers may be at war with program directors, news personnel with engineering services, and so on. In order to discover the sources of such friction so that internal communications (and profitability) can be improved, a number of media firms have contracted with outside consultants to perform a *communications audit*.[20]

An audit of a company's internal communications can reveal the sources of management rivalry and conflict and can show how and where grapevines conveying information are in use and whether those grapevines are disseminating accurate or false information. Professional audits are expensive (as much as $20,000 and above), but audits may also be performed for a smaller fee by professors and graduate students in communication and industrial psychology.

TELECOMMUNICATIONS MANAGEMENT AND LABOR RELATIONS

As we have seen, the relationship between media managers and their employees has been characterized over the years by a good deal of distrust, often outright enmity. From management's point of view, the eager supply of young, prospective employees has made it possible to keep salaries low. The profitability of many

In telecommunications, staff meetings like this one are a common means of informal communication. (courtesy of Gannett Broadcasting)

media properties created a huge schism between those at the top, earning substantial sums, and those at the bottom, perching precariously close to the minimum wage. As we have just reviewed, employee benefits have not kept pace with those in other industries.

We have seen that opportunities for women and ethnic minorities have been fewer than in other industries. Management has also been eager to automate many job functions, which has been seen (sometimes incorrectly) as a threat to job security. For these reasons, a number of intermediaries are now interposed between media employees and their employers with the goal of improving employee compensation, working conditions, benefits, and job security. Chief among them are talent agencies and trade unions.

Effective media management today often includes the important skills of contract negotiation with agencies and collective bargaining with unions. Let's examine these processes in detail.

Talent Representatives and Agencies

What do Ted Koppel, Henry Kissinger, Shaquille O'Neal, and Madonna have in common? All use the services of a personal manager or agent. Talent agencies seek to promote and protect the interests of those who appear before the public to deliver news, sing, act out dramatic roles or simply dribble a basketball.

The use of the term "talent agency" is actually a misnomer. In addition to actors, the roster of media employees likely to have personal representation includes writers, producers, directors, musicians, authors, anchors, even political figures. No longer is negotiating with agents the sole purview of Hollywood moguls. Dealing with personal managers and agencies is a task faced by many broadcast and cable program producers and station managers, particularly in the nation's largest markets, where competition in news and entertainment programming is intense and "bankable" air personalities are essential to popularity and financial success.

In return for a commission, usually in the range of 5% to 15% of a performer's gross income, an agency will perform a variety of personal services for its clients. An agency's most vital service is to provide the valuable contacts between program executives and talent. In Hollywood, agents meet with producers to "pitch" their projects and personalities. In New York or Chicago, an agency representative might meet with a news director to offer the services of one of the agency's clients for a morning news show or afternoon magazine hour.

In addition to providing valuable liaison services, agents may also assist clients in legal and financial matters, including bookkeeping, tax payments, investments, and the like. They may also handle publicity for and about their clients, and give advice about cosmetic concerns, including hairstyle, makeup, and dress.

Three major agencies dominate this area of the industry: William Morris, Creative Artists Agency (CAA), and International Creative Management (ICM). In fact, the President of CAA, Michael Ovitz, is often considered "the most powerful man in Hollywood," since he controls the business affairs of many major stars needed to make any film project a success.[21] Numerous other small or "boutique" agencies operate in New York, Los Angeles, Chicago, and other major cities.

Guilds and Unions

The level of unionization in telecommunications varies considerably by region and by type of facility. The major television networks are highly unionized, as are virtually all stations and production companies in large markets such as New York, Los Angeles, Chicago, and Washington. Unionization is less common in the south than in the industrialized northeast and is more pronounced among larger stations and cable systems than smaller ones. Canadian media are highly unionized.

There are two types of unions active in telecommunications. The first represents creative and performing personnel, sometimes referred to as *above-the-line* employees. The second represents technical and craft workers such as engineers, tape operators, camera operators, and audio technicians, who are referred to as *below-the-line, line,* or *crew* positions. While there are exceptions, in general, creative organizations are known as *guilds,* and technical associations are considered *unions.*[22]

Unlike talent agencies, which represent individual performers and artists, guilds and unions lobby for the collective interests of their constituents. Unions negotiate with management to regulate wages and hours, to establish employee benefits, and to supervise and improve working conditions.

The leading performers' guilds are the *American Federation of Television and Radio Artists* (AFTRA) and the *Screen Actors Guild* (SAG). Live or tape appearances by a performer have traditionally fallen under AFTRA's jurisdiction; film appearances are covered by SAG agreements. However, new technology has blurred the lines between TV and film, so that the current contracts for television and cable appearances are negotiated through the participation and cooperation of both guilds. AFTRA and SAG contracts set minimum wages for appearances (known as *scale*), define conditions of employment (hours, meal and rest breaks, hospitalization, etc.), and set the terms of payments to performers when a program is rerun (known as *residuals*).

Other guilds representing above-the-line personnel in telecommunications include the Director's Guild of America (DGA), the Writer's Guild (WGA), and the American Federation of Musicians (AFM). DGA's rolls boast over 8,000 members in TV and film; among WGA's membership are TV and radio newswriters, ABC and CBS graphic artists, clerks, and news researchers.

Four major labor unions represent thousands of technical and engineering workers in telecommunications.

- *The International Brotherhood of Electrical Workers* (IBEW) represents broadcast and cable technicians, including employees of CBS. IBEW also represents workers in noncommunication fields such as construction, manufacturing, and utilities.

- *The National Association of Broadcast Engineers and Technicians* (NABET) is exclusively a broadcast union. It has contracts with both NBC and ABC and with local stations in large cities, including New York, Washington, Chicago, Los Angeles, and San Francisco.
- *The International Alliance of Theatrical and Stage Employees* (IATSE) boasts over 50,000 members in over 800 locals. However, only a small percentage of locals are broadcast stations.
- The bulk of union activity in cable involves the *Communication Workers of America* (CWA). While the great majority of CWA members work in telephone and other common-carrier fields, membership of cable technicians in CWA has been increasing in recent years.

Union Organization and Operation

Understanding the process of negotiation in telecommunications requires a discussion of how unions are organized.

Like most major labor unions, telecommunications unions have a three-part structure: local union, national union, and federation. The *local* is a branch of the national union. It receives its charter from the national union and operates under the national union's constitution, bylaws, and rules. The *national union* generally negotiates the master labor agreement under which each local operates. The third level of union structure is the *federation*. Unions engaged in similar activities or representing similar industries may affiliate to form a stronger negotiating base. The AFL-CIO is a "federation of federations," representing the interests of organized labor before the President, Congress, and the courts.

An example of the three-part union structure is IBEW, which has active locals in many cities, including New York, Boston, and St. Louis. Its headquarters are in Washington, D.C., and it is affiliated with the AFL-CIO. Many performers' guilds, including AFTRA, SAG, and the American Guild of Variety Artists, are affiliated with one another and are part of the larger Associated Actors and Artists of America.

The structure of unions and the nature of telecommunication businesses make for a unique pattern of organized labor. Each network's union situation is different. For example, IBEW has been the bargaining agent for CBS technical employees since 1939. At present IBEW represents CBS camera operators, video engineers, and other technicians. On the other hand, NBC and ABC are primarily NABET shops.

Another unique facet of organized labor in telecommunications (particularly in New York and Los Angeles) is the specialization of locals by job function. For example, within IATSE on the West Coast are separate locals for animators, illustrators, still photographers, sound engineers, makeup artists and hair stylists, and cartoonists and set painters. Each of these locals requires a separate contract specifying wages, working conditions, and benefits (known in union jargon as "pension and welfare," or "P&W").

Thus, successful management in telecommunications frequently involves formal negotiations with organized labor, or collective bargaining.

Management and Collective Bargaining

Collective bargaining is the process by which union and management officials attempt to resolve conflicting interests over wages and working conditions. Two major aspects of collective bargaining are as follows:

1 *Each side needs the other.* The union needs jobs for its members at acceptable wages and working conditions; management requires an efficient work effort at minimum cost.

2 *Each side is capable of withholding something for a time which the other side needs.* The union withholds labor through sick-outs, slowdowns, or strikes; management can refuse to bend to union demands and, in some cases, even withhold or cancel employment.

Thus, collective bargaining between management and union representatives is a process requiring skillful negotiation and a fair degree of open-mindedness. Telecommunications managers and labor organizers should recognize that fruitful negotiations occur when meetings between management and union representatives are characterized by openness, honesty, and mutual respect. Progress is made when both union and management officials avoid posturing, ritualism, and role-playing, such as when union representatives stage dramatic press conferences, or management constructs and refers to elaborate flowcharts that purportedly document the negative impact of unionism on productivity. In addition, attempts should be made to handle the bargaining issues in package form rather than on the basis of a single issue at a time. For example, a technical union might agree to drop issues 3 (birthdays off) and 9 (voluntary weekend work) if management agrees to issues 15 (double time for dangerous working conditions) and 17 (tuition reimbursement plan).

Despite the best intentions of both unions and management, collective bargaining often reaches an impasse. When this occurs, both parties may agree to pursue mediation or arbitration.

In the case of *mediation,* labor and management call in a neutral third party to help resolve the impasse. Mediators may be obtained from the *Federal Mediation and Conciliation Service* (FMCS) or from state agencies. The mediator is brought in to create or propose alternative solutions to the impasse and may help to clarify the respective bargaining positions. While the intervention of a mediator often aids in resolving a collective bargaining impasse, the suggestions and actions of a mediator are not legally binding.

A disagreement between labor and management that cannot be resolved by mediation may be referred to an impartial *arbitrator,* whose decisions on contract language are final and must be accepted by both parties. This is known as *binding arbitration.*

The most frequently arbitrated items in media businesses have included management rights, jurisdictional disputes, job responsibilities, and pay rates.[23]

Management has repeatedly attempted to defend its right to change production methods in order to take advantage of technological innovations, while the unions

CHAPTER 10: PERSONNEL MANAGEMENT AND EMPLOYEE RELATIONS **289**

Potentially dangerous working conditions are one reason the telecommunications industry has been singled out for collective bargaining and other union activity. (courtesy of TCI)

have attempted to impose job-security restrictions upon those rights. In recent years the development of portable and easy-to-use equipment has led to an increasing number of job-security concerns. For example, in television news, management has argued for use of one-person (reporter-photographer) crews. The unions claim that this practice is both unsafe and unfair.

The complicated craft structure of the telecommunications industry has led to frequent arbitration to clarify the division between the jurisdiction of separate unions. Electricians at one station may belong to IBEW, at another to NABET, and at a third to IATSE. This jurisdictional scramble is getting worse as new production techniques blur the lines between various job categories.

Moreover, there is increasing confusion over job responsibilities within stations and systems. Is an engineer really needed to operate the same videotape recorder found in millions of households? Should a studio camera operator also be permitted to edit videotape? Should a radio announcer be allowed to operate the board (as in most small stations), or is a union technician required (as in Chicago, for example)? As the distinction between above-the-line positions and below-the-line positions continues to fade, issues concerning job responsibility are likely to continue to be a major reason for the need for arbitration.

In the area of pay disputes, telecommunications is unusual for two reasons. The first is that its long operating hours require the use of numerous split shifts and part-time personnel. It is frequently difficult for labor and management to develop uniform pay rates covering all situations. To remedy this problem, below-the-line personnel are generally paid on an hourly basis and talent is compensated on a fixed-salary or per-appearance basis.

The second cause for pay disputes in media is the lack of uniformity between locals. For example, television productions in New York can be as much as 30% more expensive than in California. And, of course, using union labor is more costly than hiring nonunion crews. As a result, spiraling production costs have sent production to nonunion locations and have led to an increase in the use of nonunion subcontractors at unionized facilities. Needless to say, both practices disturb guild and craft leaders and have led to arbitration disputes.

Of course, the ideal outcome of negotiations between management and labor is the development of a contract agreed upon by both parties. This binding contract is known as the *collective bargaining agreement.* In most cases, both sides wish to resolve issues and delay new negotiations for a reasonable period. For this reason, collective bargaining agreements in the media are usually multi-year contracts.

Issues in Management and Labor Relations

In this era of change, a number of critical issues pertaining to management and labor relations await resolution. They include:

- The unionization of cable
- The effect of escalating costs upon collective bargaining
- The effect of mergers and ownership changes upon contract negotiation and administration

Organizing the Cable Industry Compared with the telephone industry, the cable industry has not been a hotbed of union activity. While estimates vary, it is believed that less than 10% of cable workers are represented by a union, whereas nearly 30% of technical workers in the telephone industry and 20% of workers nationwide are union members.[24] The reasons for the lack of union representation in cable include the relative newness of the industry, its beginnings in small counties—away from metropolitan-based union activities—and ambiguity over the jurisdiction of unions over cable employees (for example, should a cable installer be regarded as an electrician, an engineer, or a telephone line installer-repairer?)

However, the explosion of cable in recent years has led to a concentrated effort by unions—especially the CWA—to organize the cable work force. This trend will no doubt increase, as cable companies join forces with the heavily unionized telephone industry to promote new interactive telecommunications services.

Rising Labor Costs The second major issue concerning both telecommunications management and telecommunications labor is the rising cost-consciousness of media management. Historically, telecommunications unions were able to negotiate comparatively high wage levels for their employees. This was in part the result of the high profitability of the electronic media, in part to the high-tech nature of the equipment (to operate and maintain media equipment required considerable training and skill), and the centrality of network and film production to union strongholds such as New York and California. However, the financial trends documented in Part Two, including rising production costs and expenses and a less impressive bottom line, have hardened management's resistance to the wage demands of labor. Many producers have fled the major production centers and are making increasing use of nonunion personnel. It is no accident that many of cable's network headquarters are located away from areas of union jurisdiction—for example, Ted Turner's WTBS and CNN, in Atlanta, and ESPN, in Bristol, Connecticut.

Much television and film production has moved to Canada, with Toronto and Montreal "standing in" for Los Angeles and New York. While Canadian crews are represented by unions, the favorable exchange rate and their lower per-diem wages make for lower costs for production executives.

Blessed with simpler, remote-operated control boards and transmitters, broadcasters have lobbied successfully at the FCC to remove requirements for full-time engineers on duty at studios and at transmitter sites. The availability of lightweight news-gathering equipment has led stations to reduce the number of crew members needed to operate studio and remote gear.

The common denominator of all these trends is a new hard-line approach to the labor movement among telecommunications managers. This tough stance is consistent with antiunion sentiment throughout U.S. industry and with prevailing public opinion, the most striking example of which was perhaps the popular support of President Reagan's decision not to give in to striking air-traffic controllers in 1981.

Ownership Changes and Labor Relations The spate of ownership changes traced in Chapter 7 has cast a long shadow over labor relations in telecommunications. Unions are concerned about how mergers and buyouts will affect negotiation and contract administration.

As media "mega-mergers" continue, union leadership will strive to protect employee job security and benefits and will try to assume work-force continuity against a backdrop of ownership and administrative change. In this volatile environment it is not unrealistic to expect an increase in job actions in radio, TV, and cable.

Can't Labor and Management Just Get Along? Is it possible for labor and management in telecommunications to have good relations—without unions? Perhaps. A study conducted for the cable industry revealed the five major errors made by media managers which spurred union activity.[25]

- Management failed to give employees adequate information about the financial status, goals, sales, and production achievements of the company.
- Management made changes in plant, equipment, or policies without advance notice or follow-up explanations to the work force.
- Management made key decisions without taking employee needs and interests into account.
- Management used pressure tactics and punitive methods to increase production and productivity.
- Management ignored or played down employee dissatisfaction within the organization.

VALUES-DRIVEN TELECOMMUNICATIONS MANAGEMENT

With all the negative information conveyed in this chapter about the rather sorry record of the industry with regard to employee treatment, compensation, and job security, it is important to note that there are media companies with an expressed commitment to the welfare of their employees at least as great as their pursuit of profit. One example is Emmis Broadcasting, the radio company headed by Jeff Smulyan. Emmis maintains its own version of "The Ten Commandments," reprinted below as Figure 10-7. The principles are listed in reverse order, like the "Late Night Top-10" made popular by David Letterman.

Another values-driven company is Bonneville, based in Salt Lake City and headed by Rodney Brady. Its mission statement is reproduced as Figure 10-8.

Note from the Emmis and Bonneville documents that each company considers its people to be as important a resource as its various products and services. Ultimately, this is the key to successful employee management and labor relations in telecommunications. To earn the loyalty and productivity of your employees, treat them as indispensable partners in your enterprise.

While this may seem simply to be a restatement of the "Golden Rule," the fact is that media management has too often ignored this credo, much to its own chagrin—in the form of contentious labor relations, work stoppages, lawsuits, and persistent bothersome regulation in the areas of EEO and affirmative action.

THE TEN COMMANDMENTS OF EMMIS

X. BE FLEXIBLE - ALWAYS KEEP AN OPEN MIND

IX. BE RATIONAL - ALWAYS TAKE A LOOK AT ALL OPTIONS

VIII. HAVE FUN - DON'T TAKE THIS TOO SERIOUSLY - BE AN UP PERSON

VII. NEVER GET TOO SMUG

VI. DON'T UNDERPRICE YOURSELF OR YOUR MEDIUM - DON'T ATTACK THE INDUSTRY - BUILD IT UP

V. BELIEVE IN YOURSELF - IF YOU THINK YOU CAN MAKE IT HAPPEN IT WILL

IV. NEVER JEOPARDIZE YOUR INTEGRITY - IF WE CAN'T WIN THE RIGHT WAY, WE'D RATHER NOT WIN

III. BE GOOD TO YOUR PEOPLE - GET THEM INTO THE GAME PLAN - GIVE THEM A PIECE OF THE PIE

II. BE PASSIONATE ABOUT WHAT YOU DO AND COMPASSIONATE ABOUT HOW YOU DO IT

I. TAKE CARE OF YOUR LISTENERS AND YOUR ADVERTISERS - THINK OF THEM AND WE'LL WIN

FIGURE 10–7 The ten commandments of Emmis. *Courtesy:* Emmis Broadcasting.

BONNEVILLE • INTERNATIONAL • CORPORATION

MISSION A·N·D COMMITMENTS

onneville International Corporation is a major national and international firm engaged in the commercial broadcast and media communications business.

We are a values-driven company composed of values-driven people.

We are committed to serving and improving individuals, communities, and society through providing quality broadcast entertainment, information, news, and values-oriented programming.

We are committed to satisfying our customers' marketing and communication needs through quality programming, creative advertising, and advanced technology.

We are committed to building our resource base and economic strength through skillful and profitable application of our human, financial and technological resources.

We are committed to providing professional and personal growth opportunities to our people; to enhancing the effectiveness, power, influence, and value of our properties; and to exerting positive leadership in the broadcast industry.

—Rodney H. Brady, President

FIGURE 10–8 Bonneville mission and commitments. *Courtesy:* Bonneville International Corporation, Rodney H. Brady, President.

SUMMARY

Personnel management in telecommunications has become increasingly important in recent years. Management tasks regarding personnel include organizational and work-force planning, recruitment and training, performance evaluation, wage and benefit structures, internal communications, and employee development.

Tables of organization are used to identify each of the units (departments) in the firm, to indicate the chain of command from top management to each unit, and to describe the relationship between individual units. Depending upon market size and type of ownership, tables of organization in telecommunications may be simple and direct or extensive and complex. However, four core units are found at virtually all facilities: general administration, sales, programming, and engineering.

Effective management of media personnel requires identification and description of jobs within the firm. Ideally, job descriptions in telecommunications should indicate the title and level of the position, its pay scale, the purpose of the job, and the specific tasks expected of the employee.

In hiring and training new employees, managers must follow equal employment opportunity and affirmative action guidelines proscribed by the EEOC and the FCC. Broadcasters and cablecasters with more than 5 full-time employees must file an EEO plan following the FCC model, which calls for specific information on how managers ensure nondiscrimination in their recruitment, training, promotion, and compensation methods.

The search for new employees is made using internal or outside sources—including job postings, trade and general advertisements, employment agencies, internships, and training programs. Prospective employees usually face a screening process which may include general aptitude and performance tests. They may also be required to submit an audition tape.

Making decisions about employee promotion and termination are among the most difficult tasks facing media managers. The most common reasons for employee termination in telecommunications include incompetence, management changes, and personality factors.

A variety of compensation methods are used in telecommunications. Technical workers generally receive hourly wages, while creative and management positions are typically salaried. Salespeople are paid in a variety of ways, from straight salary to straight commission. Employee benefits in the industry include medical benefits, life insurance, savings and investment plans, and educational services.

There has been increasing concern in the industry about employee welfare. To facilitate the physical and mental well-being of workers, more media firms are developing comprehensive internal communication systems and staff enrichment plans. Internal communications include employee handbooks, reports, and newsletters. Enrichment plans may include counseling services, recreational activities, and employee workshops.

The most difficult negotiations in telecommunications occur between management and organized labor. The communications work force is represented by two kinds of labor organizations. Creative personnel belong to guilds, such as the Writers Guild, the Directors Guild, the Screen Actors Guild and the American Federation of Television and Radio Artists. Technical and craft workers belong to unions, including the National Association of Broadcast Employees and Technicians, the International Association of Theatrical and Stage Employees, and the Communication Workers of America.

Current issues in labor relations include unionization of the cable work force, increasing cost-consciousness, the impact of new technologies on job security, and the effects of ownership changes on negotiations.

Ultimately, the contentious nature of labor relations in telecommunications can be changed only by a commitment by management to the welfare of the media work force. There is some evidence that media companies are becoming more values-driven and committed to employee participation than has been the case in the past.

NOTES

1. *Standard and Poor's Industry Surveys,* January 23, 1992, p. T-16.
2. Based on "Broadcast Industry Fringe Benefit Survey," *Broadcast Financial Journal,* Summer 1983, p. 26. At that time, the proportion was 25%. Updated by the author.
3. Ronald H. Claxton, *The Student Guide to Mass Media Internships* (Charleston, IL: Eastern Illinois University).
4. See "The Powers That Run Television," *Channels,* August 13, 1990, pp. 38-48; Kenneth Harwood, "Women in Broadcasting, 1984-1990," *Feedback,* Spring 1984, pp. 19-22; Janice Castro, "Women in Television: An Uphill Battle," *Channels,* January 1988, pp. 42-52; "Out of Focus—Out of Synch: A Report on the Film and Television Industries," (Baltimore, Md.: The National Association for Advancement of Colored People), 1991.
5. Federal Communications Commission, *47-CFR* Ch. 1 September 1, 1992, p. 361.
6. Sean Sculley, "Midterm Review Added to EEO Compliance," *Broadcasting and Cable,* June 28, 1993, p. 10.
7. See Federal Communications Commission, *1992 Broadcast and Cable Employment Report,* press release, July 6, 1993.
8. Harry Mansfield, "The Cost of Not Training," *Broadcast Financial Journal,* May-June, 1989, pp. 48-49.
9. Barbara Reising, "Tips for Successful Performance Reviews," *Broadcast/Cable Financial Journal,* January-February 1991, p. 18.
10. Simon Applebaum, "Jacks of All Trades," *Cablevision,* December 2, 1991, pp. 28-31. Simon Applebaum, "Learning to Listen," *Cablevision,* December 2, 1991, pp. 32-33.
11. Sherman P. Lawton, "Discharge of Broadcast Station Employees," *Journal of Broadcasting,* vol. 6, no. 3, Summer 1962, pp. 191-196.
12. Joseph G. Fulmer and Maryin R. Bensman, "Turnover in TV Industry Too High," *Feedback,* vol. 20, no. 1, May 1978, pp. 17-19; Thomas R. Berg, "The Phenomenon of Employee Turnover," *Broadcast Cable Financial Journal,* July-August 1990, pp. 32-38.
13. Howard M. Pardue, "Legal and Practical Implications of Employee Termination," *Broadcast Cable Financial Journal,* November-December 1990, pp. 32-33.

14 Steve McClellan, "NBC Newsmen Win in Suit Against Network," *Broadcasting and Cable,* September 20, 1993, p. 36.
15 See, for example, *1992 Radio Employee Compensation and Fringe Benefits Report* (Washington, D.C.: National Association of Broadcasters, 1992). The NAB also publishes annual salary reports for TV.
16 Data for this section are based on the NAB compensation reports cited above.
17 National trend data in this section are based on "Employee Benefits Survey," *Handbook of Labor Statistics,* Bulletin 2340 (August 1990, pp. 582–585); "Employee Benefits in Medium and Large Firms," *Statistical Abstract of the United States 1991* (Washington, D.C.: U.S. Department of Commerce, 1991), p. 420.
18 John C. Doolittle and Ellen K. Baker, *Broadcast Employee Assistance Programs: Saving Money and Careers* (Washington, D.C.: National Association of Broadcasters, Research and Planning Department, March 1984), pp. 1–2.
19 Ibid., p. 4.
20 See William L. Roberts, "Should Your Station Undergo Analysis?" *RadioActive,* April, 1984, pp. 26–27; Keith Davis, "Methods for Studying Informal Communication," *Journal of Communication,* vol. 28, no. 1, Winter 1978, pp. 112–129.
21 See, for example, "Power People," *Entertainment Weekly,* November 2, 1990, p. 19.
22 Data for this section were provided through telephone interviews with personnel at each union listed.
23 Robert Coulson, "What Has To Be Arbitrated in Broadcasting?" in Allen E. Koenig (ed.), *Broadcasting and Bargaining* (Madison, Wis.: University of Wisconsin Press, 1970), pp. 85–98.
24 See *Employment and Earnings January-March 1992,* (Washington, D.C.: U.S. Department of Commerce, 1992), p. 106; Bureau of Labor Statistics, "Compensation: Working Conditions," press release, February 10, 1992, p. 1.
25 Chuck Moozakis, "When Labor Organizers Call," *Cablevision,* May 15, 1983, p. 50.

FOR ADDITIONAL READING

Armstrong, Michael: *A Handbook of Personnel Management Practice* (3d. ed.). Englewood Cliffs, N.J.: Prentice-Hall, 1988.
Bowers, Mollie H., and David A. DeCenzo: *Essentials of Labor Relations.* Englewood Cliffs, N.J.: Prentice-Hall, 1992.
Craven, Charles B. *Can Unions Survive?* New York: New York University Press, 1993.
E. Edward Herman and Joshua L. Schwarz: *Collective Bargaining and Labor Relations* (3d. ed.). Englewood Cliffs, N.J.: Prentice-Hall, 1991.
Holley, William H.: *The Labor Relations Process* (3d. ed.). Chicago: Dryden Press, 1988.
Koenig, Allen E. (ed.): *Broadcasting and Bargaining: Labor Relations in Radio and Television.* Madison, Wis.: University of Wisconsin Press, 1970.
National Cable Television Association. *The Cable System Management Manual.* Washington, D.C.: NCTA, n.d.
_____: *A Guide to Developing Your Equal Employment Opportunity Program.* Washington, D.C.: NCTA, n.d.

PART **FOUR**

CORE DEPARTMENTS

Part Four: CORE DEPARTMENTS describes the process of telecommunications management in three key areas vital to successful operations. These are programming, sales and promotion, and audience research. Chapter 11 describes radio program management; Chapter 12 covers program management for television and cable. Chapter 13 describes the process of sales and marketing management, including promotion. Chapter 14 describes the relationship between management and audience analysis. Chapter 15 closes with a view toward the future of media management, with predictions for media in the new millenium.

11

RADIO PROGRAM MANAGEMENT

The two core departments found in all media companies—regardless of medium (radio, TV, or cable), size (large market or small), or kind (commercial or nonprofit)—are programming and sales. While the names of the departments and titles of the managers vary, the functions they perform are fairly consistent across media. We will detail the daily tasks performed by managers in each setting, beginning with the department which controls what we see and hear: the program department, and its head, the program director or PD.

The goals of this chapter are:

1 To list and define the five functions of program management.
2 To describe the process of radio programming.
3 To detail the major sources of radio program material, including networks, record companies, and other sources.
4 To describe the process of constructing a radio format.
5 To trace how radio program managers analyze their competition and evaluate their format.

Information or Entertainment?

As an overview to the task of the program manager, consider the following question: What is the goal of media programming?

One might argue that the primary function of media programming is (or should be) to educate the public. Advocates of this approach hold that radio and TV programs must keep the populace informed, must raise the level of political discourse in the country, must teach children about the evils of drugs and alcohol, and so on. Proponents maintain that even entertainment programs should be "highbrow"; classical and operatic music should be the mainstay of radio instead of popular songs; television should do more dramatic work and fewer soap operas and situation comedies. On the other hand, there are those who argue that electronic media programming should provide a diversion from the pressures of daily life in an increasingly stressed-filled environment. In this view, ideal media programming is simple-minded, innocuous, and escapist. Programming should entertain, rather than inform; soothe, rather than incite; placate, rather than stimulate. According to this view (to paraphrase TV and film producer Garry Marshall), in the education of the American public, television and radio are recess.

As a simple twist of the dial will confirm, most commercial media managers might agree with Mr. Marshall. The goal of most media programming is to entertain, not educate. Indeed, programming exists mainly to keep audiences in tow until the commercial breaks (see Chapter 15). In contrast, public broadcasting, such as National Public Radio (NPR) and Corporation for Public Broadcasting (CPB) programs, and special-interest cable services like Arts and Entertainment and The Discovery Channel seem to provide the programming which enrich and enlighten audiences.

However, to say that program managers in commercial broadcasting and cable care only about mindless entertainment would be a vast overstatement. While most see their programming goal primarily as entertainment, and while most programmers on the noncommercial side hope to inform and educate, the reality is that each does both. Even escapist fare educates: think of the lessons learned about family life from *Roseanne* or *The Simpsons,* for example. And, some of the best educational programs (*Sesame Street, Mr. Rogers*) also entertain. As stressed in Chapter 1, one of the unique facets of electronic media programming is that it provides *both* education and entertainment to its large and diverse publics. As a result, media program managers have a dual obligation and, therefore, a great responsibility.

Creativity vs. Voracity

The reality that most media programming is escapist, imitative, formulaic, and simplistic has a basis in a simple fact: there are vast amounts of time to fill. Broadcast and cable facilities (particularly in the United States) are almost always "on." Consider the task of the program director of a radio station in a large city. If the station is on 20 hours a day, the PD must fill 140 hours each week, 560 hours each month, and nearly 7000 hours each year with attractive programming. The situation is compounded in cable: the manager of a 35-channel system must provide over 300,000 hours of programming per year to subscriber households.

Is it any wonder, then, that radio stations tend to replay the same songs; that TV networks repeat their prime-time shows; that local stations air quiz shows and old situation comedies over and over again; that cable services play the same movies, as much as twice or three times daily?

The voracious appetite of the electronic media tends to divert program managers from a concern about "aesthetics" or "quality" to a preoccupation with the here-and-now. Facing the reality of an ever-ticking clock, the goal often becomes to fill the time as cheaply and quickly as possible. Under this pressure, it may be asking too much to expect "art" or "culture" from commercial radio, TV, or cable. But, surprisingly, programs of high artistic and cultural achievement do occur (just as classic automobiles, like the 1957 Chevrolet, occasionally roll off the assembly line). Radio gave us "Bob and Ray," Elvis, and the Beatles. Television spawned *Roots, M*A*S*H, All in the Family,* and *Masterpiece Theater,* among countless other classics. A relative newcomer on the scene, already cable has created such high water marks as Cable News Network, Nickelodeon, and the innovative *Larry Sanders Show.*

Of course, you might disagree with the authors' assessment of the genius of the people and programs just mentioned. That's a key point for programmers: programming involves creativity and creativity is subjective, evasive, and different from person to person. On the other hand, the fads which pervade popular culture are comparatively easy to understand, and therefore to program. Thus, there are traits common to most rock and roll records (the vocals, the instruments, the beat), game shows (the host, the contestants, the clock, the board or wheel), situation comedies (the dimwitted dad, the goofy neighbor), movies-of-the-week (the love triangle, the glamorous doctors, the fatal disease), and so forth. In radio, these similarities become a *format;* in television, they evolve into a program type or *genre.* So, if it seems that much media programming is repetitive, derivative, and formulaic, it's because it is.

THE FIVE FUNCTIONS OF PROGRAM MANAGEMENT

While it might seem that the tasks of the program manager in radio, TV, and cable are vastly different, there are great similarities. To varying degrees, each is involved in the same five basic activities. These are *personnel administration, program acquisition, program origination, program scheduling,* and *program evaluation.* Let's define each task and see how the processes operate in radio, TV, and cable.

Personnel Administration

As mentioned earlier, electronic media rely on highly creative people to develop programs and commercials which will attract and intrigue the public. Unfortunately, many creative individuals have reputations for unpredictability, incorrigibility, and a distrust of authority. This presents special difficulties for management. Nowhere is the situation more acute than in the program department. Most radio stations boast an eclectic mix of highly creative, ego-driven, and vastly different air personalities. All have their own ideas about how the station should sound. Inside television stations there are news anchors, feature reporters, announcers, graphic artists, photographers, tape editors, and other skilled and creative individuals. How does the manager get them moving in the same direction? The situation is similar at a cable network like HBO or Showtime, even in the studios of local cable channels.

Thus, a primary task of the program manager in the electronic media is personnel administration: the organization, staffing, and supervision of the program department in an electronic media firm. As we review the chain of command in the program departments of radio, TV, and cable facilities, it might be useful to review the organization charts from Chapter 10.

Program Acquisition

There are only two ways for management to fill radio and television airtime: in house, using the resources of the station or cable system alone; or out of house, by relying on program suppliers. By far, the most common source of programming in radio, television, and cable today are the outside sources: networks and syndicators. The process of obtaining programs from outside suppliers is known as *program acquisition.* Program managers spend a good deal of their time in negotiations with program sources. Such negotiations require knowledge of programming, as well as shrewd business acumen.

Program Origination

Program origination refers to programming which is produced locally by radio stations, television stations, and cable systems. Such programs are also known as "in-house" or "home-grown" productions. In radio, local origination includes live broadcasts from outside the station ("remotes"), such as concerts and news reports or sports play-by-play.

In broadcast television, the bulk of local programming is devoted to news—typically at noon, in early evening, and late-night, following prime-time programming. Many local stations produce public-affairs and interview shows on a regular basis. And, as we will see, some stations have the resources, creativity, and ambition to try other local fare, from kids' shows to soap operas to comedies.

Locally produced cable programming has its own unique vocabulary. *Public access* refers to local programming produced by individuals and groups on facilities provided free of charge by the cable system, usually as part of its franchise agreement. *Leased access* includes local programming produced by such groups on production facilities rented from the cable system, typically on an hourly basis. *Local origination* refers to cable programming produced by the cable system itself, which may range from a simple "crawl" of upcoming programs or the weather, to full-service entertainment and news services.

Program Scheduling

Scheduling refers to the strategies employed in selecting programs, including the types of programs to run, how often to run them, and the time of day they will air.

In radio, the program schedule, from the music selected to the promotion announcements, disc-jockey chatter, and commercials accepted, is known as the *format*.

The process of scheduling in television has received a great deal of attention in the press. Television programmers have developed their own colorful vocabulary to describe the scheduling process. Some shows are "skewed." Others have "topspin." Still others have "legs." Some stations seek "power programming"; others prefer to "counterprogram." Those terms, plus some others, will be defined in the next chapter.

Program Evaluation

An ongoing and increasingly vital job for the program manager is the process of program evaluation: deciding which programs to schedule; once on the air, determining which to continue and which to cancel. Radio programmers are all too familiar with the phenomenon of "burn out," when listeners tire of a record and don't want to hear it anymore. Many artists have been "one-hit wonders," famous for having only a single song make it onto radio stations. The television landscape is littered with the carcasses of short-lived series and "overnight sensations" which quickly became passé.

Assessing program performance is a critical task for management. As we will see, program managers use a range of assessment measures, from ratings to "gut reactions" in selecting and scheduling music, entertainment, even news.

Now that we understand the daunting tasks facing media program managers, let's examine each in its unique setting, beginning with radio programming. Television programming, both broadcast and cable, is examined in the next chapter.

THE PROCESS OF RADIO PROGRAMMING

Personnel

In radio, the chain of command begins with the program director, usually known by the acronym "PD." In a music-formatted station, the PD will normally hire a music director to assist in managing the station's air sound. Stations with a significant news commitment will hire a news director for this purpose. Generally, the music director and news director are lateral positions, both of which report to the PD.

Beneath the music director (MD) and news director (ND) may be a chief announcer. The chief announcer, usually one of the station's on-air personalities, will have the overall responsibility for training and evaluating the station's disc jockeys (and/or newscasters). Some stations place the supervision of disc jockeys in the hands of an operations manager (OM), whose job may also include keeping the station's program log in order so that the format is adhered to and the commercials run on time and as scheduled.

At the bottom rung in the table of organization in the program department are the air personalities themselves. This may partly explain why many disc jockeys eventually leave the airwaves for careers in sales or record promotion. Many aspire to senior management, including MD, PD, general manager (GM), and station owner. Others become program and music consultants (see below), who sell their expertise in programming to stations for a fee.

The hierarchy of the program department becomes clear upon close inspection of compensation. Nationally, the mean salary for radio program directors is about $30,000. As is the case in virtually all job categories in radio, pay is better at FM stations and in large markets. For example, the typical salary for the PD at stations in the nation's largest cities (more than 2.5 million) is in the $70,000 range. Program managers in FM stations earn more than their counterparts in AM. For example, PDs at FM music stations in the nation's top 10 markets frequently boast salaries above $80,000 per year; on the AM side, PD salaries top out in the range of $50,000–60,000.

Music directors and news directors are paid at roughly the same rate. Nationally, each position pays about $25,000 per year. Again, market size is the crucial determinate. Average pay for big-market (cities above 1 million in population) music and news directors is about $35,000. The best position for a music director is at a major-market FM music station, with a top salary approaching $50,000. For the news director, the ideal slot is at the all-news AM radio station in a top 10 market, where salaries of $60,000 per year are commonplace.

In small markets, however, news and music directors typically "double" as reporters and air personalities. The same is true for chief announcers. In these cases, the additional managerial responsibility normally coincides with bonus pay in the range of $5000 to $10,000.

The national average for salaries of on-air talent remains very low: about $20,000. Here's where market size exerts the greatest impact. Large-market air talent can earn $100,000 or more (particularly in drive-times). Indeed, the average for stations in markets above 2.5 million approaches $60,000 per annum. At the top of the salary pyramid are the kings of radio personalities, including Paul Harvey, Howard Stern, Rush Limbaugh, Don Imus, Mark and Brian, and Tom Joyner. Each earns well in excess of $1 million per year.

But most disc jockeys and talk show hosts toil in middle and small-size markets, where average annual compensation has stood at about $15,000 for over a decade.

Today, the typical program department at a radio station consists of 3 to 10 people, with the average about 5. Because of their comparatively small size, program staffs tend to be closely knit organizations. Compatibility of personality, tastes, and interests between the PD and the programming staff is often the key, especially at tightly formatted stations which program album rock, contemporary hit, or black/urban music.

Program Acquisition

Radio stations obtain outside programming from two major sources: networks and record companies.

RADIO NETWORKS

Radio networks have evolved to feature the kind of specialization found across the radio dial. There are two main classes of radio networks: short-form news/information services and longer-form music services. Table 11-1 ranks the leading short-form networks.

ABC offers five different news/information services targeted to the various radio formats, including ABC Prime, ABC Platinum, ABC Excel, ABC Genesis, and ABC Galaxy.

CBS has three: the "traditional" CBS Radio Network, a younger-oriented CBS Spectrum service, and Radio Americas, the CBS Hispanic Network. Spanish-speaking audiences are also served by the Caballero Network; African-Americans are served by the American Urban Network, formed out of a recent merger between Sheridan and the National Black Network.

Westwood One is another network radio powerhouse. Its services include two mainstays of network radio since the 1930s which it acquired in the 1980s: NBC Radio and the Mutual Broadcasting System, and its own creation, "The Source."

TABLE 11-1 Leading Radio Networks*

Rank	Network	Average listeners (000)†
1	ABC Prime	4,703
2	Westwood Mutual	2,808
3	ABC Platinum	2,463
4	ABC Genesis	2,196
5	CNN +	2,117
6	CBS Spectrum	2,078
7	Westwood/NBC	1,953
8	Unistar Super	1,664
9	CBS Radio Network	1,495
10	Unistar Power	1,277
11	Westwood/Source	1,174
12	ABC Excel	929
13	American Urban	865
14	ABC Galaxy	845

* Westwood purchased Unistar in 1994.
† People 12+, Monday–Sunday, average number of people listening
Source: Broadcasting and Cable, September 13, 1993, p. 36.

In addition to these national networks, there are dozens of regional radio network news services. Statewide news networks (like the Georgia Radio News Service, the Hawkeye Network in Iowa, and the Wisconsin Network) are common, as are sports networks centered on big-league franchises and college teams.

A second type of network service is music delivered by satellite. These operations use the relatively recent technology of satellites and digital recording to concentrate on delivering high-quality music programming to subscribing stations. These program sources tend to be long-form; that is, they provide hours of high-quality programs for local stations to use. Such services as Satellite Music Network, Superadio, United Stations, DIR Communications, and Watermark provide a range of such programming, including extensively researched music formats, live concerts, special events, and weekly music shows.

CapCities/ABC-owned Satellite Music Network is generally acknowledged as the leading provider of full-service music formats. At present, it boasts 11 different 24-hour services, from country to urban to oldies and modern rock.

A fast-growing area of network radio is syndication, whereby popular programs in one market are offered to stations in other markets. The growing popularity of talk radio (see below) has led to nationally syndicated programs featuring a diverse cast of characters, from Pat Buchanan and Jim Bohanon (Westwood One), to G. Gordon Liddy (Unistar), Pete Rose (Sports Entertainment Network) and Barry Farber (Daynet). Of course, the leaders in this form of radio networking (in notoriety as well as listeners) are Rush Limbaugh (EFM Media) and Howard Stern (Infinity).

The popularity of radio networks is underscored by one hard fact. More than 8 in 10 stations in the country boast satellite-receiving stations to import programming. And, the new technologies which marry computers with phone lines make possible even more radio networks, including the power to deliver CD-quality talk and music to program directors via their desktop PCs.

RECORD COMPANIES AND OTHER MUSIC DISTRIBUTORS

While most of America's radio stations own satellite receiving dishes to import network programs, many still rely on music distributors to provide records (and increasingly, compact discs) for local airplay. The major record labels, including Columbia, WEA, EMI, and Polygram, provide recorded music to radio stations in return for the promise of airplay. Independent labels, such as IRS, DJ International, and GRP, are also important suppliers of recorded music to radio programmers. Such labels often concentrate on new music or music not in the so-called mainstream. In recent years, independent labels have released music by post-modern rock performers, rap, and "hip-hop" acts.

Since the 1950s, radio stations have received free records for one simple reason: radio exposure is directly tied to record sales. Generally, the more airplay received by a record, the greater the sales at the record store. This relationship is particularly true in the modern music categories which appeal to youth: including rock, country, and dance music. According to record industry statistics, more than two-thirds of the records, tapes, and discs purchased in the United States each year are bought by people under 34 years of age; similarly, the great majority of record sales are accounted for by people who listen to more than 10 hours per week of radio.

However, the high costs of record production and distribution have led to fewer "freebies" provided to radio stations in recent years. Today, the record companies require proof that their product is receiving airplay. Such proof is provided by the program manager in two main ways. First, many popular music stations print a *playlist,* which indicates their most frequently played songs. Playlists are prepared monthly or even weekly, and are made available to record suppliers, distributors, and local record stores.

In addition, program managers are frequent contributors to *trade publications* and *tipsheets.* The leading trade publications—*Billboard, Cash Box, The Pulse of Radio, Radio and Records, Radio Only,* and others—publish columns, articles, and playlists which disseminate information about radio airplay of new record releases. Tipsheets, such as *The Gavin Report, Hits, Breneman Review,* and *Bobby Poe's Music Survey,* report on airplay at stations around the country, and feature articles and interviews with program and music directors about songs receiving attention at their stations.

Once programming is acquired, from either networks or record suppliers, the program manager's problem becomes: How is it all put together? This is the process of preparing the radio format.

THE FORMAT: DEVELOPING THE RADIO SCHEDULE

Because of the preponderance of radio stations (over 14,000, including noncommercials) most programmers have decided to try not to be all things to all people and instead, to be all things to only some people: those of a given age, sex, or lifestyle. This approach is known as format radio. The major formats were introduced and defined in Chapter 4. Let's review the most common radio formats, this time concentrating on the audiences which tend to align with each program type.

Table 11-2 ranks radio's 10 most listened-to formats and provides important information about the type of audience each attracts. In the jargon of the radio programmer, these descriptive audience statistics (led by age and sex) are called *demographics.*

The table ranks the top-10 radio formats, in order of popularity with American listeners. The formats are identified by the shorthand, "buzzwords," and acronyms which are used by program managers. AC is industry shorthand for Adult Contemporary; CHR for Contemporary Hit Radio. AOR stands for *Album Oriented Rock,* EZ is the industry acronym for Easy Listening; and MOR means Middle of the Road. The others should be self-explanatory (if not, recheck Chapter 4).

TABLE 11-2 Radio Format Trends

Rank	Format	(1) SHR	(2) M	(3) F	(4) 12–24	(5) 25–54	(6) 55+	(7) TSL	(8) T/O	(9) EXCL
1.	AC/Soft/Oldies	14.95	40.4	59.6	15.2	74.8	10.0	7.9	15.6	6.8
2.	AOR/Classic	14.45	69.0	31.0	39.7	58.8	1.5	8.7	14.5	8.4
3.	Country	13.72	48.7	51.6	12.5	62.8	24.7	10.6	11.9	17.7
4.	News/Talk	13.05	50.6	49.4	2.8	41.6	55.6	7.7	16.3	7.5
5.	CHR/Top 40	10.35	41.5	58.5	54.7	42.8	2.5	8.0	15.7	11.1
6.	Black/Urban	8.75	44.4	55.6	42.0	52.9	5.1	11.1	11.4	14.6
7.	BTFL/EZ	8.08	39.0	61.0	4.2	42.8	53.0	10.9	11.6	14.6
8.	Hispanic	4.07	40.5	59.5	13.9	58.5	27.6	11.3	11.1	14.0
9.	MOR/Variety	3.52	45.7	54.3	2.8	42.7	54.5	8.5	14.9	10.7
10.	Religion/Gospel	2.09	31.9	68.1	8.5	53.4	38.1	8.3	15.3	12.5

Source: *Duncan's American Radio*, Fall 1989–1992; *Billboard*, June 9, 1990, p. 12.

Format Share

The column marked SHR (column 1) indicates the percentage of people above age 12 listening at any one time to each format. To ease interpretation of this column, as you read down, ask, of every 100 radio listeners right now, how many are tuned to this format? With this particular piece of intelligence, a number of sweeping generalizations about the appeal of a given format can be made.

The top-four formats together account for more than half the radio listening in the United States at any given moment. They split the available audience fairly evenly, with each claiming about 15% of listeners. As this book went to press, Adult Contemporary (AC) led the pack, with a share of audience of 14.95%. Next was Album (AOR) and Classic Rock, with 14.45 %, Country (just under 14%), and News/Talk (13.05%).

Interestingly, those stations playing loud, aggressive urban music and those transmitting beautiful music (also known as Soft Adult Contemporary) each attract about 8 in 100 listeners at any given time. Spanish radio and MOR/Variety report national shares of about 4%. Rounding out the top-10 list is religious radio, with about 2% of the available audience.

Sex and Age Breaks

Now, examine the next five columns in Table 11-2 (columns 2–6). These identify who listens to each format by sex and age. In radio programming jargon, these data are known as sex and age breakouts, or "breaks."

As traced in Chapter 4, the goal of radio programming is to attract large audiences to sell to advertisers. Since women make the bulk of consumer purchases in the United States, it follows that the ideal radio format would be attractive to women. Moreover, since a key demographic category for advertisers (i.e., the people who buy the most products) is women between the ages of 25 and 54, the ideal radio format would boast a majority of listeners in this age group.

Now, inspect the data for AC radio. Why is it radio PDs' preferred format? Nearly 6 in 10 AC listeners are female; 3 in 4 are in the crucial 25–54 age bracket.

In fact, reading down columns 2–3 reveals that only 2 of the 10 formats "skew" (another PD term) more male than female: Album Rock and News/Talk. Indeed, Album Rock is almost a males-only club, with men accounting for nearly 7 in 10 listeners. On the other side of the coin, religious radio is a female phenomenon; particularly those in or past middle age.

In terms of age appeal, it should not be surprising that top 40, album rock, and urban music skew the youngest (with half of CHR's audience under 25) and that the majority of the audience for news/talk, beautiful music, and MOR is above the age of 55.

Loyalty Factors

Columns 7–9 in Table 11-2 provide information about listener behavior: in particular, dedication and loyalty. Column 7 presents the average *time spent listening* (TSL) for each format, expressed in hours per week. As might be expected, the so-called ethnic and lifestyle formats—black/urban, Hispanic, and country—boast the most frequent listeners. The typical listener to each of these formats tunes in about 11 hours per week; the upper third of listeners to each format averages 3 hours per day!

TSL is one characteristic which chinks the armor of the formats with the highest share. Note that AC, top 40, news/talk and even album rock have significantly lower time spent listening than many of the others. Thus, while such stations may have more overall listeners, the listeners tune in for briefer periods of time. The AC or CHR listener may be said to be more capricious, flighty, or less loyal than the urban or country fan. In other words, if these listeners hear something they don't like, they're very likely to punch the button to another station!

Further detail about listener loyalty is provided by data on format turnover and listener exclusivity. *Turnover* (T/O) refers to the number of different listener groups which contribute to a station's total audience. A station with low turnover (under 12, for example) tends to deliver the same listeners throughout the day and across the week. A station with high turnover (approaching 15 or more), delivers many different listeners, who tend to sample the station for comparatively shorter lengths of time. They may listen to one station while getting dressed in the morning; another while driving to work; still another at lunch; and a fourth at home in the evening.

Note turnover by format. Once again, the "lifestyle" and ethnic formats lead the way. Hispanic listeners are the most loyal (or, alternatively, twist their dials less); other low-turnover formats are country and black/urban. Note too that beautiful music is a low-turnover format, no doubt due in part to the all-day tune-in typical at dentists' and doctors' offices and other places of business.

News/talk is the highest-turnover format. This is because many news/talk listeners sample these stations to get current information about news, sports, and weather and then gravitate to another station for music. But near to news/talk in turnover are the popular music formats AC, CHR, and AOR, indicating listeners with busy fingers, apt to tune out at any time.

A final measure of listener loyalty is *exclusivity,* the percentage of listeners who tune in to one and only one type of radio station. The most loyal fans of all are country's; almost 20% of country fans never listen to any other format! The black and beautiful music formats are each about 15% exclusive, followed closely in listener loyalty by Hispanic (14%) and religious (12.5%). Using this measure, fans of AC, news/talk, and album rock seem the least loyal to a particular favorite station.

On the basis of this research, a program manager can make a preliminary decision about which format seems right for a given situation. However, program decisions must also be made in light of what other signals are available to the audience.

COMPETITION ANALYSIS

The evolution of a radio format cannot take place until and unless the programmer has conducted a thorough review of the other radio stations in the market. There are two phases of this research: *external* analysis and *internal* analysis.

The Bandsweep

Fortunately for radio programmers, the product and practices of their competition are laid out for all to hear on the radio. Program managers are constantly involved in assessment of the sound of their competitors. This is done by regularly scanning the dials and monitoring, live or on tape, the programming practices of the competition. This process is known as a *bandsweep* or *aircheck.* Monitoring the competition provides a number of important clues to the programmer. Most significantly, the programmer can discover the station's approach to selecting music and air talent. How that station positions itself can become immediately apparent, by the frequency it plays certain records, announces the news, conducts contests, and so on.

External analysis is not limited to listening to the station. The programmer can also get valuable information about competitors by inspecting their advertisements, such as posters, bumper stickers, and billboards. The amount and type of attention paid to the station by the local press is important, as is each competitor's policy about remote broadcasts and involvement in community events. A savvy PD can determine whether a competitor's album rock format is positioned to younger males (more "heavy metal" songs in rotation, more concert promotions, more leather in the ads, for example), whether a country competitor is skewing classic or contemporary, and so on.

Internal analysis also yields important insight to the program manager. The program manager will try to get as much information about the ownership and management of primary competitors as is possible without being an outright "spy." Here are some programming issues which can profit from this kind of intelligence:

Who owns the competition? A big group from a major city or a local entrepreneur? If it's a big group, what other properties do they have and how are they programmed? Are they trying to bring a successful formula to a new market? How much do they pay their employees? Is it perceived as a good place to work and build a career? Or could a rival steal their air talent (a common practice in the business)? How do they get and choose their music? How automated or computerized is their operation? These are just a few questions which can be asked to help position one station against another.

Puttin' On the Hits: Implementing the Format

Armed with intelligence about the format, the audience, and the competition, the next step is to move forward with the actual format itself. To do this, most programmers translate the sound of their station to a picture: by using a chart variously known as a *format wheel, hot clock,* or *sound hour. Hot clock* is the term in most common usage. Figures 11-1 and 11-2 illustrate two types of radio hot clocks.

A hot clock uses the face of a clock to determine when things will be programmed at a radio station. There are three basic elements of a station's air product: the spoken word (DJ patter, news, interviews), music, and commercial matter (i.e., spots). Of these, commercials are the most important, since they generate the revenue which keeps the station in operation. Thus, the first area to be filled in on the hot clock will be the commercial positions (sometimes labeled CM for "commercial matter"). At the majority of stations, music is next in importance, so that is usually filled in second. Program managers use their own vocabulary to refer to music.

How often a song is played on a station is known as the *music rotation.* A song in fast rotation will play frequently, as many as two or three times per hour on some contemporary hit stations. A song in light rotation may play only once a day or less.

The segment of the hour when the most popular current songs will play are plotted first on a hot clock. Most music stations call these songs "currents" or "power cuts." A color-code may be used: red is commonly used to denote these hot hits. Songs which are still popular but not necessarily at the top of the charts are often known as "recurrents." These might be coded orange or blue on the hot clock. Oldies might be programmed next, indicated in gold or yellow on the hot clock. Together, the various times color-coded for music are known as *music sweeps.* It is not uncommon for stations to program uninterrupted music sweeps of 20 or 30 minutes or more, especially when fewer commercials are sold, such as in midday and late night.

At this point, the PD will know approximately how much music will play each hour (as well as how many commercials). The remaining time will be occupied by talk elements.

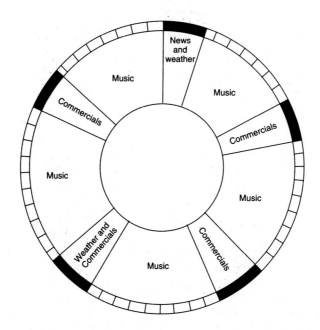

FIGURE 11-1 Radio hot clock—soft adult contemporary. *Courtesy:* CBS Radio Representatives.

News, traffic, and weather segments are plotted next. Some stations place news at :00 and :30, i.e., at the top or bottom of the hour. Others schedule news at :25 and :55. Some reserve a full 5 minutes for news; most music stations today set aside only 1½ minutes or less.

The last element to be included is disc-jockey commentary. The PD might indicate when during the hour the "jock" will read a promotional announcement ("promo"), when to make a statement about a musician or artist ("liner"), and when to give the time and temperature ("T&T"). A more flexible schedule, particularly with a witty disc jockey, might simply indicate "ad lib" or "talk" on the clock.

As strange as it may seem, news/talk stations are as rigorously scheduled as music stations. Most news/talk stations carry regularly scheduled news and feature material, from CBS Radio News to "Paul Harvey News and Comment." The PD makes sure to schedule these segments in advance, so that listeners can plan their schedules accordingly. A local news, weather, or traffic report may follow, and that report needs to be indicated on the hot clock. Open phone segments and interviews are similarly blocked out.

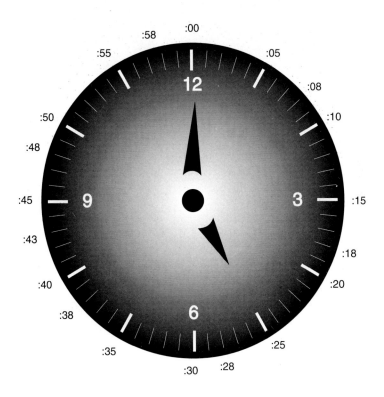

:00 CBS Network News	:20 News	:43 News
:05 News	:25 Business Reports	:45 Sports
:08 Traffic & Weather Together	:28 Traffic & Weather Together	:48 Traffic & Weather Together
:10 News	:30 News	:50 News
:15 Sports	:38 Traffic & Weather Together	:55 Business Reports
:18 Traffic & Weather Together	:40 Osgood File (Mornings)	:58 Traffic & Weather Together

FIGURE 11-2 Radio hot clock—all news. *Courtesy:* CBS Radio Representatives.

Most programmers prepare different clocks for their important broadcast time periods, known as *dayparts*. As you recall from Chapter 4, the key radio dayparts include Morning Drive (Monday to Friday 6–10 a.m.), Evening Drive (M–F 3–7 p.m.), Daytime (M–F 10 a.m. to 3 p.m.), Evening (M–F 7 p.m. to midnight), and Weekend (Saturdays 10 a.m. to 2 p.m. and Sundays from 12 noon to about 4 p.m.). Hot clocks are particularly important in planning special programming, such as commercial-free blocks, tributes to particular artists, and other events.

Formatting the news/talk station often involves plotting and promoting its various elements, such as traffic, business news, and sports. (courtesy of WBBM, Chicago)

The Computer in Radio Programming

While most programmers use the hot-clock format to get a visual fix on their air schedule (and that of their competitors), in practice hour-by-hour programming is now routinely accomplished by computer. Commercially available music rotation programs make the task of the program and music director increasingly flexible. Leading computer programs for this function include "Autoselect," "Selector," and "CSI Music." These programs allow management to produce complete program logs indicating all music, commercial, and talk elements days and even weeks in advance. Elaborate printouts give managers information about each program element. Especially important is music history: how often each song and artist in the playlist received airplay throughout a given period. When matched with ratings research, logs of telephone calls, and other feedback, music histories can identify which songs are most popular and which may be losing audience.

FIGURE 11-3 Radio programming computer printout. *Courtesy:* Burkhart/Douglas and Associates.

```
Page 1                  ===================================================
                        *                                                 *
                        *            96.1 B D A                           *
                        *        B/D&A's Modern Rock                      *
                        *    10 AM Tuesday      12-14-93                  *
                        ===================================================

            LINER                          LINER
   CART     TITLE              ARTIST                      PY    RTIME/E
   ==================================================================

   00:00    BD&A RECORDED LEGAL ID                                 :05
   --------------------------------------------------------------------
    1173    LOW                         CRACKER             93    4:35
            KEROSENE HAT
   --------------------------------------------------------------------
   04:40    BUMPER/RECORDED ELEMENT                                :05
   --------------------------------------------------------------------
     174    HUNGER STRIKE               TEMPLE OF THE DOG   92    4:00
            TEMPLE OF THE DOG
     105    UNDER THE MILKY WAY         THE CHURCH          88    4:56/F
            STARFISH
   --------------------------------------------------------------------
   13:41    "96.1 BDA BD&A's Modern Rock"......Shotgun Frontsell Next Song    :05
   --------------------------------------------------------------------
    1159    SOUL TO SQUEEZE             RED HOT CHILI PEPPERS 93  4:49/F
            CONEHEADS (OST)
   --------------------------------------------------------------------
   18:35    :20 Break: Backsell/Liner/Spots/Sweeper               3:00
   --------------------------------------------------------------------
    1134    PETS                        PORNO FOR PYROS     93    3:31/F
            PORNO FOR PYROS
   --------------------------------------------------------------------
   25:06    Listener Interactive                                   :00
   --------------------------------------------------------------------
     206    FRIDAY I'M IN LOVE          THE CURE            92    3:26/C
            WISH
   --------------------------------------------------------------------
   28:32    BUMPER/RECORDED ELEMENT                                :05
   --------------------------------------------------------------------
    1131    RUNAWAY TRAIN               SOUL ASYLUM         93    4:19
            GRAVE DANCER'S UNION
   --------------------------------------------------------------------
   32:56    **Optional Song Next.....Drop If No Time (Circle if not played)*   :00
   --------------------------------------------------------------------
    1011    DIGGING IN THE DIRT         PETER GABRIEL       92    5:16
            US
                    PRODUCED BY DANIEL LANOIS
   --------------------------------------------------------------------
   38:12    Speed Break                                            :00
   --------------------------------------------------------------------
     913    WHO'S GONNA RIDE YOUR WI    U2                  91    5:16
            ACHTUNG BABY
     280    UNBELIEVEABLE               EMF                 91    3:28
            SHUBERT DIP
   --------------------------------------------------------------------
   46:56    "96.1 BDA BD&A's Modern Rock"......Shotgun Frontsell Next Song    :05
   --------------------------------------------------------------------
    1176    CANNONBALL                  THE BREEDERS        93    3:33/C
            LAST SPLASH
```

PROGRAM EVALUATION IN RADIO

The enormous competition on the radio dial and the relative capriciousness of the listening audience combine to make the program manager involved in daily evaluation processes. Format evaluation involves testing each song on the playlist, evaluating disc jockeys and other air talent, and troubleshooting all advertisements, contests, and promotions for the station. This job is made easier (if not more costly) by engaging the services of radio programming and research consultants.

Radio Program Consultants

Major radio consulting companies include Eagle, The Research Group, Strategic Research, and Burkhart/Douglas. These companies are normally staffed by experienced radio programmers, trained researchers, and experts in promotion and marketing. They assist program and station management in the three major areas of format evaluation: music testing, air talent, and station promotion and image.

Music Evaluation

Stations utilizing a music format must evaluate each song for its popularity to its target audience. Songs which the listener tires of can cause tune-out, the greatest fear of the radio programmer. Program managers use two main procedures to test music: call-out research and auditorium tests. In each case, stations prepare tapes of their music libraries, with each song edited down to a 5- to 15-second representative sample, known as a *hook*.

In call-out research, the station will telephone listeners, play the hooks to them, and ask them to rate the song on a 1 to 5 scale (with 1 meaning "like a little," and 5 meaning "like a lot," for example). Normally, only about 10–20 hooks can be tested over the phone before listeners grow tired or irritable.

On the other hand, stations may invite as many as 200 listeners to a large auditorium (usually in a hotel) and play up to 300 or 400 hooks. But an auditorium test is costly: consultants can charge up to $30,000 per test.

Stations test music in a variety of other ways. New music is often tested on the air, with listeners asked to rate a song by calling the station (such as MTV's "Smash or Trash?"). Stations contact record stores to track sales of artists and songs. One of the reasons stations program remote broadcasts or provide talent for dances and "sock hops" is to get feedback on music (as well as indications of the appeal of their disc jockey).

Performer Evaluation

The key to a successful radio station often lies in the appeal of its announcers, especially in the critical morning and evening drive times. For this reason, a good deal of effort by management is expended in selecting and evaluating the disc jockeys. Program managers are in constant communication with each other around the country in search of new and better talent. They may use consulting services for this task, including radio "head-hunting" firms like Talentmasters, which accumulates résumés and audition tapes from air personalities throughout the United States and abroad. In addition, program directors in bigger markets report receiving dozens of applications and résumé tapes each month from prospective air personalities.

One technique which has proven especially useful in performer evaluation is the *focus group*. This is a meeting run by the station or its consulting firm, in which tapes of the station's broadcasts can be played and discussed by small groups of listeners (usually between 10 and 15 people). The focus group provides in-depth reactions to the station's sound, particularly the appeal and attractiveness of the station's on-air talent. This technique is particularly useful in assessing the rapport necessary for a successful morning team, including the host and co-host, newscaster, traffic reporter, and weathercaster.

Station Image

Focus groups are also helpful in assessing the image of the station. Every element of the station's programming, from its call letters, to its slogan, its personalities, its music, its involvement in community events, even its participation in sports (like softball) combine to create a unique station image. The station image is critical in establishing loyalty and rapport with the target demographics. In the best case, listeners become station cohorts: "walking billboards" for the station, who wear its T-shirts, slap its bumper stickers on their cars, drink from its coffee mugs, and show up at its concerts and remote broadcasts.

On the basis of focus group interviews and other research (call-letter tests, analysis of listener mail, one-on-one intensive interviews, and other means), the program manager may alter the station's sound to promote a new image. For example, recently many established album rock stations have evolved into classic rock stations and "alternative" or modern rock stations. Classic rock has gone after the older male audience for rock, with an emphasis on "supergroups" from the 1960s and 1970s (from The Doors to Led Zeppelin to Journey and Genesis). Modern rock targets the younger male—especially the one who considers himself a new music connoisseur. Thus, the rise in popularity on radio of music from The Red Hot Chili Peppers, U2, INXS, and Pearl Jam, to name a few.

Format Changeover

The most radical evaluation a PD or GM can make is a complete change of format. This occurs with greater frequency in radio today than in the past, no doubt because of the increasingly competitive nature of the business. It is not uncommon to find stations, particularly low-rated stations in major markets, which have undergone a number of format changes within 2 or 3 years. Recent trends in radio format turnover suggest that more and more stations are turning to "tried and true" formats, particularly on the FM band. In the past decade, more and more stations have moved to AC, CHR, Urban Contemporary, and the other popular formats. Country remains strong; indeed, more stations boast this format than any other. Many commercial stations have abandoned more "eclectic" formats like jazz, big-band, and classical, leaving those to be found increasingly on public radio.

The AM band has seen the most format innovation. Leading the way have been WFAN, the all-sports service in New York. There are now all-sports stations in San Diego, Chicago, Atlanta, and other major cities. Other experiments have included KBTL, an all-Beatles station in Houston, and KPAL in Little Rock, targeting a child audience.

But there have been intriguing new format experiments on FM as well. One interesting example is Metropolitan's KTWV, "The Wave" in Los Angeles. Under the banner of "new adult contemporary," the station plays a range of relaxing, frequently instrumental songs sometimes known as "new age" music. The music, ranging from pianist Liz Story to the Omaha-based supergroup Mannheim Steamroller, is said to represent "beautiful music for baby boomers." More critical listeners have labeled it audible wallpaper.

Given the vast wave of immigration in the 1980s, format growth in the 1990s is expected in Hispanic radio, on both AM and FM. A mild baby boom in mid-decade bodes well for contemporary hit and dance formats, especially in the larger urban areas.

With changing demographics, increasing competition and duopolies, and changes in public tastes, radio program management promises to be a dynamic and exciting field as the medium approaches its second century of service.

SUMMARY

Program departments at radio and TV stations and cable systems face five major tasks: personnel administration, program acquisition, program origination, program scheduling, and program evaluation. The program director, or PD, is the head of programming at a radio station. The PD may be assisted by a music director, news director, chief announcer, and/or operations director.

Radio stations rely on two sources of outside programming: networks and record companies. Satellite-delivered programming and service from independent record labels have increased in recent years.

The program director's main task involves selecting, planning, and executing the format. This process may be aided by an outside program or research consultant, and usually involves testing music, air personalities, and station image. The format is mapped using hot clocks, prepared for each major time of day, often utilizing a special computer program.

Once the format has been designed and executed, the PD is involved in evaluating the format. Modifications in music, announcers, and station promotions are commonplace. Increasingly, wholesale format turnover has been taking place, owing to the competitive nature of the radio business.

CHAPTER 11: RADIO PROGRAM MANAGEMENT

FOR ADDITIONAL READING

Baskerville, David: *Music Business Handbook and Career Guide* (4th ed.). The Sherwood Company: Los Angeles, 1985.

Carroll, Raymond L. and Donald M. Davis: *Electronic Media Programming: Strategies and Decision-Making.* New York: McGraw-Hill, 1993.

Duncan, Jr., James H.: *American Radio, Fall 1992.* Indianapolis: Duncan's American Radio, Inc., 1993.

Eastman, Susan T., Sydney W. Head, and Lewis Klein: *Broadcast/Cable Programming: Strategies and Practices 4th Edition.* Belmont, Calif.: Wadsworth, 1993.

Keith, Michael C. and Joseph M. Krause: *The Radio Station* (3d ed.). Boston: Focal Press, 1993.

——————. *Radio Programming: Consultancy and Formatics.* Boston: Focal Press, 1987.

Shemel, Sidney and M. William Krasilovsky. *The Business of Music* (6th ed.). New York: Billboard Press, 1992.

12

TELEVISION PROGRAM MANAGEMENT

Television program strategies have received a good deal of attention in the popular press. Most of us are familiar with some of the jargon of TV programming, such as "the ratings game," "sweeps weeks," "network shares," and "ratings hype." However, few members of the public understand the complex and dynamic process of scheduling the airtime on a television station or filling the many channels on a cable system. This chapter will demystify television programming and identify the tasks and strategies of the program director at television stations and cable systems.

The goals of this chapter are:
1 To identify the various tasks performed by the program manager in television
2 To discuss major methods of program acquisition
3 To describe the relationship between the program manager and the TV network
4 To detail the process of television syndication
5 To discuss local program production, scheduling, and evaluation

Like their counterparts in radio, TV program managers face five main tasks: personnel administration, program acquisition, program production, program scheduling, and program evaluation.

Program evaluation typically involves ratings. This process is described in full in Chapter 14, so we will defer that discussion until then.

For convenience, broadcast television programming will be considered first, followed by programming in the cable business.

PERSONNEL ADMINISTRATION

TV Broadcasting

The larger size and more bureaucratic nature of television stations (compared with radio) leads to more extensively staffed programming departments. A major determinate is whether or not the station is involved in local news.

About half of the more than 1400 television stations in the United States do not produce at least a half-hour of continuous local news.[1] Most of these are independent stations in large and mid-size markets; some are network affiliates in medium and small markets. In these stations, the program department tends to be a streamlined operation. The program director supervises local production (if any), including talk/interview programs, locally produced commercials, public service programs, and special projects. The program manager is also responsible for acquiring the movies, situation comedies, quiz shows, and other fare available from program suppliers. Reporting to the PD may be an assistant program director, a small number of producers, directors and a staff of production technicians.

Stations committed to local news also tend to be involved in other local production, including documentaries, magazine shows, series, even dramas and other entertainment productions. Thus, their program departments tend to have at least two and sometimes three subdivisions. The senior program manager (often a vice president of programming) may have an assistant solely involved in program acquisition: getting shows from sources outside the station. This individual might be known as program director–acquisition. A second assistant will be in charge of the station's two or three local newscasts per day, plus documentaries and special news coverage (such as political conventions). This important person is the news director. A third manager will be in charge of the remainder of local production, including public affairs programs, magazine programs, and entertainment shows. This might be the director of local programs (or program director–origination).

Today, stations which "do news" average over 100 employees each in local programming, plus another 25 in public affairs, specials and other units.[2] The typical network affiliate in a large market will have as many as 150 people involved in various aspects of programming.

Program and news directors in television are better paid than those on the radio side. Nationally, the average salary for both program directors and news directors stands at about $40,000 ($10,000 more than the typical radio PD salary). Like radio, salary is related to market size. Six-figure salaries ($100,000–$150,000) are commonplace in large markets, especially for the news director with the luck and skill it takes to lead the ratings.

On the other hand, program director salaries have been slipping in recent years. This is due to a number of factors. First, at most stations, the commitment to local programming is mainly or even exclusively to news. In addition, the cost of syndicated programming (see below) has skyrocketed, leaving decision making in the hands of the general manager or owner (not the PD).

Production personnel salaries range from entry-level in the "teens," ($13,000–$20,000) for production assistants, beginning camera operators and floor persons; to a national average in the $25,000 to $40,000 range for producers, writers, video, and audio engineers. Like the networks, anchor salaries are the highest found in the news department. The national average for news anchors (about $40,000) approaches that of their boss, the news director. Talent contracts for anchors at highly rated stations in large markets typically exceed $100,000 per year.

Cable TV

Personnel management in cable programming presents an interesting contradiction. As we have seen, cable is television of abundance, with over 30 channels of programming available to the average consumer household. The reader therefore might have the misperception that program departments at cable systems are large and elaborate.

Actually, the opposite is the case. Relatively few of the thousands of individual cable systems boast full-time programming staffs. This is because the job of selecting the channels which fill the cable system is typically the responsibility of the cable system manager. Among the major multiple system operators, such as TCI or Time-Warner, programming is typically centralized at corporate headquarters. Management executives at MSO headquarters frequently negotiate contracts with program suppliers for each of the many systems owned by the company. Once contracts with the suppliers are struck, program management for the local system operator is a simple technical matter of tuning the system's various satellite dishes to receive incoming signals.

However, many larger systems operate studios and produce programming. Others make their facilities available to local businesses, civic groups, or individuals. Channels reserved by cable operators for these purposes are known as *local-origination* or *LO* channels. The individual who supervises local production is the local origination manager, or LO manager, for short.

Salary ranges for cable program personnel are closer to radio than local television. Some reasons for this apparent inequity include the lack of union jurisdiction over cable employees, the "newness" of cable programming (especially in local news), a glut of prospective employees (as networks and local stations have downsized their staffs) and the tradition of cable as a blue-collar, technical (as opposed to "creative") medium.

Consequently, LO managers earn on average only about $25,000 per year (half that of the typical TV PD).[3] In addition to their supervisory responsibilities, most LO managers are directly involved in production as writers, producers, photographers, and editors. Many participate in local advertising sales as well.

As with broadcasting, the most labor-intensive form of local cable production is news. Increasingly, large cable systems in urban areas are launching their own news operations. Some are miniature, localized versions of Cable News Network, offering 24 hours of continuous area coverage to their communities. The pioneer of this form in the mid-1980s was Cablevision Systems' News12 service in Long Island, N.Y. Today, similar news services are available to cable subscribers in suburban Los Angeles, New York City, Washington, D.C., the Chicago area, and parts of New England.

All-news cable operations are staffed much like the news departments of major television stations and networks (or all-news radio stations). They are headed by a news director and boast up to 200 employees in positions ranging from producer to writer, reporter, editor, and photographer, many with prior experience in commercial television. As a result, salaries at all-news cable operations are on the rise and might reach parity with local TV station by mid-decade. Salaries for production technicians are in the range of $10,000–$20,000; producers, writers, and anchors can command more.

For the many thousands of cable systems without a local news channel, there is great diversity in both staff size and compensation. However, a cable system heavily involved in program production (such as local advertising, community talk shows, and area sports events) may be staffed like a small television station, with up to 10 full-time programming employees. Like the situation in news, pay is apt to be lower than in commercial television, by a factor of 50% or more.

There is some good news. As the cable business entered the mid-1990s, a survey of system operators found more than three-fourths committed to significant expansion in locally produced programming.[4] The hope for program employees is that increased production will lead to better compensation. Time will tell.

PROGRAM ACQUISITION

Program acquisition refers to the role of the program manager in obtaining programs from sources outside the station or cable system. Acquisition in television falls into two main categories: *network origination* and *program syndication*.

Network Origination

Network origination refers to programming produced for national distribution on a broadcast or cable network. In broadcast television, nearly two-thirds of the nation's TV stations receive programming from one of the four major television networks: ABC, CBS, NBC, and Fox. Similarly, the bulk of cable programming emanates from cable networks like USA, MTV, ESPN, TBS, HBO, Showtime, and literally dozens of others. Before dealing with the complexity of cable programming, let's examine the relatively ancient concept of television network affiliation.

TV Stations and Network Affiliation

Stations which receive network programming enter into contractual arrangements with their networks, known as *affiliation agreements*.

The terms of the agreement, especially the financial particulars, were presented in Chapter 5. However, it is useful to review how the relationship works, since this process is crucial to the job of the program manager. The affiliation agreement is a contract which stipulates that the local station agrees to carry or *clear* the network programs in return for a fee. At present, ABC, CBS, and NBC offer about 100 hours weekly (close to two-thirds of the typical affiliate's total airtime). The bulk of that time is in the evening, during prime time (see below), when networks provide 22 total hours.

Fox offers somewhat less (about 40 hours), with most of that (15 hours) in evening prime time. However, at the time this book went to press, Fox had planned an ambitious expansion of its network offerings.

Importantly, the local station is under no obligation to clear any or all of the network shows. This *right of first refusal* often comes into play, such as when the network offers dubious sports events (*Celebrity Water Polo*), delivers programs which local viewers might find objectionable on grounds of taste or decency, or offers programs which the local program manager feels will not attain a significant audience. Those programs which the local station refuses to clear are known as *preemptions*.

The decision not to clear a network program is taken very seriously by network executives. Often, the network has sold the program to advertisers on the basis of the number of TV households it expects to deliver. The loss of one of its affiliates, particularly in a major market like Phoenix or Cleveland, can cost a program a significant number of TV homes (and guarantee a lower rating). If enough affiliates preempt the program, the advertisers will demand a rate rebate.

Should its primary affiliate refuse to clear a network show, the network has the right to negotiate with another station in the market to carry its programming. Normally, one of the independent stations will step in to carry what it feels is an attractive show. For example, in Atlanta, for years the local CBS affiliate elected not to carry the programming offered by the network, particularly weekend sports which conflicted with its own contracts to carry local college football. In this case, a small independent station (located on Channel 69) stepped in to pick up the CBS schedule.

Affiliate Relations

As might be surmised from the above, an important component of the job of the program manager in television is *affiliate relations*—in simple terms, the process of peaceful coexistence with the network. The networks maintain an entire management team, headed by a VP for affiliate relations, charged with keeping the program flow open (else the network lose valuable viewers and, therefore, advertising revenue). At least twice a year, network program executives meet with local station program managers and general managers to share their concerns. The major meetings take place in the spring, at a star-studded affiliates meeting at which new fall shows are launched; and in the winter, at the annual meeting of the National Association of Television Program Executives (NATPE, discussed in detail, below). Throughout the year, the local program manager must make decisions about carriage of programming from the network. These decisions can be critical, since failure to clear network shows can cause friction with the network and may lead to viewer confusion if a show which appears in program listings or on-air promos fails to appear on its scheduled channel. High-profile preemptions are usually covered extensively by local newspapers, which can be eager for an opportunity to get in a "dig" against their main advertising competitor.

Fifth and Sixth Networks?

As this edition went to press, two new players had entered the business of network television programming.[5] Flushed with the success of its syndicated *Star Trek* series (including *Star Trek: The Next Generation* and *Deep Space Nine*), Paramount planned to begin a new network by offering a 5-night-a-week, 10-hour service to its group of owned stations and affiliates.

Not to be outdone, Time-Warner (in association with the Tribune group of powerful independents) planned a new Warner Brothers network (WB, for short), with initial plans for a 7-day, 15-hour weekly commitment, to be spearheaded by the company's most recognizable personality—Bugs Bunny.

Most industry observers predicted a rocky future for these ventures—with most analysts suggesting there was room in the network advertising business for only one more competitor. Tracking the fate of these new networks will be interesting in the next few years. As they say, check your local listings!

Cable Networks, MSOs, and Local System Operators Since cable systems can carry literally dozens of national program services, like the emperor with 30 wives, they tend not to boast the intimate relationship with each that "monogamous" TV stations share with their network. However, for reasons which are described below, affiliate relations is becoming an increasingly important component of cable program management.

As was detailed in Chapter 6, the two main sources of national programming for cable systems are pay services and ad-supported basic networks. Program acquisition is handled somewhat differently in each case.

Splitting the Take: Acquisition of Pay Cable Programming The pay cable business works in a manner which is very similar to the movie business. In the film world, movie distributors work on a *negotiated-split* basis with exhibitors (the movie theaters). That is, the distributor and the exhibitor sign a contract which divides the box office receipts according to some prearranged formula (50–50; 60–40 and so on).

In pay cable, the most common practice is for the cable system operator to split the revenues for pay channels with the program supplier. The most common formulas call for a split in which about two-thirds of the revenue goes to the local operator and one-third to the program supplier. For example, if a cable system charges $10.00 per month for HBO, it may keep up to $7.00 and pass $3.00 on to the pay service.

The special-event nature of many pay-per-view offerings (boxing matches, rock concerts, and so forth) has resulted in a *shared-risk* arrangement. Most of these PPV events split revenues on a 50–50 basis. Pay-per-view movie arrangements are more like those in regular pay cable and the movie business; that is, they are subject to the split negotiated by the parties, most commonly with the majority of revenue accruing to the cable system, the remainder to the PPV distributor.

Count Your Pennies: Acquisition of Basic Cable Services The variety of basic cable services, from sports (ESPN) to news (CNN), movies (American Movie Classics), and government (C-SPAN), each employs an acquisition pattern based on *cost per subscriber per month.* In each case, the cable systems negotiate a price with the supplier based upon the number of households which receive the program service each month. Most charges are in the range of 5 to 50 cents per subscriber per month.

For example, ESPN charges systems about 20 cents per subscriber per month; 25 cents or more with special packages included, like NFL Football and Major League Baseball. Superstations, like WTBS, KTVU, and WPIX, are distributed by satellite syndication companies to local cable systems for about 10 cents per subscriber per month.

Cable News Network is relatively expensive for local systems to acquire. CNN costs about 25 cents per subscriber per month. Turner Broadcasting Systems, which licenses CNN to cable systems, also offers "Headline News." This additional news service is frequently "bundled" (sold together) with CNN, for a package rate of about 30 cents per subscriber per month.

If news is costly, government talk must be considered comparatively cheap. C-SPAN I and II, the channels for Senate, House of Representatives, and other government goings-on, receive their budget from a consortium of major cable MSOs. This underwriting allows for the cost to individual cable systems to be kept to a relatively low 5 cents per subscriber per month for each channel.

Vertical Integration of Systems and Suppliers Program management in cable differs from television programming in one fundamental respect: the largest cable systems are also the owners of the major programming services. Such vertical integration (review Chapter 2) has been minimal in traditional network television, since for over 20 years, the three major networks (CBS, NBC, and ABC) were generally prohibited from owning or producing the entertainment programs appearing on their air. However, these regulations, known as Financial Interest and Syndication Rules ("Fin/syn," for short), were largely abandoned in 1994.[6]

The nation's largest cable MSO, TCI, has substantial equity interest in 10 of the top 20 cable program services, including Turner Broadcasting (CNN, TBS, TNT), The Discovery Channel, American Movie Classics, and Black Entertainment Television. The huge media conglomerate Time-Warner boasts over 5 million cable subscriber households, as well as ownership of both HBO and Cinemax. Many other cable MSOs are investing with their colleagues (or competitors) to bring programming to cable, on both a national and regional basis.

Obviously, MSOs with equity positions in cable program ventures are likely to give more prominence to their services. It is not uncommon, for example, for CNN, TNT, and American Movie Classics to boast better channel positions on TCI cable systems than on some others. Similarly, Time-Warner systems are apt to more aggressively market their owned pay services (HBO/Cinemax) than those of their primary competitors, Showtime and The Movie Channel. Some independent cable operators claim that they pay 70% to 80% more for programming than do the major MSOs.

This claim was borne out in a recent FCC survey which found that the average cost per channel for cable programming for small systems (under 1000 subscribers) was double that of the large MSO-owned systems (with 50,000 subscribers or more).[7] For their part, the MSOs claim they have the economies of scale (large subscriber counts, channel capacity, higher-grade equipment) and the attractive households (more "upscale" in income, education, and social class) which are sought by the major programmers.

Affiliate Relations: A Growing Program for Cable Earlier, it was noted that a key problem in program acquisition for broadcast television is the important relationship developed between the program supplier (the network) and the local station. The relationship between cable programmers and the local cable managers which carry their services is becoming increasingly critical. This is due to a number of factors.

Clearance and Channel Capacity Time was when there were more cable channels than there were available programming services. This is no longer the case. In a recent study virtually all cable systems limited to 35 channels were operating at full capacity; nearly 9 in 10 channels nationwide were in use.[8] Thus, simply launching a new cable service today is no guarantee of carriage. Many systems simply have no room; adding a new service would mean deleting a service or sharing a channel among two or more services. The program is compounded with newer services (like HDTV and interactive channels), which require more than one channel to operate effectively.

Deferred Compensation, Promotion, and Marketing Costs While to gain early clearance, many new basic services are offered free when launched (such as the Courtroom Television Network in 1991), charges based on subscribers typically follow at a future date. In addition, local cable operators must help promote the new channels with advertising and promotion, so that viewers will sample the service. A typical promotion campaign in cable can cost up to $1.00 per subscriber. Many cable operators are reluctant to add a new budget item to their narrowing cash flow ledger, particularly for a new service.

MSO Charge-backs To many cable MSOs, their individual systems exist mainly as a revenue stream into corporate headquarters. Overhead is deliberately controlled (through centralized billing, computerized telephone service, and so on), so that the local system operates as efficiently (and cheaply) as possible. Those local program managers with the autonomy to sign up new program services must show the cost of these services (as well as costs associated with their promotion and marketing) on their balance sheets as a "charge back" to the MSO. These expenses can erode the profit margins of the MSO, which in today's uncertain economic times is not a pleasant situation for either party.

For these and other reasons, local system operators are becoming an increasingly important factor in the cable programming process. To the cable programmer, these managers are the crucial gatekeepers who decide what channels will be available to television households in a given community.

TELEVISION SYNDICATION

Even if affiliate relations are operating at the optimum level, the television manager, both in broadcast and cable, must still program a bulk of airtime weekly. The process of program acquisition by a television station or cable system from sources other than a network is known as *syndication*.

Syndicators are television's "door to door" salespeople, who call on managers with a wide assortment of programming. Salespeople from television syndication companies pay frequent visits to television stations throughout the year. But the meeting which causes the greatest interest and likely generates the most revenue for syndicators is the annual meeting of the National Association of Television Program Executives (NATPE). At the NATPE trade show, syndicators introduce new and old program product with much bally-hoo and fanfare. Elaborate booths and screening rooms are set up. Stars mingle with program executives and their families. There are dancers, acrobats, and plenty of refreshments.

The revenue generated from television syndication more than justifies what might be perceived as the excess of NATPE. In the 1980s, the syndication business grew from less than a billion in revenues, to billings of over $5 billion annually in the early 1990s. Domestic sales account for about $3 billion; $2 billion is generated in the sales of American television programming overseas.

The types of syndicated programming one sees at NATPE fall into four main categories.

Movie Packages

Movies have been a mainstay of television stations since the medium's early days in the 1950s. They have the dual attraction of being long (90 minutes, on average, without commercials), and popular with viewers.

Generally, movies are made available to television stations after their release to theaters, pay-per-view, pay cable, videocassette, and TV networks. Thus, it is highly unlikely that a film will make its first public appearance on a broadcast channel. In addition, since there are thousands upon thousands of films available for syndication, normally movies are sold by syndicators in groups, known as *packages*. This enables the syndicator to move more product, and for the TV programmer to schedule films to consistent time periods, or *blocks*. Showing a different movie each night under the umbrella term *The Late Show* is an example of such block programming.

To a program manager, a single movie is known as a *title*. Independent television stations lease the most titles; as many as 2000 to 3000 or more may be in the station's program inventory at any one time. Affiliate stations are less reliant on movie packages, and may have as few as 50 or less titles on hand at any one time.

Syndicators use creativity in creating movie packages, often bundling popular movies (known as *A titles*) with lesser efforts and outright bad movies (known as *B titles*). Some syndicators lump films with similar themes together, such as Viacom's *Young and Reckless* and *Guts and Glory* packages and Republic's *Classic Comedy* and *Home of the Cowboys* features.

Off-Net Syndication

A second class of product licensed to local TV stations and some cable services is *off-net syndication.* Off-net syndication refers to programs sold in syndication which originally aired on a major television network. Examples of some recent network shows now in domestic syndication include *Cheers, Golden Girls, Designing Women,* and *Full House.*

Typically, programs are made available to local management after three seasons on a network (or when about 100 episodes have been completed). For this reason, some popular programs can be seen on a network as well as in reruns on a local station. Recent examples of this phenomenon incude *Roseanne, Home Improvement,* and *Married with Children.*

Like movies, off-net series are sold to stations in packages, based on a negotiated price for a week's worth of shows (5 episodes). The price paid by a station is based on the competition between stations for the product, the popularity of the program in its network run, the size of the market and other competitive factors. A successful off-net program, like *Seinfeld* or *Cheers* can cost a station in Los Angeles, Chicago, or New York $100,000 per week or more; stations in smaller markets might be able to license the same program for $10,000 per week or less.

Off-net syndication in cable   In recent years, cable services have become a prime market for off-net syndication. For example, the Lifetime cable service bought the rights to air such programs as *Cagney and Lacy* and *Moonlighting* before their sale to local television stations. Older off-net series, including many in black and white, have become stalwart programs of such cable channels as Nickelodeon (Nick at Night), TBS, WGN, Arts and Entertainment, and Discovery. Some series which have not been particularly successful in their network runs have moved to cable and continued production of new programs. Some examples of this trend include *The Days and Nights of Molly Dodd* and *China Beach.*

First-run syndication

First-run syndication refers to programming which makes its debut on local television stations. That is, these programs are not reruns of older network programs. The backbone of first-run syndication is talk and game show programming.

First-run syndication is popular with program managers for a number of reasons. First, it is fresh programming; unlike an off-net rerun, the program hasn't and personnel to aid in local production. In a few large urban systems, fully staffed *access centers* are maintained to facilitate the production of local access programming.

Leased-access refers to channels leased by cable systems to providers of special-interest programs or services. Examples include real estate listings and other local advertising, and some local sports.

At present, just over 1000 public access channels remain operational nationwide, either on a free or on a leased basis. The majority are located in larger cable systems in suburban areas, where there has been a continuing public appetite for special-interest programming, as well as a suitable economic base, including local advertising, to make such facilities viable.

Perhaps the fastest-growing category of local cable programming is *operator-originated* channels. As the term implies, these are channels programmed by the cable system for its local viewers. About 300 such channels are presently operational and range from local movie channels to full-service channels which emulate independent TV stations, to full-time news operations in the manner of Cable News Network.

The main impetus for such channels is the growing volume of local cable advertising traced in Chapter 6. In addition, as cable has become increasingly "nationalized" through superstations, sports channels, and global news services, LO channels provide systems with an important local identity.

In general, budgets for LO programming are small, ranging from about $100,000–$300,000 per year. However, Time-Warner's ATC spent well over $1 million to launch WGRC, a "station-like" channel on its Rochester, N.Y. system.[14] CableVision Systems' operation in Hauppauge, N. Y. is budgeted in the $600,000 range.

Most analysts expect cable's LO commitment to grow in coming years, after diminishing in importance in the deregulatory 1980s. Continental Cablevision, for example, supports local origination on 150 of its 170 systems and has plans for augmenting that significant commitment. Other MSOs with plans for increasing LO programming include MultiVision, Multimedia Cablevision, and Cox.[15]

PROGRAM SCHEDULING

The process of scheduling is often the most creative aspect of television program management. Choosing which shows to air and what times to air them is a critical determinate of the success or failure of a television or cable service. Volumes have been written about this process, which has produced its own vocabulary and often-contradictory "rules." This unit will introduce the major components of the scheduling process: audience composition, television time periods, and common programming strategies; for more detail, consult the list of references at the end of the chapter.

TV Audience Composition

The scheduling process requires intelligence about the nature of the television audience and how that audience changes throughout the day. As a starting point, let's examine the composition of American television households, in Figure 12–1.

About 240 million people in the United States live in households which own at least one television set. In these homes, nearly 4 out of 5 people are above the age of 18, with the majority of these being women (37% to 40%). The younger audience is made up primarily of children (2–11); teens make up only 1 in 3 underage household members.

To put these figures into context, note trends in the television households of 1970. A generation ago, there were many fewer homes with television (about 190 million), with more people living in them. Children, particularly teens, were a major force in these homes, comprising 1 in 3 residents (compared with 1 in 4 today). But then as now, adult women were the most common residents in a television household.

What programming rules of thumb emanate from these data? First and foremost, the fact remains that the most successful programming appeals to adult women: the largest audience segment and the one which purchases the majority of the products pitched by advertisers. Whereas in 1970, these women were in their child-rearing young-adult years (aged 18–35 or 18–49), today, the most sought after female audience is older: 25–49 or 25–54 are the standard demographic markers in use. Over the years, the preferred program types for women have included serial dramas ("soap operas"), movies, dramatic series, game and talk programs.

FIGURE 12-1 TV household composition, 1970 and 1990. *Source:* A.C. Nielsen Company, Inc. (Used by permission).

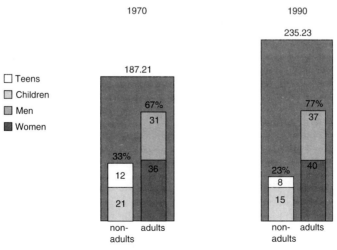

Correspondingly, adult men are next in importance to programmers. Traditionally, men have preferred action-adventure series, sports, and news, though many of the most popular programs for men are also popular with women. Indeed, an ideal TV program would appeal equally to men and women in their late thirties to early fifties. *Seinfeld* and *Home Improvement* come immediately to mind.

Secondly, while much of the historical success of television has been in its appeal to younger audiences, especially teens (who tend to be very loyal fans of situation comedies and dramatic series), the teenage audience has been in decline in recent years. Increasingly, the TV audience is either very young (under aged 12) or very old (over 50). The sheer number of small children available in the television audience creates a consistent appetite for animated characters and has returned kids' programming to the major networks (*The Simpsons* on Fox; *Family Matters* on ABC). "Kidvid," as children's programming is known in the trade, is also an important element of local programming, particularly among independent TV stations.

At the other end of the continuum is the fast-growing audience of older adults. Indeed, demographers suggest that the fastest growing segment of the population consists of people over 55 years of age. TV programmers have begun to deliver to this target audience, who have significant interest in news, movies, and talk and dramatic programming.

A third major trend is more subtly revealed by the data in Figure 12–1. In the past quarter-century, like the medium itself, the television audience has become fractionalized. Today, there are many more households than there were in 1970. More than three-fourths of these homes have more than one set. At the same time, families are smaller. There are as many single-parent homes as there are traditional units of mother-father-children. Those traditional units have fewer children. Too, many more children than in the past watch TV alone, without supervision (in so-called "latchkey" households), while parents or guardians are at work.

The upshot of these trends is that TV programming is increasingly a process of niche-marketing. Much like radio programmers, TV managers must identify their unique or desired programming niche and provide attractive or interesting shows which meets the desires of this audience subgroup. More on this phenomenon below.

TV Time Periods

As we revealed in Chapter 5, TV viewing is tied to time of day. In general, viewing grows steadily throughout the day, peaks in the evening, and declines steadily from late night to dawn. Programmers have divided the 20 or 24 hours their station is on the air into ten fairly standard time periods, also known as *dayparts*. These are charted below, organized into convenient categories of "Day," "Evening," and "Late-Night."

Daytime Programming

Early morning (6 to 9 a.m.). At its highest levels, about 1 in 4 TV homes is using television in this time period (interestingly, about the same number that is using radio, making this the most important radio daypart). The audience tends to be numerous among working adults and small children. For many reasons, this is not a good time period to reach teens. News and talk shows and cartoons are typical in this time period.

Morning (9 a.m. to noon). Viewing levels stabilize in this daypart between 20% to 25% of TV homes. These homes are characterized by the presence of non-working adults, especially women, and small children. An increasing proportion of the viewers are non-homemakers, including retirees, the self-employed, and shift-workers. Syndicated talk programs, movies, and childrens' programs have been abundant in this time period.

Afternoon or "daytime" (12 to 3 p.m.). Homes using television increase gradually through this time period, moving from 1 in 4 to 1 in 3 households. A traditional stronghold for network soap operas, this time period is also strong for movies, game shows, and childrens programming.

Evening Programming

Early fringe (4 to 6 p.m.). This time period is critical, as the audience begins to grow to its peak evening levels. As many as 40% of TV households is tuned in by the end of this block. Savvy programmers can capture viewers and hold them with popular programs in early fringe. Affiliate stations normally fill the first hour of fringe with syndicated talk, like *Donahue, Geraldo,* or *Oprah*. Other common program approaches include courtroom shows, like *People's Court* and *Divorce Court,* and recently popular situation comedies (*Cosby, Golden Girls, Cheers,* e.g.). These programs are good lead-ins to an early local news program (in the 5 p.m. hour), or to news at 6 p.m. Independent stations and many cable services use early fringe for niche programming: with cartoons, business shows, classic situation comedies, and so on.

Early evening or "early news" (6–7 p.m.). This one-hour unit is extremely important since by 7 p.m., half of the homes with TV in the United States have switched their sets on. For most network affiliates, this is the critical hour of their local newscast; independents offer alternatives through syndicated programs or movies. Cable services offer some of their strongest programs (such as CNN's *The World Today*) or offer their own news alternatives (like *Cartoon Express* on USA, and *Gilligan's Island* and *Bugs Bunny* on TNT).

Prime time access ("Prime access," or simply "Access") is 7–8 p.m. This time period, in which as many as 60% of TV households is watching, was created by the FCC to reduce network power over prime time and to stimulate local production by affiliates. While the Prime Time Access Rule (PTAR) may have been successful on the first count, it has been less so on point two. Prohibited from carrying network shows on weekdays in the 7:30 to 8:00 half-hour, most affiliates have turned to syndicated programs, like *Entertainment Tonight* or *Family Feud.* A recent strategy of affiliate programmers has been to move network news up to 6:30 p.m. and to strip popular game or situation comedies back to back in access. Thus, a station may feature *Wheel of Fortune* and *Jeopardy* back-to-back, or an hour of *Cosby* and *Cheers.*

Access time represents an opportunity for independent stations, public television, and cable systems to "steal" the TV audience an hour before prime time begins on the networks. Thus, some indies start their movie hour here. Others run off-net adventure series here. Still others go head to head with affiliates with game shows or off-network comedies. This is the home of *The MacNeil/Lehrer Newshour* on PBS.

Strategies in cable are similar. ESPN uses the access period for its strongest program, *SportsCenter.* CNN slots its business report *Moneyline,* and issue-oriented "Crossfire" in access. The Movie Channel begins its night of features promptly at 7:00 p.m.

Prime time, 8–11 p.m. (EST; 7–10 p.m. CST). Prime time is the most important segment in the television business for one simple reason: a significant majority of the American population is watching. In February, for example, as many as 7 in 10 U.S. households has its sets on in this time-period.

Prime time has been the traditional stronghold of network television. However, the changing fortunes of ABC, NBC, and CBS in recent years are perhaps best noted in the size of their audience in prime, graphically depicted in Figure 12–2.

At the beginning of the 1980s, the three commercial networks combined attracted nearly 9 in 10 TV households. Then came the VCR, cable, independent stations, Fox, even Nintendo to compete for the viewer's attention.

It would be premature to declare the networks dead in prime time. Together, ABC, NBC, and CBS still deliver more than half of all television viewers in prime time. Thus, their programming is still the most important element of the prime-time schedule. While preemptions are gaining popularity in other dayparts, with few exceptions, network prime-time programming is cleared in toto by affiliate stations.

The Fox network is now a force in prime time, with programming offered 7 days a week. And, as we have seen, the roll-out of the two new networks announced in 1993 is targeted to this important daypart.

While network television seeks a huge mass audience (with a show like *60 Minutes* pulling over 25 million households per week), the vast number of cable outlets and independent stations use prime time to showcase their best targeted or niche programs. Many independents run movies or home-team sports. This strategy is also employed by cable networks like TBS, TNT, ESPN, or USA. The Nashville Network stacks its most popular programs in prime, including *Nashville Now* and *Crook and Chase.* Sandwiched between hours of *Prime News,* CNN presents the popular *Larry King Live* talk/interview program. Amerian Movie

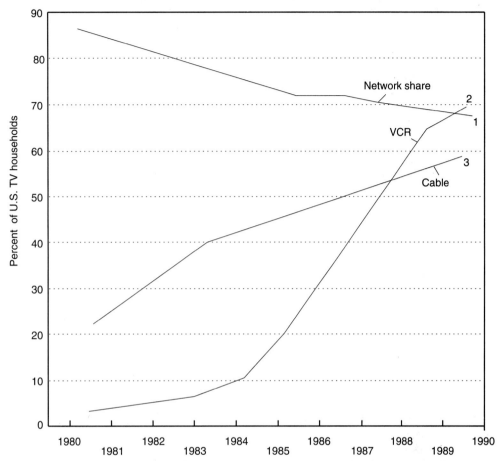

FIGURE 12-2 Prime time share, 1980–1990 (with cable and VCR proliferation). *Source:* Compiled by the author from various sources.

Classics runs a double-feature of Hollywood-era movies, and Black Entertainment Television offers 2 hours of *Video Soul.* Interestingly, viewers with nostalgia for the glory days of network dominance can revisit the 1960s and 1970s, with a prime-time block of reruns on Nickelodeon's *Nick at Night.* Its recent prime-time block included such "classics" as *Mork and Mindy, Bewitched, Get Smart, Dragnet, Alfred Hitchcock Presents,* and *Green Acres.*

Prime time is also critical to the success of pay cable services. It is here when HBO, Cinemax, Showtime, and The Movie Channel load their recent theatrical releases and special programs (like boxing on HBO).

Late-Night Programming

Late fringe (11–11:30 p.m.). This time period is local news time for affiliates, and represents a final chance for alternative programming on independents and cable services. Viewing begins to drop precipitously, from the prime-time peak of nearly 7 in 10 homes in prime to about 3 in 10 by 11:30 p.m. Normal programming strategy for independents has been to counter the "bad" news on affiliates with soothing sit-coms and fantasy science fiction. Reruns of the venerable *Honeymooners, Cheers, Taxi,* and *Star Trek* have been common at 11:00 p.m. On the cable dial, CNN offers its *CNN Sports Tonight.* ESPN counters local news with *SportsCenter.* On other cable networks, action dramas have been popular, particularly with adult themes (such as USA's *Silk Stalkings*).

Late-night (11:30 p.m. to 2:00 a.m.). While much of the country sleeps, up to 1 in 4 homes will keep its sets on, mostly looking for late-night entertainment. The king of this time period was Johnny Carson, a fixture on NBC from 1962 until succeeded by Jay Leno in 1992. In 1993, David Letterman moved to CBS to challenge the long-time dominance of *Tonight* in late night. As this book went to press, David had won the battle (with bigger ratings at 11:30 than Jay), but a newsman may have won the war. Holding its own in the time period was *Nightline,* hosted by Ted Koppel on ABC.

Other late-night alternatives include off-net syndication, movies, and "reality" programs (like *Hard Copy*). Some independent stations and cable systems move toward more "mature" programs in late-night, from R-rated movies with much nudity and gore, to syndicated programs with "adult" (or at least, raunchy) themes, like *The Love Connection.*

Overnight (2 a.m. to 6 a.m.). Less than 1 in 10 households is viewing TV in this daypart, but the numbers have been growing in recent years. Indeed, it can be argued that the most diverse and innovative programming can be found here, as stations and networks try less costly, often experimental program forms in the dead of night. Moreover, programmers are willing to sell blocks of time outright to a range of producers seeking exposure (see *paid programming,* above). Overnight news has been popular, whether provided by CNN (with *Headline News* syndicated to a number of broadcasters in this daypart) or NBC. Many stations rerun their late news in the overnight period.

Other staples of the predawn hours include movies (usually very old titles, which may explain their cult appeal at 3:30 a.m.), syndicated games, and older off-network dramas and comedies. Home shopping is also a mainstay of the late night hours.

PROGRAM STRATEGIES

Armed with the above information about viewing by daypart, there is little mystery to the art of television programming. The name of the game is popularity, as measured by the ratings. A number of strategies has evolved to achieve the lofty goal of getting TV households (or increasingly, individual viewers) to tune in to and stay with your channel. A review of these will introduce the reader to some of the jargon of the TV program director.

The Theory of Audience Flow

A long-held belief of television programmers (traditionally supported by research data) is that once viewers select a program, they tend to stay with the channel it's on throughout their viewing period. This phenomenon is variously known as "tuning inertia," or "viewer passivity." Translated into program practice, the concept is known as *audience flow*. Simply put, the idea is to match programs of similar type, which appeal to similar audiences, throughout a single or adjacent daypart so that the audience for the first program will flow into the next. For example, in network television, some evenings feature situation comedies, others movies, others adventure shows. In Fall 1993, on Monday evenings, CBS scheduled *Evening Shade, Dave's World, Murphy Brown, Love and War,* and the comedic-drama *Northern Exposure.*

Audience flow is also utilized by local stations, cable operators, and public television programmers. Many affiliate stations program syndicated talk or current affairs programming (including *A Current Affair*) just before the early news. In cable, numerous examples abound.

In 1994, for example, information-driven CNBC offered a prime-time talk block, with Vladimir Pozner's and Phil Donahue's talk show moving audiences into Tom Snyder's show and the mature-themed *Real Personal*. Nickelodeon's *Nick at Night* offered an evening bloc of sitcoms, with *Newhart* building into an hour of *Mary Tyler Moore*, followed by an hour block of *Dick Van Dyke* and *The Lucy Show.*

Audience flow is associated with two other key terms. *Lead-in* programs are those which build audience for the program to follow. *Lead-out* or "follow-on" shows are, not surprisingly, the programs designed to pick up audiences from the preceding program.

Program managers often use the audience flow concept to help programs achieve an audience. For example, TV networks will often place a new or weaker program between two hit shows, hoping the audience will flow through all three (and make successful the show in the middle). This process is known as *hammocking*.

In Fall, 1993, for example, ABC unsuccessfully hoped to build an audience for *Phenom* by placing it between *Full House* and *Roseanne* on Monday nights. CBS pinned its hopes for *Dave's World* by hammocking it between *Evening Shade* and *Murphy Brown*.

A corollary of hammocking is the more risky practice of *tent-poling*. In this case, a strong show is hammocked between two newer or weaker shows, in the hope that lead-in and follow-on audiences to the hit show will make the others successful. This is what Fox had in mind recently by initially sandwiching its hit *Married with Children* between *Roc* and *Herman's Head*. The strategy worked. By 1993, both *Roc* and *Herman's Head* were successful enough to move from Sunday to other nights (*Roc* to Tuesday; *Herman* to Thursday).

While audience flow has guided program decision making for many years, there is new evidence that viewer inertia and loyalty are dying out. The VCR, the wireless remote, and the increasingly short attention spans of viewers are combining to suggest that many fewer TV sets stay on any particular channel for a significant time. Thus, other approaches to programming are important to understand.

Power and Counter-Programming

In any given time period, TV programmers have two main choices. They may schedule a very strong program and try to garner the lion's share of available TV homes. Or, they may sacrifice the time period to a competitor with a strong program, and instead opt to pick up an underserved or specialty audience.

The first option is known as *power programming* or "challenge programming." When you power program, you try to "hit them with your best shot," even when it means you are programming the same kind of show as one or more of your competitors. For example, the 1993–94 television season showed CBS, NBC, and ABC all offering a movie of the week on Sunday evenings.

At the local level, most affiliates go head-to-head in news, often with lead-in programming with *Oprah!, Donahue,* and *Geraldo* opposite one another at 4:00 p.m.

The tamer approach, normally favored by independent stations, cable networks, and public broadcasters, is to leave the hugest audience segment to the three commercial networks and seek the "leftovers." For years, independents have countered network sit-coms and dramas in prime time with movies and local sports. Childrens' blocks have been common daytime, as an alternative to soap operas. CNN counters escapist entertainment on commercial television with prime-time news and *Larry King Live.* ESPN offers *SportsCenter* as an alternative to early and late news. MTV has countered news with *Beavis and Butt-head,* a show with arguably little or no information value.

The public broadcasting schedule has built its reputation by countering network entertainment with documentaries, nature programming, and fine arts performances. As the TV audience fractionalizes, counter-programming is becoming the norm.

Other Programming Strategies

Stripping and Checkerboarding As mentioned above in the discussion of syndicated programming, stripping is the process of scheduling syndicated shows to run in the same time period across the week, such as is commonly done with game shows (like *Family Feud*) and talk (*Regis and Kathy Lee*). A variation of this approach is *checkerboarding,* in which two or more syndicated shows are alternated in a time period, such as *Newlywed Game* on Mondays, Wednesdays, and Fridays, and *Divorce Court* on Tuesdays and Thursdays.

Stunting During ratings periods, programmers often deviate from their usual programming to attract an audience. Premiere series may be introduced as movies, running 90 minutes or more (such as the 2-hour premiere of *Twin Peaks* in 1990 on ABC). A new comedy series may run two episodes back to back; specials may preempt regular programming, shows may be moved temporarily to different days of the week, and so on. Adapted from football terminology, the process of making such temporary scheduling modifications is known as *stunting*.

The language of TV programming has many other colorful terms in its lexicon, which space precludes discussing here. The reader should consult the references at the end of the chapter for more detail about this exciting management position.

SUMMARY

Television program managers are involved in personnel administration, program acquisition, program production, and program scheduling. They are compensated at a generally higher rate than their counterparts in radio, though employees in cable programming tend to earn less than those in broadcast television.

TV programs may be acquired from a network or from syndicated sources. Network programming involves the program manager in the often-ticklish process of affiliate relations, including decisions to refuse or preempt certain programs. Syndicated programming is often costly, whether the programs are movies, talk or game shows, or sports.

Program managers may also produce their own programs. News and talk are the most common forms of original programs in both broadcast and cable television. Most managers predict an increase in local program production in upcoming years.

There are a number of theories and strategies in TV programming which emerge from the ways viewers use their sets at various times of day. Most common are the notions of audience flow, power and counter-programming, stripping, checkerboarding, and stunting.

NOTES

1 See, for example, "TV News: The Rapid Rise of Home Rule," *Newsweek,* October 17, 1988, pp. 94–96; "Local network affiliates increase news programming," *NAB Info Pak,* December 1987, p. 4.
2 This and subsequent data on salaries is based on the annual employment reports prepared by the National Association of Broadcasters. Full citations are available in the references cited in Chapter 10.
3 Based on data reported in "Salary Survey," *Cablevision,* June 7, 1993, pp. 114–120. See also Table 6–2.
4 See "Shining Up LO," *Cablevision,* January 28, 1993, pp. 23–23.
5 See "WB Backs Off After Paramount Successes," *Broadcasting and Cable,* November 15, 1993, p. 10.
6 See Joe Flint, "Network Win, Hollywood Winces as Fin/Syn Barriers Fall," *Broadcasting and Cable,* November 22, 1993, p. 6, 16.
7 See *Cable Television Business,* July 31, 1991, p. 40.
8 Tom Kurver, "To Fill or Not to Fill?" *Cablevision,* February 11, 1991, pp. 23–27.

9 See Steve McClellan, "Broadcasters, Cable: The Airing of the Green," *Broadcasting and Cable,* October 25, 1993, pp. 24–25.
10. See "Newsrooms Remain Profit Centers," *RTNDA Communicator,* April 1988, p. 52.
11 See *1992 Television Financial Report,* Washington, D.C.: National Association of Broadcasters, 1993 (various pages).
12 Harry Jessell, "Special Report: Baseball '93," *Broadcasting and Cable,* March 15, 1993, pp. 39–44.
13 See "Programming Spending Strategies: Keeping the Cash at the Station," *Broadcasting,* January 23, 1990, p. 71.
14 See "Cable does local news," *Cable TV Business,* September 15, 1990, pp. 27–29.
15 See "Local Outlook," *Cablevision,* January 21, 1991, pp. 22–28.

FOR ADDITIONAL READING

Baldwin, Thomas F., and McVoy, D. Stevens: *Cable Communications* (2d ed.). Englewood Cliffs, N.J.: Prentice Hall, 1988.
Blum, Richard A., and Lindheim, Richard D.: *Primetime: Network Television Programming.* Boston: Focal Press, 1987.
Brooks, Tim, and Marsh, Earle: *The Complete Directory to Prime Time Network TV Shows: 1946–Present* (3d ed.). New York: Ballantine, 1985.
Clift, Charles III, and Greer, Archie, (eds.): *Broadcasting Programming: The Current Perspective* (7th ed.): Washington, D.C.: University Press of America, 1981.
Dominick, Joseph, Barry L. Sherman and Gary Copeland; *Broadcasting/Cable and Beyond* (2d ed.). New York: McGraw-Hill, 1993.
Eastman, Susan T., Sydney W. Head, and Lewis Klein: *Broadcast/Cable Programming: Strategies and Practices* (4th ed.): Belmont, Calif.: Wadsworth, 1993.
Fuller, John W. *Who Watches Public Television?* Alexandria, Va.: Public Broadcasting Service, 1986.
Gitlin, Todd. *Inside Prime Time* (rev. ed.). New York: Pantheon, 1985.
Heeter, Carrie, and Greenberg, Bradley S. *Cableviewing.* Norwood, N.J.: Ablex, 1988.
Howard, Herbert H., and Kievman, Michael S. *Radio and TV Programming.* New York: Macmillan, 1983; originally published by Grid.

13

SALES AND MARKETING MANAGEMENT

Managers involved in media sales are fond of claiming that theirs is the most important department in the firm. After all, they maintain, "all other departments—programming, news, engineering, etc.—*spend* money; ours is the only one that *makes* money." Spend some time around a broadcast or cable company and the cause for such seeming self-importance is clear: the salespeople seem to be earning the most money, driving the best cars, wearing the best clothes, and so on. And, if the general job picture in telecommunications appears bleak at best, there always seem to be openings in sales and promotion. Of course, there is a downside to a career in media sales: "creative" people (like writers, directors, air personalities, and newscasters) tend to take a dim view of the sales force. Sales is a high-pressure business marked by stress and turnover. Too, for every successful sale, there are dozens or hundreds of dead-ends or "no thank you's." Working in media sales requires the thick skin necessary to take rejection, and lots of it.

This chapter examines the process of broadcast and cable sales from the point of view of the sales manager. In addition, consideration is given to the increasingly important role of marketing and promotion in a station or cable system: the art of not only selling commercials, but selling the station or the system to its audiences, suppliers, clients, communities, and other groups.

The goals of this chapter are:

1 To identify the major elements of media sales, including advertisers, agencies, representatives, radio and TV stations, and cable systems
2 To discuss the process of advertising rate setting in media companies
3 To discuss the functions of traffic and operations in sales management
4 To identify the role of the promotion department in sales and audience development
5 To identify common promotion and marketing activities in radio, TV, and cable

THE PRODUCT: SELLING TIME

Way back in Chapter 2, media businesses were distinguished from other organizations by the nature of their product. As you recall, the media tend to deal in intangible goods—like public taste, art, image, and other ephemera. This phenomenon may be most acute in sales. Unlike other forms of advertising, which sell space (on billboards or matchbook covers, in newspapers and magazines), media companies sell *time.* In fact, while most consumer-oriented businesses sell tangible goods, like cars, sneakers, or hamburgers, radio and television stations sell "air," in units of 15, 30, or 60 seconds each. And, unlike other kinds of ads, like those in newspapers, magazines, or in matchbooks, a TV or radio commercial can't be held, clipped, saved, or passed on to someone else (except on tape, of course, though many consumers try to remove commercials when recording radio or TV programs). To compound things, while space tends to remain available (some magazines seem to stay around your house for years, for example) time marches on. An unsold ad in radio or TV disappears into the ether and is gone forever. This helps explain why the sales manager in media seems less concerned with today's spots and instead, is trying to sell as far in advance as is possible: next week, next month, next season, even next year, and beyond.

Defining the Product: Some Terminology

Despite the intangible nature of time, media salespeople have borrowed terminology from other forms of selling. Unlike the general public, media salespeople almost never refer to their product as "commercials." Commercials are typically referred to as *spots,* based on their length. Thus, there are 60-spots (60 seconds in length), 30's, and increasingly, 15's and 10's. Other terms are in common use, each with a slightly different meaning.

Like the goods on the shelf in the supermarket as day dawns, the unsold commercial positions in a radio or television station's program day are considered its *inventory.* For example, a radio station on the air for 20 hours a day, which averages 10 commercial minutes per hour, will begin each day with two thousand 60-second or four thousand 30-second commercials in inventory.

Inventory is important to the sales manager in setting sales and revenue goals. If our station described in the above paragraph faces expenses of $10,000 each day for salaries, programming, rent, utilities and so forth, the sales manager knows that each 60-second commercial must be sold for not less than an average of $5.00 each; no 30-second spot for less than $2.50 each, in order for the station to break even. In order to return a nice 20% profit, a sales goal of $12,000 can be set for the day, with the lower limit for 60's set at $6.00 and 30's, $3.00. Of course, this is a very simplistic example: like listening levels, rates vary throughout the day. Managers usually set sales goals by weeks, months, and quarters (daily targets might drive them crazy). And many, many spots can go unsold. But now you should realize: each unsold spot erodes the station's bottom line, and increases the price all others must bring.

The above example also illustrates the impracticality of selling inventory one commercial at a time. For example at 10 commercial minutes per hour, this hypothetical station has 14,000 60-second spots in inventory each week. Thus, most ads are sold in *packages,* or groups of spots. In radio, an advertiser may pay $1000 and receive in return, 12 commercials per day in morning drive, 10 in evening drive, 6 each in midday and late-night, and so on. We'll detail later how such packages evolve.

A term less commonly used in radio but prevalent in broadcast and cable television is *availabilities,* or "avails." Analogous to inventory, the term refers to the unsold ads available for sale in a television station's daily schedule. The term has also come to mean the package of spots offered by the media salesperson to a potential advertiser. Thus, a TV sales executive might prepare an avail for a prospective client. For example, suppose a car dealership has decided to invest $10,000 in local television advertising. The salesperson at an affiliate TV station might prepare an avail which offers two weeks of daily spots in early and late news, plus a smattering of commercials in other dayparts, like access and late fringe. A sales representative at an independent station might structure an avail which loads a number of spots in evening movies, sports and late-night adventure programming.

Packages and avails purchased over a number of weeks or months may also be known as *flights* or campaigns. These terms come from the advertising agencies, which develop elaborate schedules for their clients, often using more than one medium, such as TV, radio, magazines, and newspapers. Thus, one may read of a 13-week flight, a 26-week flight, even a 39-week or annual campaign. However, with ad budgets tight and available media proliferating, such long-term advertising campaigns are becoming increasingly rare.

THE MAJOR PLAYERS: ADVERTISERS, AGENCIES, REPS, AND STATIONS

Now that some basic terminology has been mastered, let's examine the milieux of the media sales manager. The process of media sales is a bit more complex than it might seem, involving a number of organizations and individuals.

Major Advertisers

The flow of dollars begins with the thousands of firms which invest in advertising each year. Today, more than $130 billion is spent on the various advertising media, from newspapers, to radio, TV, cable, billboards, direct mail, even matchbook covers.[1]

The leading advertisers (known in the trade as the "major brands") are all household names, including McDonalds, Sears, Burger King, AT&T, the major airlines and automobiles, and so on. Advertising has been a surprisingly resilient industry. For a number of years, statisticians and media analysts have noticed a relationship between America's GNP and advertising spending.[2] Apparently, companies believe in advertising, both in good times and bad. In fact, from 1987 through 1990, media advertising actually outperformed GNP in rate of growth, despite generally gloomy economic conditions. After a brief advertising recession in the early 1990s, media sales rebounded by mid-decade.[3]

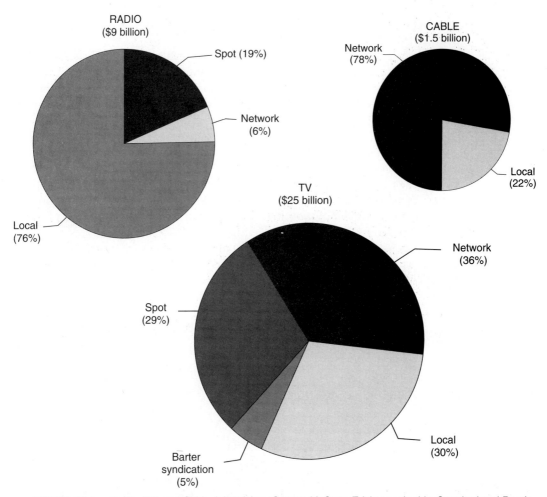

FIGURE 13-1 Radio, TV, and Cable Advertising. *Source:* McCann-Erickson, cited in *Standard and Poor's Industry Surveys,* January 1993, pp. M1–M7.

While many small businesses handle their own advertising, most advertisers contract with advertising agencies to develop and place their commercials before the public. *Media buyers* are the ad agency employees who seek to place spots where they are most likely to attract consumers of a particular product. Those who purchase radio, TV, or cable spots are "time buyers."

Dividing the Revenue Pies: Advertising Sales by Medium

Figure 13-1 illustrates how the media buyers distribute advertising revenue among the major electronic media.

TV receives about 22% of the total annual advertising volume, or more than $25 billion; radio 7%, or about $9 billion. Cable's share today is only about 2%, but cable is nearly a $2-billion dollar advertising enterprise.

TV Sales

The large TV pie ($25 billion) is divided into four slices. The largest slice (36%; approaching $8 billion) goes to the major TV networks: NBC, ABC, CBS, and Fox. In this case, media buyers at major agencies purchase time from sales executives at the networks. Most network spots are bought long in advance, such as in the spring for Fall premieres of network shows. This is known as *up-front* buying. However, the flagging performance of the major networks has caused up-front buying to diminish in recent years. Increasingly common is *scatter* buying, where agencies save a portion of their network advertising budgets to place closer to scheduled airtimes. In this way, advertisers can place spots in hit programs, or those likely to deliver the intended audience, instead of taking the risk long in advance of a network program being a success or failure.

About 1 in 20 TV ad dollars (5%; or about $1.5 billion) goes to the barter syndication market. In this case, time buyers place ads with syndicators of popular programs, like *Wheel of Fortune* and *Entertainment Tonight,* who have reserved a certain number of commercial positions for their own sale. As we traced in the previous chapter, barter syndication is a growth industry in telecommunications.

The remaining 60% of television advertising revenue, over $15 billion, goes to individual TV stations. This amount is distributed roughly equally among national/regional spot sales and local advertising sales. The *spot market,* as it is called, consists of purchases made by media buyers on a national or regional basis, on individual stations or programs known to deliver an important target audience. For example, suppose a major airline is selling trips to Florida and the Bahamas in the dead of winter. It would be a waste for the airline to spend huge amounts of money for a network commercial, in *Home Improvement* for example, which will air in many warm-weather cities, including Los Angeles, Miami, and Phoenix. Instead, the ad agency might choose to advertise on selected stations in selected "snow-belt" cities, such as Detroit, Cleveland, and Minneapolis.

Station Representative (Rep) Firms Since it is difficult for agency buyers to contact thousands of individual stations to place such ads, national and regional spot sales are normally handled by *station representative* firms. Reps maintain offices in the nation's major cities, in proximity to the leading ad agencies. Reps are paid on a commission basis, ranging from 5% to 10% in television, to up to 20% for radio.

The stations on a rep's roster usually share similar attributes (such as ownership, affiliate, or independent status, or location). ABC, CBS, and NBC have had subsidiary organizations to represent their owned stations. There is Galavision (Hispanic), Southern Spot Sales, Cable Networks, Inc., and Group W Television

Sales. Other leading rep firms include Blair; Harrington, Righter, and Parsons (HRP); Petry Television; Seltel, and Telerep.

The remaining slice of the TV revenue pie is produced by local advertising sales. These are the spots sold by station account executives to time buyers at local ad agencies or directly through local merchants. Today, local TV sales represents about $8 billion annually, or roughly 30% of all TV revenue.

Radio Sales

If drawn to scale, the radio advertising pie in Figure 13-1 would be about one-third the size of TV's, comprising about $9 billion annually in the early 1990s. Compared to TV, radio advertising is a local phenomenon. Radio networks receive only 6% of total ad revenue (about $500 million); the remaining 94% goes to stations, through national/regional spot and local advertising sales.

Radio's "Mega-reps" National/regional spot advertising accounts for about 1 in 5 radio ad dollars. Put another way, the "typical" radio station receives 20% of its revenue from this source. Spot sales are coveted by sales managers, since advertisers pay top dollar for these spots, compared with the severe discounting common with local ads. However, faced with two or three dozen stations to choose from in a given radio market, most media buyers located in major cities will allocate funds only for the most popular stations in a given market, or those that consistently deliver a large and identifiable audience segment. They will buy only "three-deep" or "four-deep," that is, they will advertise on only the top three or four rated stations in the market. This leads to aggressive promotion campaigns and considerable communication between the radio sales manager and the rep firm.

Two rep firms dominate spot business in radio. Interep is the parent company of a family of firms, including McGavern Guild, Major Market Radio, HNWH, and Torbet Radio, among others. Katz Radio Group includes such rep firms as Banner, Christal, and Eastman Radio. CBS has its own rep firm, CBS Radio Representatives, to handle spot business for its group of stations.

The competitiveness in radio is equally extreme in local sales, which accounts for more than three of every four dollars in station revenue. Local TV and radio advertising both generate annual revenues in the $8–$9 billion range. In TV, about 1100 commercial stations compete for the revenue; on average, among about six stations per market. In radio, over 10,000 commercial stations vie for the local ad dollars, as many as 30 stations or more in larger markets. This helps explain why today, most TV stations remain profitable, but more than half of all commercial radio stations are operating in the red. Severe discounting in local advertising rates is the norm in radio today, with many stations offering 30- or 60-second spots for under $10 apiece.

Cable Advertising: An Emerging Marketplace Cable is different from its major competitors in a number of ways. First, with subscriber fees as their primary revenue source, among cable managers advertising represents a comparatively small income stream (under $2 billion of more than $20 billion in total revenues).

The good news for cable advertising managers is the spectacular growth of the medium as an advertising vehicle. The volume of cable advertising grew more than tenfold between 1980 and 1990. Local advertising sales, virtually nonexistent in the early 1980s, is a common feature of virtually every cable franchise. While general advertising predictions for the remainder of the 1990s are gloomy at best, cable is the one area singled out for revenue growth, with forecasts for the volume of cable advertising reaching as high as $10 billion or more by the latter part of the decade.

There are many reasons for cable's increasing popularity as an advertising medium. As we traced in the programming chapter, the focused programming of many cable services allows advertisers to more narrowly target their campaigns. Compared with network television, cable advertising is cheap (see next section). At the local level, today most cable systems have the capability to schedule commercials properly in their channel line-up—a technical difficulty only recently solved with the development of computer-assisted advertising insertion equipment.

One consequence of this growth is considerable opportunity. Whereas, just a few years ago, virtually no cable systems had a local sales force or sales manager, today the majority of cable systems (virtually all with more than 10,000 subscribers) have at least one full-time salesperson. And, while layoffs have been commonplace at pay cable program services like HBO and Showtime, ad-supported cable networks have been expanding their sales and marketing staffs.

Cable Network Sales The bulk of cable advertising dollars (more than two in three) goes to the ad-supported cable networks, like CNN, ESPN, WTBS, USA, and so forth. In recent years, media buyers have turned increasingly to cable networks, turning this advertising segment into more than a billion-dollar annual business.

Part of the reason is cost. While a single spot in ABC's *Roseanne* costs about $300,000, an entire prime-time package on Turner Network Television (TNT) can cost less than $40,000! These smaller rates correspond to smaller ratings, however. A TNT movie will command only 2%–4% of the prime-time TV audience, while *Roseanne* will command as many as 1 in 4 TV homes in the nation. But lower aggregate numbers for cable networks is offset by demographic targeting possible in the medium. Cable networks can offer viewers with "highbrow" tastes (Arts & Entertainment, Discovery), sports fans (ESPN), teens (MTV), and so on. Network cable is one of the few bright spots in an otherwise gloomy advertising picture. In the early 1990s, this segment of the industry was growing in volume at a pace of more than 20% per year (much to the dismay of the traditional broadcast networks, ABC, NBC, and CBS).

Local Cable Sales Local cable advertising, while a growing outlet for advertisers, accounts for under 33% of total cable ad revenue, between $700 and $900

TABLE 13-1 TOP 10 CABLE INTERCONNECTS

Rank	Interconnect	Coverage area	Subscribers
1	New York Interconnect	Greater NY Metro Area	3,346,000
2	Philadelphia Cable Advertising	Philadelphia, PA	1,353,000
3	AdLink	Los Angeles, CA	1,200,000
4	Bay Cable Advertising	San Francisco/Oakland, CA	1,106,393
5	Boston Interconnect	Boston, MA	1,100,000
6	Greater Chicago Cable Interconnect	Chicago, IL	800,000
7	Tampa Bay Interconnect	Tampa, FL	773,000
8	Detroit Interconnect	Detroit, MI	765,000
9	Time-Warner CityCable Advertising	Manhattan/Brooklyn/Queens, NY	725,000
10	North Carolina Interconnect	North Carolina	702,300

Source: *Advertising Age*, Feb. 11, 1991, P. S–25.

million in the early 1990s. A major limitation of local advertising has been the problem of coverage. Whereas radio and TV stations blanket an entire community with their signals, cable advertising is restricted to the range of the individual franchise area, as narrow as a few miles or less. Advertisers seeking to blanket viewers with their messages have been reluctant to spend money in such a narrow viewing area.

To address this problem, cable systems in larger markets have been joining forces to sell advertising. Cable advertising interconnects allow advertisers to place their spots on each of the many cable systems which may serve a given community, thereby increasing their reach. "Hard" interconnects are actually wired together; a local ad designed for MTV will automatically air on the music channel on each cable system in the area at the same time. Hard interconnects require expensive equipment, often beyond the reach of mid-size and small cable systems.

More primitive (and more common in smaller communities) are "soft" interconnects, in which the various cable systems in the area agree to schedule local ads sold by the interconnect precisely where promised to the advertiser. To date, advertising interconnects have been most successful in the nation's largest markets, including New York, Philadelphia, Los Angeles, Chicago, and Detroit (see Table 13-1).

Absent interconnects, cable ads in the majority of mid-sized and smaller communities across the nation are priced like radio spots: as little as $5 or $10 each. Still, such spots are attractive to numerous local advertisers. For the price of a radio ad, they get both sound and picture. Plus, their product can take its place among the high-priced ads on the commercial networks, CNN, ESPN, even MTV!

THE MEDIA SALES WORKFORCE

With media buyers spreading $50 billion or more among radio, TV, and cable, you now realize why media sales can be a lucrative profession. A talent for sales is highly regarded and well rewarded at stations and cable systems. The next section examines the organization and compensation of the media sales staff.

The Sales Manager Media sales staffs are headed by the sales manager (SM). Typically, sales managers are the highest paid executives in media management.

Depending upon the degree of national and regional spot business in the market, radio and television stations may separate the functions of the sales manager. Where national accounts are relatively few, the sales manager will typically handle relations with the station rep and maintain liaisons with larger advertisers and agencies in the market. However, stations with more sizable spot business will generally boast both a national sales manager (NSM) and a local sales manager (LSM). As the title implies, the NSM will work with the station rep firm to increase national and regional spot business. He or she may also "pitch" the station to major brands and their advertising agencies. The local sales manager supervises the sales staff involved in local advertising. Normally, TV and radio time salespersons are known as Account Executives (AEs).

Since national and regional accounts typically bring in more revenue per spot than do local ads, GSMs tend to out-earn LSMs. In 1993, for example, the average compensation paid TV GSMs was about $90,000, with radio GSMs earning, on average, in the range of $50,000–60,000. Average earnings for LSM's in TV was $60,000 and for radio, about $50,000.[4]

The range of compensation for AEs varies widely. New AEs and trainees draw a relatively small salary, in the range of $10,000–20,000. The national median sales salary in radio and television is in the range of $40,000–50,000. However, these figures are misleading, since they sum across the range of different compensation methods used in media sales traced in Chapter 10.

RATE SETTING

One of the primary tasks of the media sales manager is the process of setting and adjusting commercial advertising rates. As the process is detailed, keep one truism in mind. As a general rule, advertising rates are negotiable. Media sales management can be viewed as a high-stakes game. The sales director who sells the most spots, at the highest cost per spot, is the winner (and usually gets the "spoils," in the form of salary, bonuses, and entry into ownership and station management).

Revenue Projections

The rate-setting process typically begins with the station or cable system manager. Senior management sets station or system revenue projections, on a monthly, quarterly, bi-annual, and annual basis. These may be self-imposed, or directed by the "home office," in the case of group ownership. Factors that enter into the setting of sales goals include the financial condition of the station, its popularity, the amount of advertising revenue anticipated in the marketplace, and other factors. While this process is understandably kept secret by most stations, it can be

Market Radio Financials
(all figures in 000's, except percentages and ratios)

	1985	1986	1987	1988	1989	1990	Δ 85–90
ESTIMATED GROSS REVENUES	$32,600	33,700	36,500	41,400	44,700	43,400	5.9%
★★★	Δ 90–91	1991	1992	1993	1994	1995	Δ 91–95
	3.0%	$44,700	46,900	49,400	52,000	54,800	5.2%

	1985	1990	1995
Revenue/Retail Sales	$4.04/1,000	$4.11/1,000	$4.00/1,000
Revenue/Captia	$25.32	$31.23	$37.37

Demographic and Economic Overview
(000's, except Retail Sales and EBI in 000,000's)

	1985	1990	Growth Rate	1990	1995	Growth Rate
MSA Population	1,288.9	1,389.7	1.5%	1,389.7	1,465.0	1.1%
Households	488.8	529.4	1.6%	529.4	559.3	1.1%
Retail Sales	8,072.0	10,567.4	5.5%	10,567.4	13,673.5	5.3%
EBI	15,097.4	18,747.2	4.4%	18,747.2	25,950.9	6.7%

FIGURE 13-2 Columbus Radio Market Overview. *Courtesy:* Broadcast Investment Analysts. Used with permission.

approximated by utilizing one of the commercially available industry guides (see Figure 13-2).

Figure 13-2 is a page from *Investing in Radio,* published by Broadcast Investment Analysts, for the Columbus, Ohio, market.[5] Similar volumes are published for TV and cable by BIA and other industry associations.

In 1994, estimated radio advertising revenue in the Columbus market was $52 million. For example's sake, let's round off market revenue to $50 million. Suppose you bought a station in Columbus. What might be your revenue goal?

In general, audience share is correlated highly with market revenue share. A station with 10% of the Columbus radio audience can expect to earn at least 10% of the advertising revenue in the market; a station with only 1% of listening will garner 1% or less in ad revenue. In practice, the highest-ranked stations usually exceed their audience share in revenue; those at the bottom of the ratings tend to underbill their audience share.

Suppose our general manager, proud of the new contemporary music format installed by the program director, anticipates a share of 4% of the radio audience. Gross advertising billings should approach $2 million annually ($50 million times .04). This translates into a quarterly projection of $500,000 in gross sales; about $170,000 per month and roughly $42,000 per week.

Typically, contemporary stations program about 12 commercial minutes per hour. Assuming only half of all 30-second spots is sold, this means the sales manager will sell about 2000 spots per week (12 spots per hour, times 24 hours, times 7 days = 2016 sold advertising spots). To meet the general manager's expectations, each spot sold would need to bring in about $21.00 in revenue ($42,000 divided by 2016). This figure is sometimes known as the "nut," the minimal amount each spot sold must bring to enable the sales manager to meet their projection.

Of course, this is a very simplistic example. In practice, most program managers expect their ratings to improve each quarter. General managers therefore expect the number of spots sold to increase, at a higher average cost per spot. In addition, in some dayparts, managers want to sell more spots (morning and evening drive, for example). At other times, the number of spots available for sale will diminish (such as in the late evening or overnight time periods). Programmers are always at odds with salespeople. The former want longer, uninterrupted periods, such as "music sweeps," the salespeople want (and need) to meet their revenue projections and the only way to do that is to sell more spots. Usually, a truce is signaled in the form of a variable guide to station rates: the rate card.

Radio rate cards Figure 13-3 is a typical radio rate card. Most stations publish these cards on a quarterly or semi-annual basis, to be distributed among potential local clients. Rates are also published in *Standard Rate and Data Service* (SRDS) an industry periodical used by advertising media buyers to place national and regional spot advertising.

Notice that the most expensive times on the rate card correspond to times of highest listening (i.e., morning and evening drive times). Chances are that the "nut," the average spot cost required to meet projections, falls somewhere between the highest and the lowest quotes on the cards.

Rate cards give sales managers and their staffs considerable flexibility in dealing with potential advertisers. Spots become cheaper when purchased in bulk. Such reductions are known as *frequency discounts* or *consecutive-week discounts*. Spots run at the same time each day cost more, and are known as *fixed-position*. At the other extreme, advertisers may pay a reduced amount for a commercial, but the station will run it whenever it has time. This is know as *run-of-schedule* (ROS), or *best-time-available* (BTA).

Space precludes a fuller discussion of the creative dynamics and unique terminology of rate card construction. For more information, consult the reading list at the end of the chapter. Of course, in the final analysis, radio rates rarely reflect those printed on a rate card: as in virtually all transactions, the sales price depends upon the price negotiated between the media salesperson and the advertiser.

Television rate guides, avails, and rankers Television rates are also set in concert with projections established by senior management. To see how this process might take place, let's examine sales data for the Rockford, Ill., TV market, illustrated in Figure 13-4.

KSON AM/FM COMBINATION

Contact your KSON AM/FM Account Executive for prevailing grid levels.

97.3 FM • 1240 AM

Class A
Monday-Sunday 3 p.m.-9 p.m.

Grid levels	Unit cost (60s or 30s)
I	400
II	340
III	290
IV	250
V	220

Class of AAA
Monday-Sunday 5 a.m.-10 a.m.

Grid levels	Unit cost (60s or 30s)
I	500
II	430
III	370
IV	320
V	270

Class B
Monday-Sunday 9 p.m.-12 midnight

Grid levels	Unit cost (60s or 30s)
I	170
II	150
III	130
IV	120
V	110

Class AA
Monday-Sunday 10 a.m.-3 p.m.

Grid levels	Unit cost (60s or 30s)
I	350
II	300
III	265
IV	230
V	200

Best times available
Monday-Sunday 5 a.m.-12 midnight

Grid levels	Unit cost (60s or 30s)
I	320
II	280
III	250
IV	220
V	190

Special Features

▲ Rollin' Radio Show
▲ Sponsorship of Local News
▲ Live Personality Ad-Libs
▲ Special Program Sponsorships

▲ Wednesday through Saturday schedules—next highest grid level.
▲ No discount for AM or FM only.

FIGURE 13-3   Radio Rate Card. *Courtesy:* Jefferson-Pilot. Used with permission.

In 1990, estimated advertising revenues for Rockford were $19,700,000. Again, let's round off, to $20 million. Checking the share summary at the bottom, we see that WIFR, channel 23, reported a share of 20 percent of the audience in February, 1990. Multiplying the advertising volume by the estimated market share, WIFR, the CBS affiliate, could expect to earn approximately $4 million in advertising revenue for the year. The independent station WQRF had a share of 8, which would lead to a revenue projection of $1.6 million. Thus, one would expect the rates on the CBS station to be considerably higher than those on the independent. And this is indeed the case. Recent editions of SRDS list typical spot costs on WIFR in the range of $300-500 in prime time; a prime-time spot on WQRF tops out at about half that amount.

The prices for inventory on TV stations vary considerably in comparison to those on radio stations. The most costly spots are those few made available to

FIGURE 13-4 Rockford Television Market Overview. *Courtesy:* Broadcast Investment Analysts. Used with permission.

Rockford Market Overview

ADI Rank: 136

Demographic and Economic Overview
(000s, except Retail Sales and EBI in $000,000s)

	1983	1988	Growth Rate	1988	1993	Growth Rate
ADI Population	415	415	0.0%	415	414	0.0%
Households	154	157	0.4%	157	161	0.5%
Retail Sales	1,781	2,548	7.4%	2,548	3,726	7.9%
EBI	4,356	5,137	3.4%	5,137	7,444	7.7%

Pop Rank #140	ADI Counties	5	White	92.5%	Avg Household	$32,720
HH Rank #136	TV Households	156	Black	6.6%	Per Capita	$12,378
RS Rank #135	ADI Cable	60%	Other	0.9%	Spanish Speaking	2.5%
EBI Rank #117	ADI VCR	75%				

Market Television Financials
(all figures in 000s, except percentages)

	1983	1984	1985	1986	1987	1988	Rate
ESTIMATED NET REVENUES	$12,300	13,300	14,300	15,400	16,500	17,600	7.4%
	1989	1990	1991	1992	1993	1994	Rate
	$18,300	19,700	20,800	22,100	23,400	24,900	6.4%

1983 Net Revenues	$ 12,300	Rev/R.S.	$ 6.89/1,000	Rev/Pop.	$ 29.56
1990 Net Revenues	$ 19,700	Rev/R.S.	$ 6.65/1,000	Rev/Pop.	$ 47.52

Rockford Competitive Overview

Calls	City of License	Ch	Visual Power (kW)	HAAT	Aff	Rep	Owner	Date Std	Date Acq	Sales Price (000)
• WREX	Rockford	13	316	710	ABC	Petry	ML Media Partners	53	8707	18,000
WTVO	Rockford	17	631	670	NBC	Young	Young Bcstg, Inc	53	8806	20,000
WIFR	Freeport	23	676	720	CBS	Katz	Benedek, A. Richard	65	8611	g
• WQRF	Rockford	39	525	570	IND	SItel	Ash, Henry A.	78	8909	2,000
WJNW	Janesville	57	1,000 cp 1,000		NOA		Tri-M Commun Ltd			

ADJACENT MARKET STATIONS

SHARE SUMMARY

	9:00 AM - 12:00 MID					PRIME TIME				
	Feb 90	Nov 89	Jul 89	May 89	Feb 90	May 89	Jul 89	Nov 89	Feb 90	May 89
• WREX	19	19	18	18	23	23	18	20		
WTVO	25	25	24	28	29	30	26	31		
WIFR	20	24	22	23	21	22	17	23		
• WQRF	8	8	7	6	7	7	8	6		
WJNW	0	0	0	0	0	0	0	0		
	4	3	6	5	2	3	8	4		
Total	76%	79%	77%	80%	82%	85%	77%	84%		
HUT %	33%	34%	30%	30%	59%	58%	45%	54%		

local stations for sale by their affiliates in prime time. In a large market, like Atlanta, a spot sold by the ABC affiliate in *Roseanne* can cost over $20,000.

Next most costly are spots sold in popular syndicated programs, such as *Wheel of Fortune*. News is costly, since it is popular with viewers and the local station controls all the inventory for sale (unlike the situation in prime time, when the network gets to sell the overwhelming majority of inventory).

Spots in daytime, early morning, and late-night are considerably cheaper, as available audiences decline. Indeed, after midnight, even network affiliates will reduce spot costs to deeply discounted rates seen on the schedules of independent stations.

This brings us to the issue of rate publication. Unlike radio, TV station rates are not widely published in publicly available guides, like rate cards and SRDS. This is due to a number of factors. First, TV program ratings tend to vary significantly from week to week and quarter to quarter. Managers do not want to be locked in when programs outperform expectations. Nor do they want to discount their rates if shows fail to meet a promised rating. Also, account executives rarely sell individual spots. More commonly, spots are packaged to fill inventory throughout the day. This requires price flexibility to meet both the audience needs of the advertiser and the revenue projections set by the sales manager.

For these reasons, TV sales executives utilize flexible yardsticks in rate setting. Most commonly used are rate guides, rankers, and avails.

Rate Guides A rate guide is similar to a rate card, but it is not published or made available to a potential advertiser. Figure 13-5 is a sample rate guide for an NBC affiliate in a major market.

The rate guide lists available programming in inventory, plus the ranges of prices account executives should use in selling advertising. For example, note that a 30-second spot in *The Today Show* should range from $225 to $450 for the third and fourth quarters of the year. The target cost for a spot in the third quarter (July-August-September) should be $300; with the onset of Christmas buying season, the station will seek a minimum spot price of $400 in the fourth quarter (October-November-December).

Two additional TV sales tools are rankers and avails, which are generated by a TV station by using special computer programs, often provided as a service by the station rep firm, which uses them to sell national and regional spot advertising.

SALES AND TRAFFIC

The success of radio and television advertising is largely dependent upon the successful scheduling of the advertising sold by account executives. Unlike print advertising, which can be verified by simply examining a newspaper or magazine, it is often difficult for a broadcast or cable client to ascertain whether the spots contracted for actually aired as scheduled.

The job of policing the commercial time sales at a broadcast station falls to the traffic department. Sales and traffic are in constant communication to handle the very complex task of commercial scheduling, a process known as *inventory*

		Submission Levels-$$ 3rd Qtr underlined and 4th Qtr in bold
	DAYTIME	
M-F 530-6AM	NBC News @ Sunrise	150 125 **100** <u>75</u> 50 25
M-F 6-7 AM	11 Alive Today	500 **450** 400 350 <u>300</u> 250
M-F 7-9AM	The Today Show	450 **400** 350 <u>300</u> 250 225
M-F 6-9 AM	Early Morning Rotation	500 450 **400** 350 <u>300</u> 250
M-F 9-12N	Morning Rotation	200 175 150 **125** <u>100</u> 75
	The Joan Rivers Show (9-10A) The Judge (10-1030A) The Judge (1030-11A) Trialwatch (11-1130A) Closer Look with Faith Daniels (1130-12N)	300 250 **200** <u>175</u> 150 (M-F/10-11A)
M-F 12N-1P	Noonday	300 275 **250** <u>200</u> 150 125
M-F 1-4 PM	Afternoon Rotation *Program Lineup Includes:* Days of Our Lives (1-2P) Another World (2-3p) Santa Barbara (3-4P)	500 450 **400** 350 <u>300</u> 250
	EARLY FRINGE	
M-F 4-430PM	The Cosby Show	500 **450** 400 350 <u>300</u> 250
M-F 4—5PM	Golden Girls	500 **450** 400 350 <u>300</u> 250
M-F 4-5PM	Cosby/Golden Girls	500 450 **400** 350 300 <u>250</u>
M-F 5-530PM	11 Alive at Five	500 450 **400** 350 <u>300</u> 250
M-F 530-6PM	Golden Girls	800 **750** 700 650 <u>600</u> 550

FIGURE 13-5 TV Program Rate Guide. *Courtesy:* Gannett Broadcasting. Used with permission.

control. Today, sophisticated computer programs assist in this process. Print-outs enable the sales force to provide proof that spots aired as promised. Such print-outs are variously known as "affidavits," "verifications," or "reconciliations."

The traffic department, through its computer, enables the sales department to provide additional services to the advertiser. Advertisers want their commercials separated from those of a major competitor. The traffic computer will ensure that ads for two auto dealers are not "back to back," or that the spot for one fast-food restaurant is not adjacent to another. This is known as *product protection* or *product separation.* As we have seen, some advertisers will pay a premium to air in fixed position, i.e., at precisely the same time each day, such as in the middle of a newscast. Advertisers sold ROS and BTA packages must be scheduled to meet the frequency of commercial placements promised by sales staffs.

The work of the traffic department is of inestimable importance to sales management. Traffic is essential to manage inventory and to provide verification to advertisers that ads they purchased actually ran (and, it is hoped, delivered the desired audience to the advertiser as a result).

Sales promotions are often prepared by a TV station in association with its rep firm, such as this one for professional football. (courtesy of Fox Broadcasting and Petry Television)

Interestingly, 4 in 10 cable marketing executives reported annual income above $50,000 compared with "only" 1 in 4 of their counterparts in radio and TV.

The message from these data is clear. While job opportunities in other areas of telecommunications management are on the decline, promotion is an area of growth and opportunity. Like sales and programming before it, the promotions department is becoming an avenue into senior broadcast and cable management.

The Promotion/Marketing Director Increasingly, promotion managers, typically known as promotion or marketing directors, are being compensated at a rate comparable to other senior broadcast and cable executives. Nationally, the average salary for TV promotions directors is in the $50,000 range. As with all other broadcast jobs, market size is the best predictor of executive compensation. In the nation's top 10 markets, television marketing managers average over $100,000 per year. Radio and cable pay less than TV stations, with the average salary for a radio promotions director and a cable marketing executive approaching $30,000.

Like program managers, most promotions managers wear multiple hats. Most also write, produce, execute, and evaluate their planned promotional activities. This brings us to the tasks of the media marketing manager.

Types of Media Promotion Eastman and Klein (listed at the end of the chapter) have identified three main types of media promotion activities: *audience promotion, sales promotion,* and *public relations.*

Audience Promotion

As might be expected, audience promotion is targeted to viewers and listeners. The goal of audience promotion is twofold: to attract and to maintain an audience, as reflected in ratings.

Getting an audience to listen to a new station, watch a new show, or subscribe to a new cable service is sometimes called "acquisitive" promotion. The intent of such promotions is to encourage *sampling.* That is, the promotions manager is trying to get a listener or viewer to sample the service, in the hope that he or she will be pleased and become a regular consumer. Huge billboard campaigns touting a new morning team would be an example of an acquisitive radio promotion. Another example would be a free weekend offered by a pay movie channel such as The Disney Channel or Cinemax. The massive ad campaigns unveiled by the major television networks each fall are yet another example of such promotions.

Retentive promotions are audience promotions which encourage repeat viewing and listening. Usually, they offer some reward or incentive for continued viewing and listening. Elaborate contests which offer as prizes the kinds of products or services identified with target audience represent such promotions. For example, Viacom's MTV offers its core audience opportunities to win concert tickets, backstage passes, even limo rides with its rock stars. Many radio stations retain listeners with contests urging continued listening in other dayparts. The morning team might announce, for example, the award of a major prize sometime in the afternoon.

Sales Promotions

Sales promotions are targeted toward the broadcast station or cable system's professional affiliations and clients, including advertising agencies, rep firms, and potential advertisers. In broadcasting, the goal of sales promotion is to sell advertising time. The cable system has the additional goal of selling its service to potential subscribers. Thus, most cable systems undertake elaborate sales promotions, often in cooperation with cable program services, to urge potential customers to subscribe. Such campaigns may include bill-stuffers, point-of-purchase displays, billboards; indeed an entire arsenal of promotional efforts.

Public Relations

PR promotion has as its goal the promotion of the image of media facility in its community. Participation in food drives, Christmas programs, and other civic and philanthropic ventures is an example of public relations promotion. Also in this category would fall tours of the station or system for local groups, including schoolchildren, participation in parades, awards banquets for community leaders, and so on. While the intent of PR promotion is the creation of a reputation of the media firm as a "good citizen," such promotions have the added benefit of putting the company's call letters and logo before many important groups, including potential advertisers and consumers, i.e., viewers and listeners.

COMMON PROMOTIONAL AND MARKETING ACTIVITIES

While these general types of promotion help delimit the field, we close this section with an inventory of the specific tasks performed by media marketing managers.

Table 13-3 ranks the most common promotional and marketing engaged in by the three main types of electronic media firms.

Radio Promotion Activities Not surprisingly, nearly all radio promotion managers are involved in some type of sales promotion: everything from rate cards to brochures touting a new disc-jockey tandem, to elaborate presentations aimed at potential advertisers. PR promotions are next in importance to radio, with over 9 in 10 station promo departments involved in community projects and arranging personal appearances by station personalities. Other prominent radio promo activities include on-air contests, image campaigns, and large-scale events promotion like concerts, picnics, and raft races.

TABLE 13-3 MEDIA PROMOTION/MARKETING ACTIVITIES

Overall rank		Percent of stations/systems		
		TV	Radio	Cable
1	Community involvement projects	90%	94%	74%
2	Sales promotions	89%	97%	78%
3	Personal appearances by station personalities	86%	94%	59%
4	On-air contests	79%	93%	67%
5	Topical news or sports promotions	76%	88%	59%
6	Overall image campaigns	70%	84%	74%
7	News image campaigns	62%	34%	30%
8	Concerts or large events	59%	84%	37%
9	Point-of purchase promotions	58%	68%	74%
10	Outside promotions consultant/production cos.	57%	47%	63%
11	Outside advertising agencies	37%	47%	67%
12	Promotions of air personalities	19%	65%	19%

Source: BPME Image, July/August, 1991, p. 16.

KIIS-FM (Los Angeles) morning air personality Rick Dees tars-and-feathers Ted the Tax Man one recent April 15. Promotions like this help keep radio stations "top of mind" with their listeners. (courtesy of Gannett Broadcasting)

TV Promotion Activities TV promotion managers report community involvement projects as their primary promotional activity. Sales promotions, especially the preparation of avails (see above) rank second in frequency. Like radio, personal appearances by station talent is important, as are on-air contests. In TV, topical news and sports promotions are commonplace, reflecting the importance of local news to many TV stations. It is not uncommon today for TV stations to design full-scale ad campaigns solely to tout their weatherperson or sports anchor. Image, especially news image, is critically important to TV marketing managers.

Cable Promotion Activities Cable promotion is slightly different from that in broadcasting. As discussed above, sales promotion, targeted to current and potential subscribers and advertisers, leads the pack among tasks performed by cable promotions managers. Its history of marginal public service is reflected in the commitment of more than 7 in 10 cable marketers to community projects and overall image campaigns. Much of the cable marketing manager's time is spent with outside promotion units and production companies, i.e., the major program suppliers (cable networks, superstations, etc.). But, like their counterparts in radio and TV, increasingly, cable marketing executives are involved in such PR activities as arranging personal appearances and promoting on-air talent. Image promotion is becoming increasingly critical to cable, particularly as possible competitors lurk below ground (the telephone companies) and in the air (direct broadcast satellite firms).

Public Broadcast Promotion

It is fitting that this chapter closes with public broadcasting, since it is an area which merges both the sales and the marketing functions of media management. While there is considerable diversity in the titles given to public radio and television sales/marketing directors, they share a common goal: raising money in a difficult economic and political climate.

During its growth period of the 1960s and 1970s, the primary funding sources for public broadcasting were federal and state moneys. In fact, as recently as 1975, government funding accounted for nearly $7 in $10 allocated for public broadcasting. Foundations and other philanthropic organizations provided about 15% of public broadcasting dollars. Viewers and listeners provided about 12% and private enterprise, primarily through corporate underwriting of particular programs or series (such as Texaco's Metropolitan Opera broadcasts) provided the remaining funding.

In the 1990s, funding sources are precisely the reverse. Today, audiences and businesses provide the lion's share of public broadcast operating income. Federal and state support is on the decline, as is foundation/philanthropic giving.

This dramatic shift is due to political, economic, and competitive factors. On the political side, since the election of Ronald Reagan in 1980, the executive branch (through its conservative fiscal policies and appointments) has applied a free-market model to many governmental agencies, including the Federal Communications Commission, Corporation for Public Broadcasting, and National Public Radio. It has not helped matters that many on Capitol Hill have viewed public broadcasting as a haven for those of leftist or at least liberal leanings. It has also not helped that audits of public broadcasting have revealed considerable fiscal mismanagement over the years. The result has been diminishing funds from federal sources for public radio and television.

The general economic climate is another factor. Federal deficits and state revenue shortfalls are major news, as is a general public sentiment in favor of lower taxes. Many public programs have suffered budget cuts or been eliminated altogether. Against a backdrop of closing hospitals, day-care, and drug treatment centers, it has been difficult to argue for continuing support of public broadcasting.

Too, public broadcasting was founded in part to provide choice and variety in a limited broadcast environment. In the 1950s and 1960s, public television emerged as a noncommercial alternative to the three networks. Similarly, public radio grew predominantly as a soft-spoken, typically classical alternative to the popular music and screaming DJs found on commercial radio.

Today, the number of TV and radio stations, program types, and formats has proliferated to a point where many stations are attractive, commercially supported competitors to public broadcasting. Lifetime, Bravo, The Discovery Channel, Arts and Entertainment, Disney, and other channels provide the documentary, drama, and children's programming once seen only on public TV. On the radio side, commercially supported jazz, "new age," even classical stations operate side by side with noncommercial stations. And, some markets have as many as five noncommercial stations competing with one another for dwindling dollars.

The result has been the rise of the sales/marketing manager in noncommercial radio and TV, usually known as the underwriting manager. Like their commercial counterparts, underwriting directors are charged with raising revenue. How that job is done is considerably different, however.

Legal Restrictions on Station Underwriting

What sets noncommercial broadcasting apart from its commercial counterpart is clear: public stations are prohibited from selling advertising. However, recent FCC rules changes have made it easier for public stations to engage in on- and off-air sales and marketing activities.

The rise in noncommercial underwriting was facilitated by two FCC rulings. According to its *Second Report and Order,* noncommercial stations were no longer bound to identify corporate sponsors by name only.[9] Instead, stations could prepare on-air announcements which acknowledged the name, location, and brief description of business underwriters. A year later, stations were authorized to sell airtime outright to outside organizations, provided they were of a nonprofit nature.[10] More lenient underwriting rules soon followed.

Today, some critics argue that the line between commercial sponsorship and corporate underwriting has blurred to a degree that a qualitative difference is indecipherable. As a practical matter, the tasks of the commercial marketing manager and the non-commercial underwriting manager are beginning to merge. Let's survey those tasks.

Promotion and Marketing Activities in Public Broadcasting

On-Air Promotions Many on-air promotional activities of public radio and TV are identical to those in commercial broadcasting. These activities include promoting programs, developing a consistent on-air "look" in graphics (from logo slides to station break announcements) and set design, promoting major personalities and other activities.

However, some on-air promotions occur in public radio and TV which are unheard of on the commercial bands, at least so far.

Pledge Drives Sometimes caustically called "Beg-A-Thons," most public broadcast operations engage in marathon on-air fund-raising events. Three months are common for these events: March, August, and December. Public TV pledge drives typically last for 10 days to 2 weeks, overlapping 2 weekends for maximum viewer exposure. Normally, prime-time (8 p.m. to 11 p.m.) is the focus of these events. Popular programs such as *Nova, Masterpiece Theater,* and *Frontline,* along with well-received specials (such as *The Civil War*) are often rerun during pledge weeks, to maximize ratings as well as to focus viewer attention on the best attributes for public broadcasting.

The on-air fund-raiser is a staple of public broadcasting promotional activity. Gale Zucker/Stock, Boston.

Public radio pledge drives focus on other "prime times," especially weekday and weekend mornings, evenings, and weekend nights. These correspond to the most attractive public radio programs, such as NPR's *Morning Edition, All Things Considered, Weekend Edition, Jazz Alive,* and others.

Auctions Related to pledge drives, some public stations suspend programming to raise funds by auctioning items of value to viewers or listeners. These items, which range from the small (T-shirts, mugs, and the like) to the significant (vacation travel, VCRs, large appliances) are made available free or at reduced cost by station underwriters and other groups interested in preserving public broadcasting. Stations then auction the items on-air to the highest bidder, with proceeds entering the station budget. Auctions have diminished in frequency in recent years, as they are logistically difficult to schedule and they tend to be a tune-out to viewers expecting the regular program schedule.

Even though they are relatively infrequent, planning pledge drives and auctions, with their complex logistics (obtaining programming, securing telephone lines, recruiting volunteers, etc.) occupies a considerable amount of the public marketing manager's time.

Off-air promotions These activities truly set noncommercial marketers apart from their commercial counterparts. Most stations are involved in one form or another with the following off-air promotions.

Membership magazines One benefit offered viewers and listeners for pledging funds is a station newsletter or magazine. Usually sent on a monthly basis, such guides feature program schedules, personality profiles, and other articles of interest to a public station's audience. Many carry advertisements touting the products and services of the station's underwriters, as an added bonus to them.

Direct mail/telemarketing Many large public stations, especially those in large TV markets, are involved extensively with direct mail and telemarketing. Computer-generated, personalized letters are mailed to potential subscribers, sometimes with samples of the station's program guide. Because public stations must collect the money pledged by audience members during pledge drives, most stations are also involved in mailing annual dues statements, reminders, and other attempts to solicit funds.

Other off-air promotional activities emulate those at commercial stations, including advertising (on billboards, buses, and in newspapers, for example), community relations activities, and the like.

PROFESSIONAL SALES AND MARKETING ORGANIZATIONS

A range of industry associations assist media sales and marketing managers with the complexities of their tasks. On the radio side, the Radio Advertising Bureau (RAB) provides sales and marketing assistance to radio account executives. A major goal of the RAB is to attract advertising to radio from other media, especially newspapers. At the outset of this chapter, we mentioned that radio's share of total advertising expenditures is under 7%. The RAB is determined to increase that share through marketing and promotion of the sound medium. To do this, the RAB prepares on-air announcements, advertiser profiles, case studies of radio success stories and other materials, for use by sales executives at member stations. Each year, at its Managing Sales Conference, over 1000 radio sales executives convene to increase advertiser awareness and use of their medium.

The TV counterpart of the RAB is the Television Advertising Bureau (TVB). Like the RAB, TVB vigorously promotes television advertising, particularly for networks and their affiliated stations. With the decline of network advertising and the rise of barter syndication, TVB has set its sights on increasing spot and local advertising, the bread and butter for TV stations.

Membership in TVB has been declining in recent years, owing to ownership changes, station budget cutting, and other factors traced throughout the text. Today, about 400 stations (roughly 1 in 3) belong to the association, down from over 500 stations (nearly half of the station universe at the time) in the early 1980s.

On the marketing side, two organizations serve the professional needs of the growing number of promotion executives in broadcasting and cable. For 35 years, the Broadcast Promotion and Marketing Executives association (BPME) has itself promoted the interests of broadcast promotion managers. Its Gold Medallion Awards program annually recognizes the outstanding promotion and marketing campaigns in radio, TV, and cable. With offices in Los Angeles, BPME maintains a library and resource center for marketing executives. It offers occasional satellite teleconferences, a monthly magazine (*BPME Image*), an important annual meeting, and other services to its members in the United States and over 20 other countries.

The Cable Television Administration and Marketing Society (CTAM) has grown significantly in recent years, reflecting cable's increasing emphasis on marketing and advertising sales. Located in Washington, D.C., like BPME, CTAM maintains a resource center for cable marketers, including a file of success stories in such important areas as customer service, image building, subscriber retention, public relations, and program promotion. At its annual meeting, CTAM presents its awards for the best cable marketing each year.

SUMMARY

Sales and promotion are critical functions of telecommunications management. Broadcast stations and cable systems sell time, also known as spots, inventory, availabilities, and other unique terminology. Media time sales are complex negotiations involving advertisers, their agencies, station representative firms, and the sales staffs of individual radio and television stations.

Among the major electronic media firms, the bulk of advertising expenditures goes to television, followed by radio and cable, a relatively new ad medium. There are three distinct types of advertising sales: network, national/regional spot, and local time sales.

The promotion and marketing functions in telecommunications firms are becoming increasingly important, as reflected by growing staffs and increased compensation. Broadcast and cable marketing executives are involved in a range of activities, from promoting on-air talent, to designing contests, concerts, and other on- and off-air activities.

Marketing is also increasingly important to public broadcast facilities, which face severe shortfalls in government and foundation funding.

NOTES

1 Data on gross advertising sales for each medium (radio, TV, and cable) are drawn from *Standard and Poor's Industry Surveys,* January, 1993, pp. M1-M6.
2 See "Advertising Revenue's Great Expectations," *Broadcasting,* April 15, 1991, p. 82.
3 See "Media revenue up," *Broadcast Sales Training Executive Summary,* July 26, 1993, p. 1.
4 Data are extracted from *1993 Radio Employee Compensation and Fringe Benefits Report* and *1993 Television Employee Compensation and Fringe Benefits Report,* (Washington, D.C.: National Association of Broadcasters, 1993).
5 BIA Publications, Inc.,: Washington, D.C., 1991. Used with permission.
6 Stan Sossa and Joan Voukides, "The State of Our Art," *BPME Image,* July/August, 1991, pp. 16–19.
7 *Ibid.*

8 *Ibid.* See also, *NAB Employee Compensation Report(s), op. cit.*
9 *FCC Reports,* 86, 2nd, 1981, pp. 155.
10 "Memorandum Opinion and Order," *FCC Reports,* 90, 2nd, 1990, p. 99.

FOR ADDITIONAL READING

Bergendorff, Fred L., Smith, Charles H., Webster, Lance, *Broadcast Advertising & Promotion.* New York: Hastings House Publishers, Inc. 1983.

Eastman, Susan Tyler, Klein, Robert A., *Promotion & Marketing for Broadcasting & Cable.* Second Ed. Prospect Heights, IL: Waveland Press, Inc. 1991.

Haines, William John, *The Effect of FCC Rulings on Underwriting Policy in College Radio.* Unpublished Thesis. Athens, GA: 1986.

Heighton, Elizabeth J. and Cunningham, Don R. *Advertising in the Broadcast and Cable Media.* Second Edition. Belmont, CA: Wadsworth Publishing Company. 1984.

Kaatz, Ronald B. *Cable Advertiser's Handbook* Second Ed. Lincolnwood, IL: Crain Books. 1985.

Kossen, Stan. *Creative Selling Today.* San Francisco, CA: Canfield Press. 1977.

O'Donnell, Lewis B., Hausman, Carl and Benoit, Philip. *Radio Station Operations—Management and Employee Perspective.* Belmont, CA: Wadsworth Publishing Co. 1989.

1990 Radio Employee Compensation & Fringe Benefits Report, Radio Financial Report and Television Report. (Washington, DC: National Association of Broadcasters. 1993).

Schulberg, Bob. *Radio Advertising: The Authoritative Handbook.* Lincolnwood, IL: NTC Publishing Group. 1989.

Warner, Charles and Buchman, Joseph. *Broadcast and Cable Selling.* Second revised edition. Belmont, CA: Wadsworth Publishing Company. 1992.

Zeigler, Sherilyn K. and Howard, Herbert H. *Broadcast Advertising.* Second Ed. Columbus, OH: Grid Publishing, Inc. 1984.

14

AUDIENCE ANALYSIS

Telecommunications management is becoming increasingly reliant on research. Chapter 8 stressed how market research is crucial to the establishment of a media facility. Chapter 9 documented the importance of research in financial analysis and planning. The concern of this chapter is research into the media audience: its size and composition, viewing and listening habits, leisure and lifestyle interests, and other factors.

An understanding of the uses and misuses of audience research is critical to this generation of media managers. In broadcasting, ongoing and comprehensive audience analysis is demanded by advertisers—for gauging the impact of their commercial messages—and by programmers—for gauging the popularity and appeal of their shows. Such research drives the industry: high viewing and listening levels translate into greater advertising revenues for stations and networks.

Managers of cable networks and local systems need similar information to maintain profitable operations. Their dependence on subscriber fees for revenue requires system operators to undertake research into viewer motivations and preferences: what programs and services attract and retain subscriber households?

Research on the emerging new information and transmission systems must cover two areas: consumer acceptance (will the product be used, by whom, and how?) and market impact (how will the new service affect existing program and delivery systems?).

This chapter describes the major types of research used by media managers in sales and programming.

Specifically, the goals of this chapter are:

1. To discuss the two major forms of audience research in telecommunications: demographics and psychographics
2. To describe the major sources of audience research, including national research services, consulting firms, and in-house research activities
3. To discuss the role of research in sales and promotion at broadcast stations and cable facilities
4. To describe the uses of research in programming, including music and format testing in radio, and program and performer testing in television
5. To discuss new developments in and approaches to audience research
6. To discuss issues facing management in the use of research, including accuracy, costs and benefits, and legal and ethical concerns

FORMS OF MEDIA RESEARCH

Demographics

Media managers and advertisers require information on the number and characteristics of viewers, listeners, or subscribers. Such quantitative information is known as demographics. Demographic information is also known as *ratings,* or *head-count,* data. In the radio business, demographic data are usually presented for individual listeners; television and cable typically use demographics based on households.

The most common demographic categories, or *breaks,* in use in media research are based upon age and sex. Children 2 to 11 are called tots; 12- to 17-year-olds are teens; and adults are divided into two groups—18- to 34-year-olds and 35- to 54-year olds. In media, senior citizenship begins at 55. Adults can be further subdivided, for example, into the 18–24 and 25–49 categories, among others.

Thus, radio program directors might speak of "skewing" to a female-teen audience, or a television news director may seek to increase his share of 35- to 54-year-old males (M 35–54). The operator of a cable music channel may pinpoint her service to 18- to 34-year-old males (M 18–34), while another might target a wider group, men and women 18 to 49 years old (M-W 18–49).

Other demographic categories involve ethnic background and socioeconomic status (SES). Research on the ethnic background of audiences provides useful information to the media manager. For example, it is known that black households watch more television than white households.[1] Similarly, Spanish-surnamed listeners report extremely high radio listening levels, and represent an attractive, loyal audience to potential advertisers.[2]

There are a range of indicators in use for the socioeconomic status of audience members. Level of education is one useful bit of demographic information. Managers of noncommercial broadcasting facilities recognize that many of their audience members have a college degree and this is reflected in their programming of opera, ballet, and dramatic performances.[3] The cable business has known for years that subscribers to pay, multipay, and interactive services are likely to be better educated than the basic-only household, and this knowledge has affected their program offerings, which include foreign films, stock and news reports, and other services.[4] And the fact that education is highly correlated with income has not been lost on advertisers, who try to place advertisements for big-ticket items such as luxury automobiles, home entertainment products, and computers in programs attracting a more educated audience, such as CBS's *60 Minutes.*

Other indices of socioeconomic status include job category (blue or white collar) and home ownership. Such information is readily available from U.S. Census data, as well as from private research sources.

In aggregate, demographic data provide a picture of the size and makeup of the media audience. Telecommunications demographics are the equivalent of newspaper and magazine circulation figures. However, demographics do not explain *why* people listen to one station over another or drop one premium cable service in favor of another. Nor do they link media use to the broad range of personal activities, including sport and hobby pursuits and religious and political beliefs. That is the concern of psychographic research.

Psychographics

Psychographics is the broad term applied to research designed to provide qualitative information about media audiences. Psychographic research seeks to identify and describe audiences in terms of psychological dimensions, such as leisure and lifestyle activities, personal interests, opinions, needs, values, and personality traits.[5] Psychographic techniques may also be used to analyze the strengths or weaknesses of individual stations and their competitors by measuring audience perceptions of them.

One primary application of psychographic research involves what are known as *segmentation* studies. These identify audiences in terms of specific subgroups linked by various demographic and lifestyle characteristics. Identifying such subgroups helps in the development and testing of programming and advertising. For example, a study of the subscribers to a large cable system might break down the audience into four distinct groups. Segment 1 might be comprised of "innovators"—upper-income, well-educated households sharing an interest in foreign and classic films. This group might be the target for an advertising campaign for a specialized premium service. A second subscriber segment might be comprised of "weekend athletes." This group, residing mainly in middle-class areas, may share an interest in televised sports, yardwork, jogging, and recreational pursuits. They might be the target for advertising for sports and health channels. The last two audience segments might be "passive homebodies" and "swinging singles," and each would require its own programming and marketing plan.

Another application of qualitative research is *program testing*. As detailed below, radio program managers use call-out and focus-group research to test records, formats, and personalities. Television managers use similar methods to aid in their decision making about news and entertainment programs. Program testing may take place in preproduction stages, in an attempt to gauge audience reaction before making a significant budgetary commitment. In the absence of completed programs, researchers use photographs, animation, storyboards, and rough cuts. Program research may also take place during production—to assist in plot, theme, or character development—and after the completion of a project—to ascertain viewer responses.

Other applications of qualitative research include ascertainment and image studies. For 30 years, the FCC required broadcasters to formally *ascertain* the problems, needs, and interests of their communities. Broadcast managers were instructed to systematically poll community leaders and "average citizens," produce a list of community problems, and develop programs designed to respond to them. Although deregulation efforts have virtually eliminated such formal ascertainment requirements, many managers continue the ascertainment process in many informal ways. Such research provides invaluable information on the social and political climate in the community as well as on the perception of the station by community leaders and residents.

Image studies are specifically designed to discover how the public perceives a station or cable system. Image can often mean the difference between financial success or disaster. A television station can spend millions of dollars on news equipment and personnel, but if it is not perceived as a "news leader," the expense may be wasted. Similarly, in recent years, many cable systems have learned the lesson that despite a plethora of programming services, brusque customer service representatives or technicians may create an image of the company as aloof, greedy, and insensitive. Regularly conducted image studies keep the station or cable system apprised of the current social and political climate within a community and of the community's perception of the media operation within that climate. As such, they can be of inestimable value to management, sales, and programming personnel.

SOURCES OF AUDIENCE RESEARCH

There are four main sources of research available to the media manager. They are national research companies, industry groups, specialized research and consultation companies, and in-house research services.

National Research Companies

The leading firms providing ratings information to the broadcasting and cable industry are *A.C. Nielsen* and *The Arbitron Company*. Nielsen provides research services primarily in the area of television, including broadcast TV and cable. After providing ratings services for both radio and television for many years, Arbitron exited the TV ratings business in 1994. It remains the major company supplying ratings to the radio business.

A.C. Nielsen Company   A.C. Nielsen is the world's largest market research company, with offices in over 20 nations.[6] While its ratings of the electronic media provide only a small portion of its company earnings, *Nielsens* have become the most important set of television ratings. Nielsen services used by the television industry include the following.

National Nielsen Services   Nielsen offers three major measures of national viewing to television programs: the Nielsen Television Index (NTI), Nielsen Syndicated Services (NSS), and Nielsen Home Video Index (NHI).[7] The NTI provides data on ratings of programs produced by the major networks, particularly in prime time. The NTI is the source of most of the ratings for TV shows listed in the major newspapers. Figure 14-1 is a page from a Nielsen NTI report.

The NSS provides information about the program ratings for syndicated shows, such as those discussed in Chapter 12 (including *Wheel of Fortune* and *Oprah*).

FIGURE 14-1 A page from a Nielsen Television Index (NTI) report. Used with permission.

Nielsen NATIONAL TV AUDIENCE ESTIMATES — EVE. WED. NOV. 18,

TIME	7:00	7:15	7:30	7:45	8:00	8:15	8:30	8:45	9:00	9:15	9:30	9:45	10:00	10:15	10:30	10:45
HUT	58.1	58.9	59.2	60.9	64.0	66.3	67.1	68.6	68.6	69.4	68.7	68.1	64.5	63.3	61.8	59.1

ABC TV — HOME IMPROVEMENT / DOOGIE HOWSER, M.D. / JACKSONS' AMERICAN DREAM (PAE)

					8:00	8:15	8:30	8:45	9:00	9:15	9:30	9:45	10:00	10:15	10:30	10:45
HHLD AUDIENCE% & (000)					17.4 16,200		13.8 12,850		23.9 22,250							
TA%, AVG. AUD. 1/2 HR %					20.5		16.1		32.7	22.1*		24.2*		24.9*		24.2*	
SHARE AUDIENCE %					27		20		36	32*		35*		39*		40*	
AVG. AUD. BY 1/4 HR %					16.7	18.1	13.8	13.8	21.4	22.8	23.9	24.6	24.9	25.0	25.3	23.2

CBS TV — IN THE HEAT OF THE NIGHT / 48 HOURS — PORN IN THE U.S.A.

	8:00	8:15	8:30	8:45	9:00	9:15	9:30	9:45	10:00	10:15	10:30	10:45
HHLD AUDIENCE% & (000)	13.1 12,200								11.6 10,800			
TA%, AVG. AUD. 1/2 HR %	20.6	12.0*	12.7*		13.6*		14.1*	17.0	11.7*		11.4*	
SHARE AUDIENCE %	19	18*	19*		20*		20*	19	18*		19*	
AVG. AUD. BY 1/4 HR %	11.8	12.3	12.6	12.9	13.5	13.7	14.1	14.0	11.8	11.7	11.3	11.4

NBC TV — UNSOLVED MYSTERIES / SEINFELD / MAD ABOUT YOU / LAW AND ORDER

	8:00	8:15	8:30	8:45	9:00	9:15	9:30	9:45	10:00	10:15	10:30	10:45
HHLD AUDIENCE% & (000)	12.2 11,360				13.0 12,100		10.3 9,590		10.6 9,870			
TA%, AVG. AUD. 1/2 HR %	17.7	10.9*		13.5*	15.0		11.6		13.8	10.5*		10.6*
SHARE AUDIENCE %	18	17*		20*	19		15		17	16*		17*
AVG. AUD. BY 1/4 HR %	10.7	11.1	13.0	14.0	12.9	13.1	9.9		10.3	10.6	10.7	10.6

FOX TV — BEVERLY HILLS, 90210 / MELROSE PLACE

	8:00	8:15	8:30	8:45	9:00	9:15	9:30	9:45
HHLD AUDIENCE% & (000)	11.0 10,240				6.6 6,140			
TA%, AVG. AUD. 1/2 HR %	14.1	10.2*		11.8*	9.0	6.7*		6.4*
SHARE AUDIENCE %	17	16*		17*	10	10*		9*
AVG. AUD. BY 1/4 HR %	9.7	10.8	11.7	12.0	6.9	6.6	6.7	6.2

INDEPENDENTS (INCLUDING SUPERSTATIONS EXCEPT TBS)

	7:00	7:15	7:30	7:45	8:00	8:15	8:30	8:45	9:00	9:15	9:30	9:45	10:00	10:15	10:30	10:45
AVERAGE AUDIENCE	16.2 (+F)		15.8 (+F)		5.8		6.0		6.0		6.1		9.0 (+F)		7.4 (+F)	
SHARE AUDIENCE %	28		26		9		9		9		9		14		12	

PBS

	7:00	7:30	8:00	8:30	9:00	9:30	10:00	10:30
AVERAGE AUDIENCE	2.1	2.3	3.3	4.3	3.7	3.9	2.5	2.3
SHARE AUDIENCE %	4	4	5	6	5	6	4	4

CABLE ORIG. (INCLUDING TBS)

	7:00	7:30	8:00	8:30	9:00	9:30	10:00	10:30
AVERAGE AUDIENCE	10.8 (+F)	12.1 (+F)	11.2	11.7	11.9	11.7	10.9 (+F)	8.7 (+F)
SHARE AUDIENCE %	18	20	17	17	17	17	17	14

PAY SERVICES

	7:00	7:30	8:00	8:30	9:00	9:30	10:00	10:30
AVERAGE AUDIENCE	1.5	1.5	1.5	1.9	2.0	1.5	2.5	3.2
SHARE AUDIENCE %	3	3	2	3	3	2	4	5

U.S. TV Households: 93,100,000

For explanation of symbols, See page

A-7   For SPANISH LANGUAGE TELEVISION audience estimates, see the Nielsen Hispanic Television Index (NHTI) TV Audience Report

The Nielsen Peoplemeter. Note how the remote control has buttons corresponding to each member of the household, as well as to visitors. (Courtesy of Nielsen Media Research)

The NHI provides ratings information about program popularity among the major cable television networks (including USA Network, MTV, CNN, and ESPN). The NHI also provides data about household use of VCRs to play prerecorded tapes or to record programs off the air.

The national Nielsens are produced through the use of the Nielsen Peoplemeter, a device resembling a cable box with a remote control. Over 4000 homes in the United States comprise a sample which, theoretically at least, represents TV viewing in America's more than 94 million TV homes. The people meter is attached to each set in the participating household. Viewers push buttons assigned to them to track their viewing activity. The data are transmitted to the Nielsen Company and processed overnight so that ratings information can be used by programmers and advertisers by the beginning of business the following day.

Local Nielsen Services As we have learned, there are over 200 individual television markets in the United States. In each of those markets, at least four times a year, Nielsen produces reports on local television viewing. These four visits each last about a month, and are known as "sweeps." Sweeps periods are critical for local stations to evaluate viewing to their local news and syndicated programming, as well as to track the performance of the network programs they may be carrying.

Sweeps make use of the diary method. Samples of TV households in each market (usually between a few hundred and a few thousand) are sent diaries in which they record their viewing throughout the day. Completed diaries are mailed back to Nielsen and a summary report—Nielsen's "Viewers in Profile" (VIP) is prepared within 3 weeks of the completion of the sweeps period. Figure 14-2 is a sample page from a VIP report.

Local Metered Markets By 1995, Nielsen expects to have each of the top 50 television markets equipped with meters to enable overnight ratings for use by local stations, cable systems and advertisers. This service, known as Nielsen's Metered Market Service, does not use the comparatively sophisticated people meter. Instead, a small box attached to each sample household (known as a storage instantaneous audimeter) records whether the TV set is on or off, and if on, to which station it is tuned. This information is stored in a microprocessor, then fed overnight by phone line to Nielsen's computers. By the following morning, subscribers to the metered service receive information on a Nielsen computer terminal about a program's overall rating and share (see below). Demographic data are not provided, so metered markets also rely on the diaries distributed during the sweeps period for this information.

The Arbitron Company The Arbitron Company, a subsidiary of Control Data Corporation, dominates the field of radio ratings. Like Nielsen, Arbitron provides a range of services for its clients, who include radio sales and programming executives.

Arbitron Local Market Reports Like Nielsen in the television market, Arbitron provides radio managers with periodic reports on radio listening in their communities. These reports, issued as many as six times a year in the major radio markets (but only twice in some smaller communities), are known as local market reports. Figure 14-3 is a sample page from an Arbitron radio local market report.

Arbitron local market reports use the diary method to assess radio listening. In some counties in a market, as few as a dozen diaries may represent listening during a given period. In larger areas, a few thousand diaries might be placed and returned to reflect radio listening in a large metropolitan county.

Other National Services While Nielsen and Arbitron are the clear leaders in national ratings, two other services deserve mention. Ratings of the national radio networks are provided by *Statistical Research Inc.* in its RADAR report (Radio's All-Dimension Audience Research). It is the RADAR report which is watched closely by producers of *Rush Limbaugh, Larry King, Paul Harvey News and Comment,* and other network radio fare.

In 1992, a competitor to Arbitron in the radio field was introduced. Strategic Radio Research began issuing reports in the major radio markets, known as AccuRatings.[9] Rather than rely on diaries, AccuRatings uses a sample of radio listeners contacted by telephone to produce its reports. At the time this book went to press, the future of smaller AccuRatings against the giant Arbitron Company was unclear. However, a previous competitor, Birch Radio (which also relied on telephone polling), had gone out of business in 1991.

FIGURE 14-2 A page from a Nielsen VIP Report. Used with permission.

ATLANTA, GA

WK1 7/08-7/14 WK2 7/15-7/21 WK3 7/22-7/28 WK4 7/29-8/04

PROGRAM AVERAGES

METRO HH RTG	METRO HH SHR	STATION DAY PROGRAM	DMA HOUSEHOLD RATINGS WEEKS 1	2	3	4	MULTI-WEEK AVG RTG	MULTI-WEEK AVG SHR	HUT	DMA RATINGS PERSONS 2+	18+	12-24	12-34	18-34	18-49	21-49	25-54	35+	35-64	50+	WOMEN 18+	12-24	18-34	18-49	21-49	25-49	25-54	50+	WKG	MEN 18+	18-34	18-49	21-49	25-49	25-54	TNS 12-17	CHILD 2-11	6-11
1	2		3	4	5	6	7	8	14	15	16	17	18	19	20	21	22	23	24	25	26	27	28	29	30	31	32	33	34	35	36	37	38	39	40	41	42	43
2		*R.S.E. THRESHOLDS 25+%*	*3*	*3*	*3*	*3*	*2*			*1*	*1*	*4*	*3*	*2*	*2*	*2*	*2*	*2*	*2*	*2*	*2*	*7*	*4*	*2*	*2*	*3*	*2*	*5*	*3*	*2*	*5*	*3*	*3*	*3*	*3*	*10*	*6*	*9*
1		*(1 S.E.) 4 WK AVG 50+%*	*1*	*1*	*1*	*1*	*LT*			*LT*	*LT*	*1*	*1*	*1*	*LT*	*1*	*LT*	*LT*	*1*	*1*	*LT*	*2*	*1*	*1*	*1*	*1*	*1*	*1*	*1*	*LT*	*1*	*1*	*1*	*1*	*1*	*3*	*2*	*2*
		11:00PM																																				
10	20	WAGA MON EYEWITNSS NWS	12	11	8	10	10	19	53	6	7	4	4	5	5	5	6	8	7	11	7	4	6	6	5	6	6	11	6	6	4	4	4	4	5	2	1	2
11	19	TUE EYEWITNSS NWS		11	10		10	18	57	6	7	6	4	5	5	5	5	8	7	11	9	12	7	7	5	5	6	13	7	5	2	3	4	4	4	3	2	3
14	25	WED EYEWITNSS NWS	13	11	11	15	13	24	53	7	8	4	5	6	7	7	8	10	9	10	10	5	8	10	10	11	11	12	8	5	4	5	5	5	5	3	4	6
12	23	THU EYEWITNSS NWS	10	11	11	14	11	22	52	7	8	4	6	6	7	6	7	9	7	12	9	5	7	8	7	8	8	12	7	7	5	5	5	6	7	4	2	4
11	21	FRI EYEWITNSS NWS	9	9	11	10	10	20	50	7	7	8	7	5	5	5	6	8	7	12	8	3	4	6	6	7	8	12	7	6	6	5	5	4	4	11	1	2
12	*22*	*AV5 EYEWITNSS NWS*	*11*	*11*	*10*	*12*	*11*	*21*	*53*	*8*	*7*	*5*	*5*	*6*	*6*	*6*	*7*	*9*	*8*	*11*	*9*	*5*	*6*	*7*	*7*	*8*	*8*	*12*	*7*	*6*	*5*	*5*	*5*	*5*	*5*	*5*	*2*	*3*
12	24	SAT EYEWITNSS NWS	14	11	11	11	12	24	49	7	8	4	4	4	6	6	8	11	11	15	10	4	4	7	7	8	9	15	8	7	4	5	5	5	6	4	2	2
16	30	SUN EYEWITNSS NWS	16	15	14	14	15	28	52	8	10	5	7	8	7	7	8	12	11	18	12	5	10	9	10	10	11	19	10	8	5	5	4	4	6	4	2	2
12	*23*	*AV7 EYEWITNSS NWS*	*12*	*11*	*11*	*12*	*12*	*22*	*52*	*7*	*8*	*5*	*5*	*6*	*6*	*6*	*7*	*10*	*8*	*13*	*9*	*5*	*7*	*8*	*7*	*8*	*9*	*13*	*8*	*7*	*5*	*5*	*5*	*5*	*5*	*5*	*2*	*3*
3	6	WATL MON MASH	3		4	3	3	6	54	2	2	4	3	3	2	2	2	1	1	1	2	4	3	2	2	2	2	1	1	2	3	2	2	2	2	4		
3	6	TUE MASH	4	3	4	3	3	6	56	2	2	2	3	3	3	3	3	2	2	1	3	3	4	3	4	3	3	1	4	2	2	2	2	2	2	2		
4	7	WED MASH	4	3	4	4	4	7	53	2	3	6	3	3	3	2	2	2	3	2	2	5	3	3	3	2	2	1	3	3	2	2	2	2	3	5		
5	9	THU MASH	5	4	5	3	4	8	52	2	2	3	2	2	3	3	3	2	3	1	2	3	2	3	3	3	3	1	3	2	3	2	2	2	2	1		
4	8	FRI MASH	5	5	3	3	4	8	50	2	3	6	5	6	3	2	2	1	1	1	3	6	7	4	4	3	3		3	2	4	3	1	1	1	2		
4	*7*	*AV5 MASH*	*4*	*4*	*4*	*3*	*4*	*7*	*53*	*2*	*2*	*4*	*3*	*3*	*3*	*3*	*2*	*2*	*2*	*1*	*3*	*4*	*4*	*3*	*3*	*3*	*3*	*1*	*3*	*2*	*3*	*2*	*2*	*2*	*2*	*3*		
3	6	SAT COMIC-LIVE-FOX	2	3	2	3	2	5	45	1	1	4	3	3	2	2	1				1	3	3	2	2	1	1			1	3	2	2			1	1	2
6	16	SUN SUN MOV	6	6	6	5	6	15	37	2	3	4	4	5	3	3	2	1	1	1	3	7	7	4	4	2	2		3	2	3	2	2	2	2			
6	*16*	*NOR SUN MOV*	*6*	*6*	*6*	*5*	*6*	*15*	*38*	*2*	*2*	*3*	*4*	*5*	*3*	*3*	*2*	*1*	*1*	*2*	*3*	*7*	*6*	*4*	*4*	*2*	*2*		*3*	*2*	*3*	*2*	*2*	*2*	*2*			
3	5	WGNX MON STAR DS9-AS R	2	3	2	2	2	4	48	1	1	2	1	2	2	2	2	1	1		1	1		1	1	1	1		1	2	3	3	3	3	2	1		
3	6	TUE STAR GEN-AS R	4	2	3	2	3	5	51	1	1	2	2	2	2	2	1	1	2		1	1	1	2	1	2	1		1	2	2	2	2	2	1	2		
4	8	WED STAR TK-GENRTN	4	2	3	3	3	7	48	2	2	2	2	2	2	2	2	2	2	1	1		1	2	2	2	2		2	3	3	3	3	3	2	1	1	1
4	8	THU STAR TK-GENRTN	4	3	3	3	3	7	47	2	2	2	2	3	3	3	3	2	2	1	2	1	2	2	2	3	2		1	3	4	3	3	3	3	1		
3	6	FRI STAR TK-GENRTN	3	3	3	2	3	6	46	1	2	2	2	2	2	2	2	1	2		1	2	1	1	1	1	1		1	2	3	3	3	3	2	1	1	1
4	*7*	*AV3 STAR TK-GENRTN*	*4*	*3*	*3*	*3*	*3*	*7*	*47*	*2*	*2*	*2*	*2*	*2*	*2*	*2*	*2*	*2*	*2*	*1*	*1*	*1*	*1*	*2*	*2*	*2*	*2*		*1*	*3*	*3*	*3*	*3*	*3*	*2*	*1*		*1*
3	6	SAT APOLLO CMDY HR	4	2	2	3	3	6	45	1	1	1	1	1	1	1	1	1		1	1		1	1	1	1	1	1	1	1	1	1	1	1	1	1		
3	5	SUN TWILIGHT ZONE	3	4	1	1	2	5	52	1	1				1	1	2	1	2	1	1		1	1	1	2	2	1		1					1		1	2
13	25	WSB MON CH2 NWS NTBEAT	10	13	11	16	13	24	53	7	8	5	6	6	7	7	8	10	10	13	10	5	8	9	9	10	10	14	9	6	4	5	5	5	5	4	4	3
11	19	TUE CH2 NWS NTBEAT	14	8	10		11	19	57	6	7	4	4	4	6	6	7	8	9	9	7	5	6	7	7	8	8	9	7	6	3	5	5	5	5	4	2	1
10	19	WED CH2 NWS NTBEAT	12	11	8	8	10	19	53	6	7	4	4	5	6	6	6	8	8	9	8	4	6	8	8	8	8	10	8	5	4	4	4	4	4	2	1	1
13	24	THU CH2 NWS NTBEAT	11	14	12	12	12	24	52	8	9	7	7	5	7	7	8	11	10	13	10	5	6	8	8	9	10	14	9	7	5	6	6	7	7	11	3	4
14	27	FRI CH2 NWS NTBEAT	12	14	13	11	13	25	50	7	9	3	5	5	7	8	9	11	11	13	10	5	8	9	10	10	11	13	8	7	3	5	6	7	7	3	2	2
12	*23*	*AV5 CH2 NWS NTBEAT*	*12*	*12*	*11*	*12*	*12*	*22*	*53*	*7*	*8*	*5*	*5*	*5*	*7*	*7*	*8*	*10*	*10*	*11*	*9*	*5*	*7*	*8*	*9*	*9*	*9*	*12*	*8*	*6*	*4*	*5*	*5*	*5*	*6*	*5*	*2*	*2*
12	24	SAT CH2 NW NTBT SA	11	9	12	12	11	23	49	6	8	2	4	4	7	7	8	10	10	11	9	3	6	8	9	10	10	11	9	6	2	5	5	6	7	2	2	3
12	22	SUN CH2 NW NTBT SU	9	12	13	10	11	21	52	7	8	8	7	6	7	8	8	9	9	9	9	4	7	8	8	9	9	10	8	7	6	6	7	6	6	10	3	3
2	5	WTBS SUN NETWORK EARTH		2	1	3	2	4	52	1	1			1	1	1	1	1	1	1			1	1	1	1	1			1		1	1	1	1			
<<		WTLK MON OPPRTNTY UPDTE	<<	<<			<<		51																													
<<		MON SPCL ED PYCHK			<<	<<	<<		55																													
<<		TUE SPCL ED PYCHK	<<	1	<<	<<	<<		56																													
<<		WED DIXIE SPEEDWAY	<<	<<	<<	<<	<<		48																													
<<		THU OPPRTNTY UPDTE	<<		<<		<<		52																													
<<		THU SPCL ED PYCHK				<<	<<		51											1								1										
<<		THU UNFAIR ADVNTGE		<<			<<		48																													
<<		FRI HOOKED-PHONICS	<<				<<		50																													
<<		FRI SPCL ED PYCHK		<<	<<	<<	<<		49																													
<<		*AV2 OPPRTNTY UPDTE*	*<<*	*<<*	*<<*		*<<*		*52*																													
<<		*AV4 SPCL ED PYCHK*	*<<*	*<<*	*<<*	*<<*	*<<*		*53*																													
1	2	SAT WCW PRO WRSTLN	2	1	<<	1	1	2	45																													
<<		SUN OPPRTNTY UPDTE				<<	<<		48																													
<<		*AV3 OPPRTNTY UPDTE*	*<<*	*<<*	*<<*	*<<*	*<<*		*51*																													
3	5	WVEU MON STUDS	2	2	3	2	2	4	53	1	1	2	3	2	1	1	1				1	3	2	1	1	1	1		1	1	2	1	1	1	1	3	2	3
3	5	TUE STUDS	3	2	2	2	2	4	56	2	2	5	4	4	3	2	2	1	1		1	6	3	2	1	1	1		1	2	5	3	2	2	2	4	1	2
2	4	WED STUDS	2	2	1	3	2	4	53	1	2	2	3	3	2	2	2				2	4	4	3	3	3	2		1	1	3	1	2	1	1	2		
3	6	THU STUDS	2	3	3	2	3	5	52	2	1	2	3	3	2	1	1				1	4	3	2	1	2	2			1	2	1	1	1	1	4	5	8
3	5	FRI STUDS	2	3	2	2	2	4	50	1	1	4	2	2	1						1	5	2	1					1	1	2	1	1	1	1	2		
3	*5*	*AV5 STUDS*	*2*	*2*	*2*	*2*	*2*	*4*	*53*	*1*	*1*	*3*	*3*	*3*	*2*	*1*	*1*				*1*	*5*	*3*	*2*	*1*	*1*	*1*		*1*	*1*	*3*	*2*	*1*	*1*	*1*	*3*	*1*	*3*
2	4	SAT STUDS	2	2	2	1	2	4	49	1	1	4	2	1	1	1	1	1	1		1	6	1	1	1	1	1			1		1	1	1	1	6		
<<		SUN JACK VAN IMPE	<<	<<	1	<<	<<		52																			1										
3	*5*	*AV6 STUDS*	*2*	*2*	*2*	*2*	*2*	*4*	*52*	*1*	*1*	*3*	*3*	*3*	*2*	*1*	*1*				*1*	*5*	*3*	*2*	*1*	*1*	*1*		*1*	*1*	*3*	*2*	*1*	*1*	*1*	*4*	*1*	*2*
8	16	WXIA MON 11-ALIVE NW-LT	11	8	8	9	9	17	53	5	6	2	4	4	6	6	7	7	7	6	6	3	4	6	6	7	6	7	5	6	4	6	6	7	7	3	3	3
9	15	TUE 11-ALIVE NW-LT	8	7	5	13	8	14	56	4	5	3	3	4	5	5	5	7	7	7	6	2	4	5	5	6	6	8	5	5	3	5	5	4	4	2	1	1
8	14	WED 11-ALIVE NW-LT	10	8	7	8	8	15	53	4	5	1	3	3	3	3	4	6	4	10	5	1	3	3	3	3	4	10	3	5	4	3	3	4	3		1	
7	13	THU 11-ALIVE NW-LT	8	8	6	7	7	14	52	4	5	2	3	3	4	4	5	5	5	5	5	3	4	5	4	5	5	5	4	4	3	4	4	5	5	2	1	2
7	13	FRI 11-ALIVE NW-LT	5	6	7	9	7	14	50	4	5	4	4	4	4	4	4	5	4	6	5	4	5	5	4	4	4	6	4	4	3	4	4	4	4	3	1	1
8	*14*	*AV5 11-ALIVE NW-LT*	*8*	*7*	*7*	*9*	*8*	*15*	*53*	*4*	*5*	*2*	*3*	*4*	*4*	*5*	*5*	*6*	*6*	*7*	*5*	*3*	*4*	*5*	*5*	*5*	*5*	*7*	*4*	*5*	*3*	*4*	*5*	*5*	*5*	*2*	*1*	*1*
7	13	SAT 11-ALIVE NW-LT	7	6	7	7	7	14	49	4	5	4	3	3	4	4	4	6	6	7	5	4	4	5	5	4	4	7	4	5	3	4	4	3	4	2	1	1
7	14	SUN 11-ALIVE NW-LT	6	11	7	8	8	15	52	4	5	2	2	2	4	4	4	7	7	10	5	2	2	4	4	4	4	9	6	5	2	4	4	4	4	1	1	1
7	*14*	*AV7 11-ALIVE NW-LT*	*8*	*8*	*7*	*9*	*8*	*15*	*52*	*4*	*5*	*2*	*3*	*3*	*4*	*4*	*5*	*6*	*6*	*7*	*5*	*3*	*4*	*4*	*4*	*5*	*5*	*7*	*4*	*5*	*3*	*4*	*4*	*4*	*4*	*2*	*1*	*1*
		11:30PM																																				
9	20	WAGA MON LOVE CONNECTN	8	8	7	9	8	19	43	4	5	4	4	3	4	4	5	5	5	6	5	4	4	5	5	5	6	7	3	4	3	3	3	3	4	5	1	2
12	26	TUE EYEWITNSS NW L				13	13	28	44	6	6	9	6	3	5	5	6	8	6	10	9	15	6	7	7	7	9	15	6	3		2	3	3	3	16		
7	14	TUE EYEWT NW SP RP		7			7	14	47	4	4	6	5	5	4	4	3	3	3	3	5	12	8	6	5	4	4	4	6	1	1	2	2	2	2	5	4	
9	19	TUE LOVE CONNECTN			8		8	17	45	4	5	8	4	4	5	3	3	5	4	6	6	12	6	6	4	4	5	6	3	4	3	3	3	2	2	3	2	
9	21	WED LOVE CONNECTN	7	9	5	10	8	19	42	4	5	4	4	5	5	5	5	5	5	5	6	4	6	6	6	7	7	7	6	3	4	3	3	3	3	3	2	3
11	25	THU LOVE CONNECTN	11	8	11	11	10	24	42	6	6	7	7	6	6	5	6	6	6	8	8	7	7	7	7	7	8	9	7	5	6	4	4	4	4	8	2	3
12	27	FRI LOVE CONNECTN	10	11	12	9	11	25	42	5	6	6	5	4	4	4	5	7	6	10	8	7	5	6	6	6	7	12	7	4	3	3	2	2	3	8	2	4
10	*23*	*AV5 LOVE CONNECTN*	*9*	*9*	*9*	*10*	*9*	*21*	*42*	*5*	*5*	*5*	*5*	*5*	*5*	*4*	*5*	*6*	*5*	*7*	*7*	*6*	*6*	*6*	*6*	*6*	*7*	*9*	*5*	*4*	*4*	*3*	*3*	*3*	*3*	*6*	*2*	*3*
10	24	SAT IN HEAT OF NTE	11	9	8	9	9	24	39	5	6	3	3	4	5	5	5	6	6	8	7	2	5	6	6	6	7	9	7	4	4	3	3	3	3		4	6
10	22	SUN SPRTS UNLMTD	9	10	8	9	9	21	42	5	6	4	4	5	4	4	5	7	6	11	7	4	5	5	5	5	6	10	7	5	4	3	2	2	4	4	1	2
1	2		3	4	5	6	7	8	14	15	16	17	18	19	20	21	22	23	24	25	26	27	28	29	30	31	32	33	34	35	36	37	38	39	40	41	42	43

See Program Index for complete details of program start time, duration and weeks of telecast.

JULY 1993 66

FIGURE 14-3 A page from an Arbitron Local Market Report. Used with permission.

Metro Audience Trends*
PERSONS 12+

	MONDAY-SUNDAY 6AM-MID					MONDAY-FRIDAY 6AM-10AM				
	WINTER 92	SPRING 92	SUMMER 92	FALL 92	WINTER 93	WINTER 92	SPRING 92	SUMMER 92	FALL 92	WINTER 93
KAAY										
SHARE	.6	.4	.6	**	.4	.6	.3	.3	**	.8
AQH(00)	4	3	4	**	3	6	3	3	**	8
CUME RTG	3.0	2.5	1.4	**	2.1	1.2	1.3	.7	**	1.0
KARN										
SHARE	7.2	9.1	7.9	8.7	10.8	9.6	12.4	9.3	9.4	13.6
AQH(00)	48	62	54	58	74	101	123	87	93	138
CUME RTG	16.9	17.6	15.8	15.8	19.6	9.3	11.7	10.4	9.8	12.2
KBIS										
SHARE	.7	1.2	1.3	.6	1.2	.7	1.3	2.1	.6	2.0
AQH(00)	5	8	9	4	8	7	13	20	6	20
CUME RTG	3.5	4.4	4.9	4.2	3.9	1.6	1.7	2.5	1.7	1.8
KDDK										
SHARE	**	**	5.0	9.8	8.2	**	**	5.3	10.5	8.4
AQH(00)	**	**	34	65	56	**	**	49	104	85
CUME RTG	**	**	12.8	22.5	21.6	**	**	7.7	14.7	14.1
KEZQ										
SHARE	6.3	6.6	4.7	3.0	3.5	6.5	7.3	5.4	2.9	3.7
AQH(00)	42	45	32	20	24	68	73	50	29	38
CUME RTG	14.3	14.4	10.2	6.3	8.3	7.6	7.9	5.9	3.3	4.8
KGKO										
SHARE	.7	1.0	1.3	1.4	.3	.3	.6	.8	.6	
AQH(00)	5	7	9	9	2	3	6	7	6	
CUME RTG	1.5	2.5	2.7	3.3	1.0	.7	1.4	1.6	1.9	.2
+KGKO-FM										
KAKI-FM										
SHARE	.9	.3	.1	.2	1.2	.6	.7	.2	.3	1.0
AQH(00)	6	2	1	1	8	6	7	2	3	10
CUME RTG	1.8	2.3	1.5	2.4	2.7	.9	1.0	.6	.9	1.0
KHLT										
SHARE	2.8	2.0	2.9	3.0	4.5	2.7	1.9	3.2	2.3	4.1
AQH(00)	19	14	20	20	31	28	19	30	23	42
CUME RTG	9.5	9.3	10.3	8.8	13.2	4.9	3.7	5.0	4.4	7.4
KIPR										
SHARE	9.4	10.1	9.3	10.2	8.9	9.8	11.2	8.9	9.8	8.3
AQH(00)	63	69	64	68	61	103	111	83	97	84
CUME RTG	15.4	15.6	15.7	16.2	17.4	10.2	10.7	9.9	10.9	11.6
KITA										
SHARE	1.8	1.0	1.2	2.1	1.2	1.8	1.3	1.0	3.2	.2
AQH(00)	12	7	8	14	8	19	13	9	32	2
CUME RTG	4.3	3.7	4.3	5.5	4.0	1.2	1.4	1.1	2.5	1.0
KKYK										
SHARE	3.9	4.2	3.6	3.8	5.1	2.8	4.4	3.3	3.8	3.3
AQH(00)	26	29	25	25	35	29	44	31	38	34
CUME RTG	13.2	14.8	12.4	13.9	14.4	5.9	7.6	7.2	7.6	6.8
KLRG										
SHARE	1.9	2.5	2.2	1.7	2.2	2.0	2.3	1.9	1.7	1.8
AQH(00)	13	17	15	11	15	21	23	18	17	18
CUME RTG	4.3	5.2	3.4	3.9	7.5	2.5	3.1	2.2	2.5	3.2
KMJX										
SHARE	10.0	11.3	10.2	8.7	9.8	10.7	10.8	9.4	10.3	10.0
AQH(00)	67	77	70	58	67	113	107	88	102	101
CUME RTG	17.7	18.7	18.5	16.4	15.8	12.0	11.3	12.7	11.4	9.4
KMTL										
SHARE	.6	.9	1.3	**	.4	.4	1.1	1.4	**	.5
AQH(00)	4	6	9	**	3	4	11	13	**	5
CUME RTG	1.7	1.7	1.2	**	2.5	1.0	1.2	.7	**	1.2
KMZX										
SHARE	3.3	2.5	1.3	3.0	2.2	2.3	1.5	.5	1.3	1.2
AQH(00)	22	17	9	20	15	24	15	5	13	12
CUME RTG	6.8	6.6	4.5	5.8	5.1	3.6	2.2	1.4	2.9	2.1
KOLL										
SHARE	4.6	3.8	5.8	3.2	3.5	4.5	2.9	5.5	3.6	3.5
AQH(00)	31	26	40	21	24	47	29	51	36	36
CUME RTG	11.4	13.0	13.9	10.5	13.3	6.5	6.6	7.6	5.1	6.7
KSSN										
SHARE	25.7	20.9	22.5	19.1	18.4	26.8	19.6	23.7	20.1	19.7
AQH(00)	172	143	154	127	126	282	195	221	200	200
CUME RTG	37.7	36.2	39.9	41.2	36.5	26.4	24.2	25.3	26.6	22.5
KURB										
SHARE									.1	.2
AQH(00)									1	2
CUME RTG	.8	.1	.2	.8	.4	.2		.2	.3	.3
KURB-FM										
SHARE	6.7	6.0	6.9	7.2	7.1	7.2	6.7	8.5	8.7	8.6
AQH(00)	45	41	47	48	49	76	67	79	86	87
CUME RTG	20.2	18.9	18.6	17.5	19.4	11.8	11.2	10.5	11.3	11.4
KYFX										
SHARE	**	**	**	.5	1.5	**	**	**	.6	1.0
AQH(00)	**	**	**	3	10	**	**	**	6	10
CUME RTG	**	**	**	3.3	3.5	**	**	**	1.5	1.9
KYTN										
SHARE	**	.7	**	1.2	.6	**	.8	**	1.1	.8
AQH(00)	**	5	**	8	4	**	8	**	11	8
CUME RTG	**	1.8	**	3.2	2.7	**	1.3	**	1.8	1.8

Footnote Symbols: ** Station(s) not reported this survey. + Station(s) reported with different call letters in prior surveys - see Page 5B.

ARBITRON

LITTLE ROCK   6   WINTER 1993

* See page iv Restrictions On Use Of Report for restrictions on the use of Trends data.

Industry Research Services

Audience research is available to media managers from many industry groups, including information offices, advertising agencies and station representatives, networks, group owners, and MSOs.

Promotion and Marketing Firms Audience information is maintained by a variety of industry organizations involved in advertising, promotion, and public relations. Studies of radio audiences are maintained by the *Radio Advertising Bureau* (RAB). Its equivalents in television and cable—the *Television Advertising Bureau* (TVB) and *Cable Advertising Bureau* (CAB)—compile demographic, psychographic, and consumer purchase information for use by members in promotion and sales development.

Audience research in broadcasting is a routine function performed by the leading advertising agencies. In addition to making their own analyses and interpretations of national ratings like Nielsen and Arbitron, most major advertising agencies carry out their own audience and industry research. Common areas investigated by agency researchers include revenue and ratings projections for broadcasting and cable, the impact and effects of new communications technologies, changing audience habits and interests, attitudes and opinions of media management and sales executives, and new forms of advertising in media (products, appeals, commercial lengths, approaches, etc.). Ad agencies particularly regarded for their audience research include J. Walter Thompson, BBDO, and McCann-Erickson.

Station representation firms are equally committed to audience analysis. Like their counterparts at the ad agencies, rep research analysts track trends in media management and sales, projections for the new media (such as VCRs and home computers), and consumer tastes and preferences. In television, rep firms have their own program departments which develop and maintain analyses of program performance and audience characteristics.[10]

Networks, Group Owners, and MSOs A consequence of the growing trend toward group ownership of telecommunications properties is the rise of the research department at network, group owner, and MSO headquarters.

Each of the major networks has a research staff concerned with both demographics and psychographics. As might be expected, network research departments are interested in testing new programs and concepts for their fall schedules and the performances of network shows in the two hundred or so individual markets in which they are aired. However, in addition to such standard industry research, network research departments sponsor and conduct inquiries on the social impact of the medium (for example, the effects of violence on viewers and the use of television by children). Like agencies and reps, network researchers investigate trends in audience attitudes and opinions and produce projections on the spread and impact of new communications technologies.

Research departments are also found at the major cable networks. For example, the research department at Turner Broadcasting Company monitors the performance of WTBS and Cable News Network nationwide and at all the affiliated cable systems. Research staffs are also found at the pay cable networks, including HBO and Showtime.

Most group broadcasters and cable MSOs maintain research departments. Group and MSO researchers are primarily involved in providing services to their individual stations and systems. Among broadcast groups, this may include interpretation and analysis of ratings reports; research on local programming, including music and news on radio; television news, and magazine-style shows; image studies; and analyses of promotional campaigns. Research personnel at cable MSOs look into subscriber motivation and satisfaction, program preferences, and customer service. They trace reasons for nonsubscription and disconnection, and, like group researchers, they may be involved in developing advertising campaigns.

Independent Research and Consultation Companies

The number of research companies offering audience, program, and image studies to media managers has increased rapidly in recent years. In 1980 about 50 research companies were listed in *Broadcasting Yearbook*. Today, more than 100 research companies are listed in *Broadcasting and Cable Marketplace*. Some research companies with a substantial client list among radio and television stations include Audience Research and Development (ARD), Bruskin Associates, FMR Associates, MediaMark, and Paragon Research.

In-House Research

For many years, media managers scrupulously avoided conducting their own research. Reasons for their reluctance included the aura and mystery surrounding the gathering and production of national ratings, a generalized fear of numbers and computations (sometimes referred to as "math anxiety"), and the lack of personnel with sufficient training or opportunity to conduct local research. The situation has changed markedly in recent years. Many local stations and cable systems have staff members with training and experience in marketing, advertising, or audience research gained in college or on the job. In addition, the presence of the microcomputer in the station or head end streamlines the processes of data collection, analysis, and report generation.

For these reasons, in-house research is a rapidly growing area of audience analysis. The National Association of Broadcasters has been particularly supportive of this trend. In cooperation with ratings experts from the academic world and the industry, the NAB has issued a series of reports designed to assist broadcast managers in their own research efforts. To date, manuals on a survey research, call-outs, and focus groups have been issued (see the suggested readings at the end of the chapter). In addition, the NAB has spearheaded the development of computer programs to be used by station personnel in sales and program analyses of ratings data. Such programs, available on low-cost floppy disks or published in newsletters, enable managers to conduct their own custom-tailored ratings research that is used to supplement the data provided by the national services.

RESEARCH AND SALES DEVELOPMENT

A popular misconception about the use of audience research is that ratings are of primary interest to programmers. The fact is, however, that, while program directors, producers, and writers follow the ratings assiduously, the main motive for conducting audience research is *advertising sales.* The primary purpose of audience research is to measure and describe telecommunications audiences so that advertisers can plan commercial campaigns that match products with consumers in an efficient and cost-effective manner.

There are two general classes of ratings of particular interest to sales personnel at stations, networks, agencies, and reps: (1) measures of audience size and (2) methods of media planning.

Measures of Audience Size

Rating and Share The basic measures of audience in broadcasting are rating and share. *Rating* refers to the percentage of people or households in an area tuned to a specific station, program, or network. For example, if, in a Nielsen sample of 1000 homes, 250 households were tuned to the ABC network, the rating for ABC during that time period would be (250 ÷ 1000), or 25%. For ease of reporting, the percentage sign is dropped in the ratings book. *Share* refers to the number of people or households tuned to a particular station program or network as correlated with the number of sets in use. Continuing the above example, if only 750 of the sampled households were actually watching television in the time period covered, ABC's share would be (250 ÷ 750) = 33%. Since there are always more sets in a market than there are sets in use, the share figure is always higher than the rating.

The ratings books produced by Nielsen and Arbitron report rating and share in a variety of ways. Ratings and shares are provided for different *dayparts* (such as morning and evening drive times in radio, and prime, access, and fringe time in television). In addition, ratings are provided for three different market areas.

The total area in which viewing or listening takes place in a market is known as the *total survey area,* or TSA. Nielsen uses *designated market area* (DMA) to indicate the primary area of television use. Unlike TSAs, DMAs do not overlap; each county in the United States belongs to only *one* DMA. The *metro survey area* (MSA, or *metro*) includes the area nearest to the station's transmitter, and generally corresponds to the United States Office of Management and Budget's *standard metropolitan statistical area* (SMSA).

Since TSA figures represent the largest number of households or persons, many broadcasters use these in their promotion and sales efforts. However, advertising agencies rely predominantly on DMA and metro data in making their buying decisions, since these areas represent the location of most retail sales in the marketplace.

Cume and Average Quarter Hour Advertisers are interested in the total number of people or households exposed to their commercial messages; this is known in the industry as *reach*. In broadcast research, the most common estimates of reach are cume and average quarter hour (AQH) figures.

Cume, short for cumulative audience, refers to the number of different people in a given demographic group tuned to a particular station for at least 5 minutes during a specified time period. In larger markets, cumes are usually reported in thousands in the ratings book, so that advertisers and programmers add three zeros to the figures in the cume columns. For example, a radio station with an M 25–49 cume of 150 in morning drive time is delivering 150,000 different adult males during its morning programming.

Average quarter hour figures indicate the average number of persons or households in a given demographic group tuned to a specific channel for at least 5 minutes during a 15-minute time segment. Extending the above example, if the radio station has an AQH estimate of 75 in the M 25–49, M-F 6a–10a category, it is delivering, on average, 75,000 men aged 25 to 49 in any quarter-hour during the time period from 6 to 10 a.m. Monday to Friday. There are a number of other calculations made to describe the size and characteristics of audiences on the basis of ratings data. These include, among others, time spent listening (TSL), audience turnover, exclusive cume, and percent recycling. Further information on these figures is available from sources listed at the end of the chapter.

Cost per Thousand and Gross Ratings Points

The primary estimates used by advertisers to budget and measure the effectiveness of their campaigns are cost per thousand (CPM) and gross ratings points (GRP).

Cost per Thousand The CPM calculation is used to compare advertising outlets and to budget advertising campaigns. In the electronic media, the CPM is calculated by dividing the cost of an advertisement by the size of the audience (in thousands). For example, if the cost of a 30-second advertisement in a television news show is $150 and the show reaches 30,000 people, the CPM is (150 ÷ 30) = $5.

The reliance of advertisers on CPM in making their time-buying decisions is the primary reason for the intense competition between local stations traced in Chapters 4 and 5. The most popular programs attract the largest audiences and provide the lowest CPM figures for advertisers, in spite of extremely high spot costs. Programs and stations with lower ratings face an unfortunate bind. Poor ratings mean high CPMs, and spot costs must be reduced in order to lure advertisers. The end result is the high turnover of air personalities and frequent changes in format in radio. In television, poor ratings can lead to musical chairs played by news anchors, as they shift from market to market, as well as the cancellation of network and local programs.

Gross Ratings Points Most broadcast advertisers buy commercial *packages* rather than individual spots. GRPs are a measure of the impact of an advertising campaign over the length of its run. The GRP is calculated by summing the ratings for each program or time period in which an advertisement is run. For example, assume that an advertiser seeking men 18 to 34 has purchased a single spot in the programs listed in Table 14-1. In this example, the schedule achieved about 95 gross ratings points. While it might seem that 95 GRPs translates into a campaign reaching 95% of the target audience, keep in mind that there is a high degree of audience duplication from program to program.

Advertising campaigns are planned by media buyers to reach a given number of GRPs, such as 160 or 180. Usually, the station that can deliver those GRPs at the lowest CPM will get the contract.

Other ratings of interest to salespeople and media buyers include *cost per ratings point* (CPP) and *gross impressions* (GI). Fuller discussions of these and other figures of particular importance to sales can be found in the sources listed at the end of the chapter.

RESEARCH AND PROGRAMMING

While ratings data are of primary importance to sales personnel, programmers also track ratings, shares, and cumes and use them to aid their decision making. In radio, a decline in shares in drive times inevitably leads to changes in air personalities or music. In television, the costs of syndicated programs are directly tied to ratings performance on the network or in other markets. Thus most programmers have more than a passing knowledge of the above statistics.

Remember, however, that ratings are "post hoc." That is, they are based only on past performance (last month, last quarter, last year, last network run, and so forth). The program manager faces the unenviable task of predicting the future with some accuracy. The network program executive must choose among dozens of program proposals and decide which one is likely to be a hit. The radio

TABLE 14–1   GROSS RATINGS POINTS

Program	Rating (M 18–34)
WAAA *Nightly News*	25.3
Entertainment Buzz	22.0
Inside Scoop	29.0
Sportsview	18.8
	GRP: 95.1

programmer must select the songs for his playlist. Which should he choose and how long should he play them? The pay-cable operator must buy or produce feature films to satiate her audience's appetite for uncut movies. Will a box-office failure translate into cable success? Will audiences pay to watch first-run, made-for-cable films?

Questions like these have led to the use of extensive qualitative research by program managers in telecommunications. The next segment traces major trends in such research in radio, television, and cable operations.

Radio Programming Research

Radio programmers have three major areas of concern with regard to the personalities they employ and the music they play:

- Familiarity (recognition)
- Popularity (like or dislike)
- Fatigue, or burnout

To gather data in these areas, three forms of qualitative audience analysis are used to supplement Arbitron data: focus groups, call-outs, and auditorium tests.

Focus Groups Derived from the field of market and product testing, focus-group research involves conducting round-table discussions with selected audience subgroups in order to obtain information about programs, personalities, commercials, promotions, etc. Focus groups generally include 10 to 12 individuals selected on the basis of an important characteristic or characteristics (most frequently, these are target demographic segments such as "country listeners, 18–34" or "adult nonlisteners to WAAA"). They are led by a discussion leader (called the *moderator* or *facilitator*). The sessions last from 1 to 3 hours and usually include sample programs played to encourage audience response. Although group members generally fill out a short questionnaire about their radio-use habits, the most important element of a focus-group session is discussion. Discussions are recorded on cassette audio or videotape for evaluation by program personnel. In radio research, focus groups are used in the following areas:

- Evaluation of music and format preferences
- Evaluation of nonmusical elements (news, public affairs, etc.)
- Evaluation of talent (announcers, disc jockeys)
- Evaluation of station image
- Promotional research

Focus-group research has become very popular in the radio business. Stations pay anywhere from $3000 to $5000 for each group session, depending on the type of group recruited, the facilities desired, and other factors. Advantages of the method include the relative speed with which audience information can be obtained, the fact that groups can be customized to meet the needs of individual radio stations, and their comparatively low cost.

However, managers contemplating focus research should keep some limitations of the method in mind. For one thing, targeted small groups are *not* necessarily representative of the public at large. Poor selection methods can lead to a range of group biases and threaten the usefulness of the sessions. In addition, a focus group hinges on the abilities of the moderator. A badly trained moderator may generate little information of use or may, at worst, generate misinformation.

Call-Out Research Call-out research is a widely used method of measuring audience reaction to music. A call-out is a telephone survey targeted to samples of station listeners (or nonlisteners). Call-outs conducted in-house by radio stations usually use phone lists of known station listeners, including contest respondents or letter writers. Those conducted by independent research companies may utilize random selection procedures.

To test a variety of different songs, as well as to avoid needlessly long and tedious telephone calls, researchers play only a 5- to 15-second segment of each song to be tested. This short, highly identifiable portion of a song is known as a *hook*. To assess the audience's reaction to each hook, a simple measurement scale is used. For example, after hearing a hook, the listener might be asked to rate it on a scale of 1 to 5, with 1 meaning "hated the song" and 5 "liked it a lot." About 25 to 50 songs can be rated in this manner before respondents get bored or tired.

Call-out research has the advantages of low cost and quick turnaround. Ongoing or occasional call-outs secure important feedback about the popularity of records, the lifeblood of music stations.

However, like all research approaches, the call-out method has its limitations. Phone lists constructed from such sources as names of contest winners are not necessarily representative of the audience at large. In addition, using station personnel to conduct the calls introduces the possibility of bias. Finally, hooks work best with known music selections. The call-out method does not gather much useful information about new, or "breaking," musical selections.

Auditorium Tests Auditorium testing is a cross between the focus-group and call-out methods. Here, hooks are played to an auditorium filled with 50 to 200 target listeners. Auditorium tests enable programmers to test more selections (as many as 200 to 300) than is possible in call-out research. Since there is face-to-face contact between the researcher and the audience, it is possible to gauge listeners' reactions through their nonverbal communication, including facial gestures, mannerisms, body movements, and so forth. In addition, if a selection elicits an extreme reaction in either direction, it is possible for the moderator to ask questions and stimulate discussion.

Auditorium tests have their limitations as well. For one, conducting such research usually exceeds the resources of the station. Expensive outside research consultants must be used. The costs for a single auditorium test can range from $5000 to $50,000. And, like focus groups and call-outs, auditorium studies are only as useful as the samples selected and the abilities of the moderators and researchers will permit.

Television Programming Research

Like their counterparts in the sound medium, television program managers seek qualitative data on their audiences. Common areas of television program testing include:

- Performer evaluation
- News evaluation
- Program and concept testing

Performer Evaluation The success of much television fare, from news to soap opera, is tied directly to the popularity of its personalities. There are two indicators of performer popularity of major interest to television producers: *familiarity* and *likability*. In theory, all other factors being equal, audiences will respond to programs more favorably if the programs feature personalities they both know and like. "Ideal" television personalities fitting this mold include Pat Sajak, Oprah Winfrey, and Peter Jennings. Of course, there are exceptions. Personalities with high familiarity and low likability can be successes (for example, famous sportscaster Howard Cosell and the aforementioned Howard Stern). And, with proper grooming and promotion, a highly likable personality with low recognition can blossom into a media superstar (for example, John Tesh).

Market Evaluations, Inc., has achieved considerable reknown for its performer evaluation studies. Using a nationwide sample, it produces a widely known assessment of performers known as *TV-Q*. TV-Q provides estimates of familiarity and likability of hundreds of individual performers, across more than seventeen categories of entertainment from general variety to comedy and news.[11]

TV-Q figures are used by the national networks and pay television services to aid in their program casting and purchasing decisions. Local broadcasters and cablecasters can gauge performer popularity by utilizing the same techniques as radio research: focus groups, call-outs, and auditorium tests. This most frequently occurs in the important battleground of television news.

In fact, in one recent survey, Peter Jennings of ABC led all network newscasters in TV-Q, far ahead of competitors Dan Rather and Tom Brokaw.[12] Rather's low likability may have contributed to CBS's decision to team him with Connie Chung.

News Evaluation The importance of news as a revenue source to television broadcasters, especially local affiliates, has led to a flurry of research activity. Firms such as Frank N. Magid Associates, Audience Research and Development, and Reymer and Associates have become well known for their news consulting research. Using the entire marketing research arsenal from marketing surveys to focus groups, news consultants provide in-depth analyses of television newscasts. They make suggestions about newscaster delivery and style, set design, and even the makeup and hairstyles of anchors, reporters, meteorologists, and sportscasters.

Critics of news evaluation argue that it has compromised the ethics and trivialized the practice of journalism. They contend that news directors who use such techniques are more concerned with the "look" of their shows than their editorial content. Reporters and anchors have been known to cringe when the consultant arrived, feeling, perhaps, like a rookie football player afraid of being cut from the team.

Despite their unpopularity in some quarters, consultants are an important part of the television news environment. News is a hotly competitive, capital-intensive enterprise, and consultants provide valuable information for managers to use in protecting their investment.

Program and Concept Testing   The failure rate of network television programs is normally above 90%. Production costs and labor demands are enormous. And competition from new sources, notably independents, cable television, and VCRs, has made the television audience increasingly selective and fickle. For these and other reasons, conducting audience research has become a standard part of the television programming process.

Program research takes place in each major phase of television production. Before production, when a program is in the idea stage, program managers engage in *concept* testing. In lieu of a program excerpt, target audiences assembled in interview, focus-group, or auditorium settings are asked to comment on a short program or program description. Sometimes photographs or storyboards are used. Although such materials are sketchy at best and can never communicate the essence of a program, audience reaction to them can provide some insight into potentially popular plot lines, settings, or characters.

Producers and other program managers frequently expose partially completed programs to potential audiences. Such films or television programs are usually in a semifinal stage known as a *rough cut.* Audience reactions to rough cuts enable program executives to fine-tune problem areas in action, dialogue, or characterization; to select among alternative endings; and to prepare final touches such as music, sound effects, and canned laughter.

The final version of a film or television program is often screened before preview audiences. These screenings are particularly useful for scheduling and promotion purposes. Audience responses can aid the selection of scenes and segments to be used in advertising campaigns and can help isolate the specific factors that may contribute to a program's success.

NEW DIRECTIONS IN AUDIENCE RESEARCH

The dual phenomena of audience fragmentation and the multiplicity of channels available to the media consumer have created a need for more innovative and sophisticated approaches to audience research. Major additions and modifications are being made in the way audience research data are being collected, analyzed, and reported. The goal of these new approaches is to merge demographic and psychographic research and produce as comprehensive a picture of media audiences as possible.

Consumer Psychographics and Geodemographics

One consequence of the computer age has been the voluminous amount of data available about households and individuals. Using sophisticated analytical procedures and powerful computers, it is possible to link media use to a full range of demographic and psychographic variables. A number of market research companies now provide research services of this type.

One example is Stanford Research Institute's VALS II approach (VALS = Values and Life Styles).[13] The VALS model taps the opinions, attitudes, and leisure pursuits of people, which can be matched with their media habits. In the VALS II typology, people fall into three basic groups: principle-oriented, status-oriented, and action-oriented. Under this approach, it is possible to develop a program or advertising strategy designed to reach each kind of person in the audience.

Other psychographic and geodemographic approaches which have received attention from media buyers and programmers in recent years include PRIZM, introduced by Claritas Corporation, and ClusterPlus, developed by Simmons and Donnelly Information Services.

Psychobiological Research

A research method that has achieved a good deal of notoriety is the use of measures of physiological functions as a means of obtaining data on responses to program and commercial material. The assumption underlying such research is that exposure to media stimuli produces measurable and interpretable responses among audiences. If this concept seems farfetched, recall the last time an exciting chase scene caused your heart to race, or a passionate screen kiss elevated your blood pressure.

Advertisers, market researchers, and academic investigators have focused on a number of physiological measures of audience response, including heart and brain activity (EKG and EEG), galvanic skin response (GSR), or electrodermal conductance, and dilation of pupils, or pupillometry. When properly administered, tests producing psychobiological data have been useful in determining when and how audiences demonstrate attention to and recall of programming and commercial material.[14]

"Passive" Peoplemeters

One of the concerns frequently expressed by programmers is that the peoplemeter currently employed by Nielsen requires considerable attention and activity on the part of the viewer. The meter "takes attendance"—that is, viewers must regularly punch in their identifying number in order to be counted in the demographic analysis. Visitors to the home must do the same.

To rectify these problems, in development are so-called "passive" peoplemeters which will be able to track specific viewers without their conscious awareness (and active participation).[15] One system uses a program of photographic recognition to scan the area in front of the TV set at regular intervals to "see" who is watching. There have been some reliability problems with this approach (as a large dog might be confused with a household member, potentially inflating ratings for reruns of *Lassie*). Continuing improvements in computer visual and voice recognition may alleviate these concerns in coming years.

"Portable" Peoplemeters

An enduring criticism of the diary method of measuring audience activity, especially radio listening, is the fact that it tends to underreport away-from-home estimates. Typically, we cannot also write in a diary as we listen to a radio in the car, while exercising, or while jogging. One answer to this problem is the proposed "pocket peoplemeter," announced by Arbitron in 1993.[16] Under the system envisioned by Arbitron, listeners would carry a small device which would pick up an inaudible code transmitted by each radio station in the area. When that station is tuned in (in the car, for example), the meter would automatically note the time and channel.

As this edition went to press, there were many concerns raised about the new system. For one, it was unclear whether the device would be sensitive enough to decode listening on small headset receivers. Also, there were concerns that some radio stations might jam or drown out the audio signal of their competitors in order to rig their ratings (not unlike the days when the former Soviet Union jammed the incoming signals of Radio Free Europe).

ISSUES FOR MANAGEMENT CONCERNING AUDIENCE ANALYSIS

There is no doubt that audience research is of central concern to media managers. However, in gathering and using research, telecommunications managers should keep in mind some limitations of research processes.

Accuracy

The most common concern raised about audience research in telecommunications concerns its accuracy. In statistical terms, any test is defined by two characteristics: its validity and its reliability. A valid test actually measures what it is supposed to measure; a reliable test achieves the same or similar results each time it is administered.

The issue of validity and reliability is of central importance to media management using audience research techniques. Managers must keep in mind that *no* method of reporting radio and television use is flawless; all of them involve techniques which alter the "normal" environment in which viewing or listening takes place.

As the ratings services are quick to point out, ratings are only *estimates* based on *samples,* and are always subject to error. Unfortunately, ratings are often treated as gospel by sales and program managers.

There are two basic ways in which telecommunications managers can help to achieve accuracy and reliability in research procedures: (1) education and (2) quality control.

First, managers must become familiar with typical research procedures. Such knowledge can be obtained in a number of ways. Marketing and audience research procedures are taught at most universities, often in night classes under continuing education programs. In addition, many industry groups offer short courses and training seminars in the analysis and interpretation of research data. The motivated manager can also attain research knowledge through self-study. Research manuals are available from the ratings services themselves, as well as from industry and academic sources. The bibliography at the end of this chapter is a good starting point for self-study in audience analysis.

Next, managers must be vitally concerned with the issue of quality control. Since advertising, programming, and promotion policies are developed on the basis of audience measurements, bad research is not just useless, it can cause fatal damage to a media company. Thus research firms must be carefully monitored to assure reliability and validity.

An industry group committed to the goal of accurate research is the *Electronic Media Ratings Council* (EMRC). Made up of research specialists from stations, cable operations, and advertising agencies, the EMRC regularly scrutinizes the procedures of the ratings companies. If the company's gathering and reporting methods meet EMRC criteria, it gives its seal of approval, known as *EMRC Certification*.

Of course, it is impractical for the EMRC to evaluate the practices of the dozens of smaller research companies which service the telecommunications industry. It is the task of the media manager to gauge the reputability of research done by such companies. Before contracting with a research firm, managers should carefully examine its client list and track record. Research firms which refuse to divulge the names of its other clients, do not provide examples of its efforts, or fail to provide references should be avoided.

Management's involvement in research should not end with the signing of the contract. Continuous close monitoring not only keeps the research firm on its toes, it may also provide useful information to management. In this regard, many clients of Arbitron routinely travel to its Laurel, Maryland, headquarters to inspect the "raw" diaries used in its radio surveys. They can check whether station listening has been correctly attributed and can examine review material not reported in the book (for example, open-ended comments on programming, personalities, and special events).

Cost-Effectiveness

Like most management services, research carries a high price tag. In the major markets, the cost of individual market reports alone can exceed $100,000. Extra research services, such as focus-group studies and call-outs, add additional expenses. Managers must carefully weigh the cost of research against its presumed benefits.

There is growing evidence that media managers are beginning to question whether the insights provided by research justify the high costs.

Cost-cutting at television stations and advertising agencies led to their dropping Arbitron's TV service in favor of Nielsen's, leading directly to Arbitron's decision to exit the TV ratings business.[17] Arbitron benefited in radio, however, by the decision of many leading radio managers to drop the Birch service, which ultimately sealed its demise.

Some managers (for example, many radio executives in small markets) have elected not to subscribe to any professional research service. Instead, their programming and sales efforts are determined by instinct, intuition, and experience. Many of the facilities guided by such managers are highly successful, in terms of both economics and community service.

The point is that research is not, and should not be, perceived as a substitute for managerial initiative and expertise. While sophisticated analytical tools describe audiences and their behaviors in new, more detailed ways, they only aid, not replace, decision making. Managers should keep in mind that among the most rigorously researched and market-tested products in history was the ill-fated Edsel automobile.

Ethics

The processes and practices of audience research raise ethical concerns for telecommunications management. Such issues include "hyping," concealment and deception, voluntary participation and informed consent, and invasion of privacy.

Hyping and Ratings Distortion Ratings are supposed to reflect radio and television listening patterns in a "pure" state. To gain an accurate assessment of audience behavior in a community, stations should program normal schedules and not call attention to the fact that the ratings process is under way. In reality, however, networks and stations make significant changes in their programming and promotional strategies during ratings periods, a process known as *hyping*. Hyping may include scheduling unusual or blockbuster programs (like television miniseries or radio music specials). Common hypes during ratings time are major promotions, such as the million-dollar giveaway.

While a degree of hyping has become standard industry practice, a related and more serious ethical problem is *ratings distortion*. Ratings distortion refers to station programming and promotion activity designed specifically to increase diary reporting or meter use. Encouraging people to write down their viewing or listening as a means of entering a contest is one example of such distortion. Because this practice may cause sample households to record more viewing and listening than they actually engage in, it can result in biased, or distorted, results.

Guidelines against hyping and ratings distortion have been adopted by both the Federal Trade Commission and the FCC. In addition, if such activities are observed by the ratings companies during a survey period, they will be noted in detail in the front of the market report. Serious violations can void the results, lead to a new survey, and invite a lawsuit from a ratings company.

One recent attempt to improve accuracy actually involved hyping. Facing declining response rates from diary keepers, Arbitron engaged broadcasters in Atlanta to air an unprecedented series of on-air announcements urging listeners with diaries to fill out and return them.[18] While there was evidence that the announcements increased the response rate, the EMRC temporarily suspended its accreditation of the ratings book on the grounds that the announcements amounted to a "live" test of a new sampling methodology.

Concealment and Deception On occasion, researchers conceal information from participants in audience research. For example, a radio station doing call-outs seeking to find out which is the market's favorite music station should not announce itself as the source of the calls. Similarly, a focus group designed to rate a single female news anchor should include other female news anchors to disguise its true purpose.

While concealment and deception are often necessary if accurate, unbiased information is to be obtained, managers should follow a simple, but important, guideline when such techniques are used. They should guarantee that respondents are fully *debriefed* about the intent and uses of the research. During the debriefing process, which typically comes immediately after the telephone survey, interview, or focus group, respondents should be invited to ask any question they might have about the research. On the rare occasion in which a respondent wants a copy of the results of the research, one should be provided. Taking the time to fully debrief research participants helps maintain their faith in the sponsor of the research and can avoid an unfortunate circumstance, such as a lawsuit.

Voluntary Participation and Informed Consent Individuals involved in audience research should be free to terminate their participation at any time. In addition, their participation should be *voluntary,* not coerced. For example, requiring people to answer a battery of questions in order to claim a prize is unethical. Similarly, asking for "5 minutes" of respondents' time when the interviewer knows the survey will last 25 minutes violates basic ethical standards.

Voluntary participation and termination at will are, of course, particularly important in laboratory-type research testing physiological responses. With such procedures, there is the possibility that audience members will experience physical, even emotional, discomfort. However, even such seemingly innocuous methods as telephone interviews or mail surveys must meet basic ethical criteria. Telephone interviewers should avoid hard-sell tactics. If a respondent hangs up, he or she should not be repeatedly called back. Mail surveys should not be designed to resemble census forms, tax returns, social security questionnaires, or other official government documents.

Related to the proviso of voluntary participation is that of *informed consent*. Simply stated, that means that people should base their decision to participate in a research study on sufficient prior knowledge of the study's procedures. For example, they should be told that a focus group session might last 3 hours, that they will be watching three different newscasts, and that they may be called in for a follow-up interview at a later date. The fact that their responses are being observed and recorded on tape should also be noted.

Invasion of Privacy   On occasion, radio stations have been known to dispatch employees to parking lots and repair shops to check the dial positions of car radios (and, if necessary, to reset them!). A few years ago, a market research firm announced that it had developed new mobile scanning equipment making it possible to station a van outside a person's home and monitor the television and radio use inside the household. Today, addressable, two-way cable technology makes it feasible for cable managers to obtain a detailed log of all household viewing without the subscriber's knowledge.

The new passive meters in development suggest that "big brother" really will be watching us, as ratings companies will know where, when and what we are watching and listening to, possibly without our own knowledge that we are being monitored.

While the need for detailed audience information and simple curiosity might lead the media manager to utilize such techniques, all are in clear violation of the principles of research ethics. Participation in such research is not voluntary, nor has informed consent been obtained. And, perhaps most importantly, such research fails to respect the basic right to privacy that is essential to a democratic society.

Managers should exercise extreme care in sponsoring and participating in research which crosses the border between objective inquiry and covert surveillance. At stake is the credibility of their company specifically and the telecommunications profession in general.

There is concern that ours is becoming an overresearched society and that the glut of telephone calls, surveys, and interviews conducted by marketing research companies is becoming tiresome to the consumer. Such oversaturation carries with it great risk, for tired, annoyed respondents are liable to provide false or misleading data, sometimes merely to irritate the researcher. Media managers must keep in mind that a primary reason people watch television and listen to radio is to be alone—to avoid or escape outside pressures, including face-to-face contacts and telephone calls! Thus, despite the thirst for knowledge of audiences, it is prudent for managers occasionally to leave their audience alone—and it is ethical to respect individuals' privacy at all times.

SUMMARY

Telecommunications management is becoming increasingly reliant on audience research in programming and sales development.

Audience research takes two major forms. Research which describes audiences on the basis of size, age, income, education, and similar factors is known as demographics. Psychographic research focuses on the needs, interests, attitudes, and lifestyle pursuits of media audiences.

There are four major sources of audience research available to media management: national research companies, industry groups, consulting research firms, and in-house research activities. The Nielsen rating service is the leader in the television industry; Arbitron dominates the field of radio ratings.

The two research areas of major interest to media sales personnel are estimates of audience size and estimates of campaign cost. Key indicators of audience size are rating, share, cume, and AQH. Major media planning indicators are CPM and GRP measures.

In addition to national Nielsen and Arbitron reports, program managers use a variety of special types of audience analysis. Radio programmers conduct focus-group, call-out, and auditorium tests designed to assess audience perceptions of a station's personalities and music playlists. Television managers use similar methods in the evaluation of news, movies, and promotional strategies.

New developments in audience analysis include more sophisticated metering and geodemographic and physiological measurement. Both are potentially valuable aids to further understanding of the impact and effectiveness of media programming and advertising.

However, managers must carefully analyze the cost-effectiveness and accuracy of audience research, follow sound ethical practices, and recognize that research is an aid to, and not a substitute for, informed decision making.

NOTES

1 See, for example, J. Michaelson, "TV Adjusts Its Mirror," *Los Angeles Times,* September 22, 1991, p. 5; M. Freeman, "Producers Say Black Viewing Numbers Simply Don't Add Up," *Broadcasting and Cable,* September 13, 1993, pp. 26–27.

2 See, for example, "Hispanics: Last Frontier for Marketing," *Broadcasting and Cable,* September 6, 1993, p. 54.

3 See Ronald E. Frank and Marshall G. Greenberg, *Audiences for Public Television* (Beverly Hills, Calif.: Sage, 1982), pp. 79–99; Charles Clark, "The Issues: Public Broadcasting," *CQ Researcher,* September 18, 1992, pp. 811–818.

4 See Dean M. Krugman, "Evaluating the Audiences of the New Media," *Journal of Advertising,* vol. 14, no. 4, 1985, pp. 21–27; Carrie Heeter and Bradley S. Greenberg, *Cableviewing* (Norwood, N.J.: Ablex, 1988).

5 See William D. Wells "Psychographics: A Critical Review," *Journal of Marketing Research,* vol. 12, no. 2, May 1975, pp. 197; Rebecca Piirto, *Beyond Mind Games: The Marketing Power of Psychographics* (Ithaca, NY: American Demographic Press, 1991).

6 Scott Hume, "Nielsen vs IRI: Battle of the Research Titans," *Advertising Age,* October 12, 1992, pp. 1, 50.

7 "Nielsen Media Research: The Quality Behind the Numbers," *Nielsen Media Research News,* n.d., pp. 1–7. The author is indebted to Maria Zimman of Nielsen Media Research for contributions to this section.

8 See "RADAR 46—Network Radio, Fall 1992," *Broadcasting and Cable,* March 8, 1993, p. 33.
9 See "Accuratings More Stable than Arbitron," *Billboard,* August 28, 1993, p. 85; Peter Viles, "Accuratings, Arbitron Vary Widely by Format," *Broadcasting,* January 4, 1993, p. 55.
10 Richard A. Bompane, "Station Representatives' Role in Programming," in Susan Tyler Eastman, Sydney W. Head, and Lewis Klein (eds.), *Broadcast/Cable Programming,* 2d. ed. (Belmont, Calif.: Wadsworth, 1985), pp. 86–88.
11 Roger D. Wimmer and Joseph R. Dominick, *Mass Media Research: An Introduction* (Belmont, Calif.: Wadsworth, 1983), p. 290.
12 "Jennings Tops Q Scorers," *Broadcasting,* August 19, 1991, p. 35.
13 See Martha F. Riche "Psychographics for the 1990s," *American Demographics,* July 1989, pp. 24–31.
14 See James E. Fletcher, "Physiological Responses to the Media," in Joseph R. Dominick and James E. Fletcher (eds.), *Broadcasting Research Methods* (Newton, Mass.: Allyn & Bacon, 1985), pp. 89–106.
15 Erik Larsen, "Watching Americans Watch TV," *The Atlantic,* March 1992, pp. 66–80.
16 "People Meter Tunes to Young Listeners," *Billboard,* August 29, 1993, p. 85.
17 See "Stations Opting for Only One Ratings Service," *Broadcasting,* June 3, 1991, p. 39; Bill Carter, "Arbitron Closing down National Ratings Services," *The New York Times,* September 3, 1993, p. D6.
18 Peter Viles, "Thumbs Down on Atlanta Book," *Broadcasting,* July 5, 1993, p. 24.

FOR ADDITIONAL READING

Beville, Hugh Malcolm, Jr.: *Audience Ratings: Radio, Television, and Cable,* rev. ed. Hillsdale, N.J.: Lawrence Erlbaum Associates, 1988.
Buzzard, Karen: *Electronic Media Ratings: Turning Audiences into Dollars and Sense.* Boston: Focal Press, 1992.
Bower, Robert T.: *The Changing Television Audience in America.* New York: Columbia University Press, 1985.
Dominick, Joseph R., and James E. Fletcher: *Broadcasting Research Methods.* Newton, Mass.: Allyn & Bacon, 1985.
Fletcher, James E.: *Handbook of Radio and TV Broadcasting: Research Procedures in Audience, Programs and Revenues.* New York: Van Nostrand, 1981.
————: *Squeezing Profits Out of Ratings: A Manual for Radio Managers, Sales Managers and Programmers.* Washington, D.C.: National Association of Broadcasters, 1985.
Fletcher, James E., and Roger D. Wimmer: *Call-Out Research in Managing Radio Stations.* Washington, D.C.: National Association of Broadcasters, 1982.
————: *Focus Group Interviews in Radio Research.* (Washington, D.C.: National Association of Broadcasters, 1981).
Frank, Ronald E., and Marshall G. Greenberg: *Audiences for Public Television.* Beverly Hills, Calif.: Sage, 1982.
Hiber, Jhan: *Hibernetics: A Guide to Radio Ratings and Research.* Los Angeles, Calif.: R&R Books, 1984.
Piirto, Rebecca: *Beyond Mind Games: The Marketing Power of Psychographics.* Ithaca, N.Y.: American Demographic Press, 1991.

Reagan, Joey (ed.): *Applications of Research to Media Industries.* Dubuque, Iowa: Kendall/Hunt, 1992.
Webster, James G.: *Ratings Analysis: Theory and Practice.* (Hillsdale, N.J.: Lawrence Erlbaum Associates, 1991).
Wimmer, Roger D., and Joseph R. Dominick: *Mass Media Research: An Introduction,* 4th ed. Belmont, Calif.: Wadsworth, 1994.

15

THE FUTURE OF TELECOMMUNICATIONS MANAGEMENT

As we have seen, the telecommunications environment is rife with new technologies, new services, new investors, and new alliances. Government leaders speak of "the telecommunications infrastructure," or "the information superhighway." As the leading media companies get bigger, receiving devices get smaller—with hand-held "personal digital assistants" (PDAs) coming on line, which emulate the form and function of *Star-Trek*'s "communicator" device. The typewriter, computer, telephone, and TV set seem to be merging into a single in-home information/entertainment device. As cable and telephone companies begin to merge, new satellite firms literally "fly over" them, offering voice, video, and data services. All this is targeting a public which still seems to have difficulty removing the flashing "12:00" from VCRs. What does it all mean?

As has been pointed out, forecasting in a field as volatile as telecommunications is particularly difficult. (After all, an English lord once surmised that the telephone would be a commercial failure because there was an ample supply of cheap messenger boys in England.)[1] Despite the perils and pitfalls of prediction, the prospect of continued technological, political, economic, and social change places forecasting at the heart of telecommunications management functions.

This chapter speculates about how telecommunications is likely to "shake out" in the near future, with particular emphasis on management.

Specifically, the goal of this chapter is to examine issues in media management raised by continued economic, technological, and social change. Four main areas of impact are assessed:

1 Globalization and vertical integration of media companies
2 Technological convergence of media services and devices
3 The outlook for media corporations with respect to the global economy
4 Media consumption in an era of sociopolitical restructuring

Against this backdrop, the text concludes with some tentative predictions for successful media management in the 1990s and beyond.

The Global Village: Utopian Community or Pricey Theme Park?

Three decades ago, media theorist Marshall McLuhan speculated that modern telecommunications technology would create a "global village," a community of the world's population linked through telecommunications technology: most notably television and computers.[2] While some observers predicted a utopia—a world in which interactive, global communications would eliminate regional, ethnic, and linguistic barriers—McLuhan himself was more guarded:

> The global village absolutely insures maximum disagreement on all points. It never occurred to me that uniformity and tranquillity were the properties of the global village. It has more spite and envy...
>
> The village is not the place to find ideal peace and harmony. Exact opposite.[3]

Those of us who watch world events unfold live on CNN, or communicate to friends we've never met on an international computer bulletin board, know that the "new world order" seems particularly chaotic and disordered. But there is now no doubt that the global village has evolved from theorist's conception to reality. And, as an English professor and literary critic, McLuhan may have underestimated one aspect of the communications revolution: its reliance on international capitalism and global business conditions. One truism of media management for the future is that it will increasingly be an international and multinational activity.

GLOBALIZATION AND VERTICAL INTEGRATION

In Chapter 7, we examined patterns of telecommunications ownership and found that the media landscape is now dominated by a handful of huge multinational corporations, including such giants as Sony and Matsushita of Japan, Bertelsmann of Germany, Australia's News Corporation (parent of Fox), and Time-Warner and Disney of the United States. Once mainly a domestic activity, telecommunications management is now and will continue to be an international activity.

Tied to this phenomenon are the blurring distinctions between hardware and software firms and between the production, distribution, and exhibition phases pointed out in the first chapter. Today, many media firms which based their fortunes on hardware (the aforementioned Sony and Matsushita, for example, and the regional Bell operating companies) are hurrying to gain a toehold on the programming (software) side of the business. At the same time, software giants (such as television networks, movie studios, and computer program designers) are seeking alliances with manufacturers to develop new "platforms" on which to showcase their wares (CD-ROM, hand-held games, arcades, and so on).

Regulations in place for over a century designed to control the size and influence of American corporations are being streamlined or eliminated altogether as the United States struggles to retain its leadership in the fields of

CHAPTER 15: THE FUTURE OF TELECOMMUNICATIONS MANAGEMENT   **407**

CNN in Japan. The globalization of operations is one of the important trends for telecommunications management. (courtesy of Cable News Network)

entertainment and telephony. As discussed in Chapter 7, it is reasonable to expect further deregulation in this area. Indeed, this is precisely the goal of executive-level policy announced by Vice President Al Gore, and echoed by FCC chair Reed Hundt.[4]

The upshot of globalization and vertical integration of telecommunications is that the media manager of the mid and late 1990s will truly be a citizen of the world. The major media corporations of the twenty-first century will hold properties around the globe, and will be involved in all its activities—from telephony, to news, information, and entertainment; and all its forms—from traditional radio and TV receivers to computers, personal digital assistants, cellular phones, even virtual-reality environments whose design is yet to be developed.

One way to appreciate the dual phenomena of globalization and vertical integration is to revisit Figure 1-1 (page 10) and recast it as Figure 15-1.

The media environment of the future is no longer the "flat earth" depicted at the beginning of the book. The world is round: rotating on the twin axes of product and function. The horizontal axis of Figure 15-1 accounts for the various functions performed by media: from providing information (computer bits and news broadcasts) to purveying pure escapist entertainment (such as *Beavis and Butt-head* and *Mortal Kombat*).

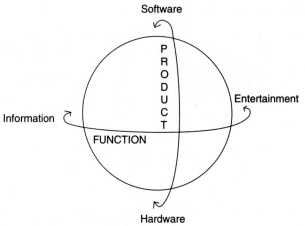

FIGURE 15-1 The world of modern telecommunications.

The vertical axis comprises the range of media products available now and on the drawing board for the immediate future. Included here are such traditional display devices as the century-old telephone, the comparatively ancient methods of communication like radio, movies, and over-the-air television, and the newer technologies, from interactive TV to CD-ROM computing.

TECHNOLOGICAL CONVERGENCE: ONE STOP SHOPPING, AND VIEWING AND LISTENING, TOO

Another quiet revolution is taking place along the vertical axis depicted in Figure 15-1—the technological convergence of discrete communication devices into single-source digital communications instruments for in-home and away-from-home consumption.[5]

In-home Technological Convergence

As the current system of mass communication evolves into the next century, it is clear that devices once considered separate items will merge into a single entity. At home, we will see some combination of the telephone, television, VCR, stereo system, personal computer, video game system, and home computer. While such a device might take any form, the most likely configuration will involve a large screen, surrounded by twin speakers, with an input/output device from which to issue our commands (imagine a real smart remote control). With this array, we will be able to watch sports events, movies, listen to music, order pizza, and take telephone calls (with full motion video) from friends and relatives, all without having to leave the sofa.

Away-from-Home Technological Convergence

The new technologies in development will merge the range of portable telecommunications devices as well. The personal stereo receiver ("Walkman"), the pager, the cellular phone, the camcorder, and the laptop computer will merge into a single communications device allowing individuals the ability to communicate with anyone else in the world, regardless of where each party happens to be—in a car, on a beach, or aboard a transcontinental airliner.

Some aspects of this convergence have already been demonstrated. In 1993, Apple introduced the Newton computer: a combination business diary, portable computer, and fax machine (albeit to disappointing initial results).[6] In a similar vein, cable companies (and their new partners, the telcos) are working on personal communication devices (the new generation of cellular phones).[7]

By early in the next century, we have the potential to be in touch, with full audio, video, (and the old standby, text), with literally anyone else in the world—so long as each of us has access to and can afford the new telecommunications products. This brings us to our next set of concerns for media management in the future: economics and social change.

ECONOMICS AND MEDIA MANAGEMENT

In forecasting the future of media management, it is useful to restate some of the economic trends traced in Chapters 4 through 6. As we have seen, historically, the development of the electronic media generally paralleled strong economic conditions, allowing for accelerated diffusion, profits, and growth. Recently, new media developments have occurred in an uncertain national and global economy. Management can no longer look ahead with unbridled optimism about the prospects for success. More than ever, they need to keep a sharp eye on prevailing economic conditions, both in the United States and abroad.

One of the great attractions of the broadcasting business has been its track record in keeping its profit and operating income margins ahead of the prevailing inflation rates. That is, in "real dollars," it has been an ever-growing income-generating machine.

However, getting into a telecommunications business is no longer a guarantee of economic success (if, indeed, it ever was). With the power of the networks seemingly slipping, with increasing competition commonplace in cable and the radio business, and with audiences having new viewing and listening options, media management faces an era of continuing cost-consciousness.

The go-go 1980s have been supplanted by the cautious 1990s—even among the normally aggressive, entrepreneurial media businesses. While merger mania continues to occur, it appears that the new alliances are more concerned with the long-term future of the industry, rather than short-term profit-making (and profit taking). As we saw in Chapters 4 and 5, the pace of radio and television sales has slipped. Cable companies are consolidating, rather than expanding (or are joining forces with telephone companies). The technological possibilities documented above may not come to pass simply because management lacks the venture capital to bring the new devices from the drawing board to the marketplace. Few of the futurists have tapped the pulse of the public. Do we really want and need these new communication modes? Can we afford them?

It is a fact of life for media managers that they are mainly in the entertainment business, competing for a share of the disposable or discretionary income of consumers, *not* the money they spend on necessities, such as food, shelter, and clothing. Thus, despite the giddy optimism of press agents and R&D specialists, there is doubt whether large numbers of consumers either want or will pay for the new hardware and services appearing on the media horizon.

In addition, some observers, particularly social scientists, contend that new media services are creating two classes of media consumers. The high cost and complexity of operations of the new equipment mean that only a small group of citizens will be able to equip their homes with the latest, most sophisticated media gadgetry. This wealthy, technologically literate group will be hotly pursued by programmers, manufacturers, and advertisers. The remainder of society will be "media poor"—lower-income, less-sought-after audiences without the resources to obtain the maximum benefits from the new technologies. This group may be cut off from the most significant advantages of the new media, including expanded news and cultural information, interactive capabilities, and so forth.[8]

The potential for such a development has serious implications for media management. A strength of the mass media system in the United States has traditionally been its accessibility to all segments of the population.

A basic goal of regulatory policy has been to make telecommunications services available in all regions, to all potential users. If the ability to obtain premium programming such as movies and selected sports events is limited to a select few, serious questions are likely to be raised in Congress and the courts. Lack of media access for low-income, minority, and other less-privileged social groups could conceivably lead to a new era of protest, perhaps even to demonstrations and violence.

Telecommunications in the United States prospered as long as its basis in the free enterprise system was not at odds with democratic principles such as the right to equal access and free speech.

To that end, Vice President Gore has placed the concept of universal service at the core of the administration's pledge to deregulate media industries so they can develop the new technologies described in the previous section.[9] While the plan calls for the new media to be funded by private capital, the Vice President specifically warns against the creation of "communication haves and have-nots." Full participation of Congress, industry, public-interest and consumer groups, and state and local governments will be sought to protect the public interest in the communications system of the next century.

CHAPTER 15: THE FUTURE OF TELECOMMUNICATIONS MANAGEMENT   411

Al Gore, one architect of the electronic superhighway. As Senator and Vice President, Gore has advocated availability of new media services to all segments of the population to avoid creating a situation of "media rich and poor." (Courtesy of Cable News Network)

TELECOMMUNICATIONS MANAGEMENT AND SOCIAL CHANGE

A theme that has been repeatedly stressed in the foregoing chapters has been the impact of social change upon media management. In this closing chapter, we will reexamine some demographic and political shifts and speculate on their potential impact on the management of electronic media.

The Aging of America

The population in the United States is aging. The baby boom of the post-World War II period is over. This fact, combined with the improvement of health care and services to the aging, means that media audiences of the future will be older.[10]

The impact of these trends has already been felt by media businesses. In radio, many beautiful music stations have switched from a diet of stringed instrumentals to a rotation of mellow pop tunes (their older audiences would rather relax with the Beatles or Kenny Loggins than with Mantovani or 101 Strings). Some album rock stations now eschew modern music (Pearl Jam and Smashing Pumpkins, for example) in favor of the "classic" rock approach (The Who, The Grateful Dead, and so on).

In network television, Angela Lansbury's long run on *Murder She Wrote,* Andy Griffith's *Matlock,* and Carroll O'Connor's *In the Heat of the Night* demonstrate that the young audience is not the only one sought by programmers. On the cable side, Viacom has successfully repackaged popular programming from the 1950s to the 1970s on prime time as Nickelodeon's *Nick at Night*. As this book went to press, there were plans to make this nostalgic block of reruns into a 24-hour, full-service network.

As the median age of the population continues to climb, further impact upon the media businesses can be anticipated. Advertising in the health field will increase, and products and services associated with money management, leisure, and retirement will become a staple of commercial messages.

In television, for example, news and information services are likely to be in greater demand. Entertainment programming can be expected to emphasize adult themes. Managers can expect a continued appetite for nostalgia—shows from the 1960s and 1970s are likely to be resurrected through syndication or given new life in first-run "reunion" shows.

There will be an ironic shift in media management. Businesses traditionally run by older managers seeking young audiences will be headed by young mavericks trying to capture mature audiences. It remains to be seen whether such a shift will result in better, more publicly responsive and responsible services.

The Multiethnic, Multicultural Audience

Not only will the media audience of the future be older, it will be more heterogeneous. As census figures reveal, the population is becoming less white: the number of African, Hispanic, Asian, and Middle-Eastern Americans continues to increase, particularly in metropolitan areas.[11] Unlike waves of prior immigration, there is little pressure upon these immigrant groups to blend into the melting pot of American culture; indeed, most attempt to preserve their history, language, and customs.

This trend will have important consequences for telecommunications management in coming years. As their numbers have increased, minority groups have gained political and economic clout. Managers can no longer merely pay lip service to their ethnic audiences by programming little but talking-head public affairs shows. They will have to develop and promote programs of interest to their diverse communities.

In radio, the number of formats targeted to specific ethnic groups is likely to increase, particularly in the urban metros in the Northeast and Southwest. In television, expect a growing number of programs offered on a multilingual basis, either dubbed or with captions provided. A model for the future may be found in Switzerland, where programs are regularly broadcast in German, with English and French subtitles. For example, it is not unreasonable to predict that television programs of the future in Los Angeles will be transmitted in Spanish, with English and Vietnamese subtitles.

The commitment to multicultural service will extend beyond programming, into management structures themselves. The male, lily-white corporate telecommunications boardroom is likely to yield to a management team featuring increasing numbers of women and members of minority groups. This realignment is bound to have important consequences for the way radio, television, and cable are programmed, promoted, and managed. We can expect a greater sensitivity to the needs of minority and female audiences, reflected in both information and entertainment programming. It is possible that persistent stereotypes about these subgroups will give way to new, perhaps more accurate perceptions, as actual members of the groups being portrayed, rather than observers, create their media characters and images.

For years, researchers have cited the media as the great instrument of social identity and cohesion. Traditionally, our telecommunications system has communicated to ourselves and to the world what made us all uniquely "American" in spite of our vast geographical and cultural differences. Ironically, it is likely that the telecommunications system of the future will instead communicate how we *differ.*

The International Audience

Traditionally, media companies made the bulk of their profits in the domestic market—from American viewers and listeners. But, as we have seen, in many ways domestic telecommunications is a mature business. Today, most Americans have telephones with answering machines, TVs and radios, VCRs, even cable TV. This has caused both domestic and international media companies to look outside the United States for their growth in the years ahead.

The maturing American market has coincided with political change. Countries whose media have been controlled by central governments (and have suffered as a result in the number and kinds of services offered) have begun the slow and painful creep toward democracy. Media companies have been at the forefront of developing the telecommunications infrastructure in these new democracies, including such services as cable TV in Prague, cellular telephones in Moscow, and commercial radio in the former East Germany.

At the same time, many Western democracies with a traditionally paternalistic media system (Great Britain, France, and Canada, to name a few) are facing cutbacks in government-sponsored media and are opening their telecommunications services to private investment. Just now, in the 1990s, are many of these countries getting commercial radio and TV stations, cable TV, and telephone services like "call waiting," and "caller ID." American media companies are actively involved in this new international consumer marketplace.

In sum, these trends suggest that today's media hardware and software cannot be targeted solely for the U.S. consumer. This puts considerable pressure on media managers to understand cultural and regional differences, as well as the nuances of policy and government in other countries. More than a few media managers admit to culture shock when trying to introduce their domestic services to new international audiences.

THE BOTTOM LINE: SUCCESSFUL MANAGEMENT
FOR THE NEW MILLENIUM

Against this backdrop of technological and social change, it is important to close with an inside view of the future of media management. That is, what will be the elements of success for you, the next generation of media managers?

A partial glimpse into the unknown may have been provided in a survey of radio management conducted for the National Association of Broadcasters.[12] The study was performed by McKinsey and Company, creators of the best-selling book *In Search of Excellence: Lessons from America's Best-Run Companies*.[13] The McKinsey staff spent 3 weeks visiting 11 of the country's top-performing stations. Data gathered during those visits were compared with a sample of "typical" stations.

Results underscore many of the main points raised throughout the text. Stations which outperformed their competitors in profit, programming, personnel relations, and public affairs shared the following attributes.

* *They targeted their programming to well-defined audience segments.*

This result suggests the continuing fragmentation of the media audience. It highlights the importance of audience research and underscores the value of sound public relations and promotions policies. In short, it illustrates that the successful media ventures of the 1990s and beyond, like those of the 1960s and the 1930s, will *know* and *respond* to the interests, needs, and expectations of their audiences.

* *They shared the belief that quality and service are as important as making a profit.*

Time and again, stations with a strong bottom-line position seemed to be the most active in civic affairs, the most open to listener criticisms and suggestions, and the most likely to originate local programming. This finding has important implications for the next generation of management. There is a growing concern that relaxed ownership restrictions and increasing profits will bring a new class of carpetbaggers into the media marketplace, replacing spirited public servants with greedy capitalists. McKinsey and Company's findings suggest that such a belief is probably incorrect. As always, being a good broadcaster or cablecaster also means being a good citizen and neighbor.

* *They allowed staff to share in decision-making procedures and in the financial rewards of the station.*

Successful stations involved all levels of management in decision making. Equally important, profitable stations shared their bounty with their employees through increased salaries, benefits, and bonuses.

As stated many times, media employees have too often been treated as expendable items and the telecommunications industry has been rife with employee turnover and labor strife. Yet this situation can be avoided by creating an atmos-

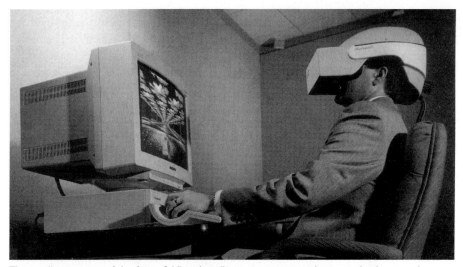

The media manager of the future? Virtual reality systems personal communications services, and other new developments portend change for telecommunications management in the new millennium. (Courtesy of Intel)

phere of shared goals, backed up with shared profits. It is unrealistic to expect the media businesses to wipe out generations of staff rivalry and friction overnight. However, the challenge to the media manager of the year 2000 will be to treat employees with increased respect and concern.

• *They made sufficient capital reinvestments in order to maintain their market position.*

As the adage states, "You've got to spend money to make money." That old saw will remain true in coming years. Managers who respond affirmatively to change by upgrading programming and production facilities, by diversifying into new enterprises, and by taking chances with new products and ideas are likely to be the ones that prosper. Too many managers have stood pat, sticking their heads in the sand as both their businesses and audiences changed. In this group fall the radio station owners who thought "TV will never catch on," the AM station operators who said "Nobody will listen to FM," and the television broadcasters who argued "No one would pay to watch television" on cable. Clearly there is little room for such executive shortsightedness in the current telecommunications environment. Like the earliest days of the electronic media in the 1920s, the times are ripe for entrepreneurs with vision and new ideas. The difference is that this time they will need *a lot* of venture capital!

SUMMARY

The future telecommunications manager faces many technological, political, and social changes. Continued merger and coventure activity guarantees that the media marketplace of the future will be led by vertically integrated, multinational corporations active in both hardware and software.

Discrete telecommunications technologies—including telephone, movies, radio, TV, and cable—are merging into single-source devices. This phenomenon will occur both at home and away from home, creating a climate of change among media managers.

Economic and political changes will also affect media management. Domestic media companies face a saturated media marketplace and will look increasingly to international consumers to fund telecommunications services.

At home, demographic changes that will have an impact on management include the steadily rising median age of the audience and the increase in the number of minority audiences. These and other trends will cause management to rethink its programming, advertising, and promotional strategies.

Against this backdrop of change, the factors that portend success for management include careful audience research leading to targeted programming, increased concern about quality and service, including staff in decision making and profit sharing, and a commitment to reinvest capital in new products and new ideas.

NOTES

1 Cited in Joseph Pelton, "The Future of Telecommunications: A Delphi Survey," *Journal of Communication,* Winter 1981, p. 177.
2 The concept is introduced and explained in Marshall McLuhan, *Understanding Media: The Extensions of Man* (New York: McGraw-Hill, 1964).
3 Gerald Emanuel Stean (ed.), *McLuhan: Hot and Cool* (New York: Signet Books, 1967), p. 272.
4 Al Gore, speech to the National Press Club, Washington, D.C., December 21, 1993; Vincente Pasdeloup, "New FCC Chairman: I'll Use Reregulation to Boost Economy," *Cable World,* November 29, 1993, pp. 1, 164.
5 See "Your Digital Future," *Business Week,* September 7, 1993, pp. 56–64. Philip Elmer-DeWitt, "Take a Trip into the Electronic Superhighway," *Time,* April 12, 1993, pp. 50–58.
6 See "Big Hit or Scully's Folly?" *Fortune,* July 26, 1993, p. 52.
7 See "FCC Acts on Frequencies, Phones, TVs; Migration Policy Explored for PCS at 2GHz Bands," *Electronic News,* September 21, 1992, p. 2.
8 These arguments are forcefully made in Mike Holderness, "Down and Out in the Global Village," *New Scientist,* May 8, 1993, pp. 36–41.
9 See "Gore Unveils Tele Plan," *USA TODAY Update,* December 22, 1993, p. 1.
10 See Kenneth J. Doka, "When Gray Is Golden: Business in an Aging America," *The Futurist,* July–August, 1992, pp. 16–21; Ken Dychtwald, *Age Wave: The Challenges and Opportunities of an Aging America* (Los Angeles: J.P. Tarcher, 1989).
11 See Dinker I. Patel, "Asian Americans: A Growing Force," *Journal of State Government,* March–April, 1988, pp. 71–77; Leslie Whitaker, "Dancing to the Latino Beat: Hispanic Media Reach a Vast Audience but Lag with Advertisers," *Time,* October 23, 1989, p. 114.

12 Bernadette McGuire, "NAB Survey of Top Performing Radio Stations," *Radio: In Search of Excellence* (Washington, D.C.: National Association of Broadcasters, 1985), pp. 51–60.
13 Thomas J. Peters and Robert H. Waterman, Jr., *In Search of Excellence* (New York: Harper and Row, 1982).

FOR ADDITIONAL READING

Abramson, Jeffrey B.: *The Electronic Commonwealth: The Impact of New Media Technologies on Democratic Politics.* New York: Basic Books, 1988.

Brand, Stewart: *The Media Lab: Inventing the Future at MIT.* New York: Viking, 1987.

Brody, E. W.: *Communication Tomorrow: New Audiences, New Technologies, New Media.* New York: Praeger, 1990.

Madu, Christian N. (ed.): *Management of the New Technologies for Global Competitiveness.* Westport, Conn.: Quorum Books, 1993.

McPhail, Thomas L.: *Electronic Colonialism: The Future of International Broadcasting and Communication.* Newbury Park, Calif.; Sage Publications, 1987.

Media at the Millenium: Report of the First Fellows Symposium on the Future of Media and Media Studies. New York: Freedom Forum Media Studies Center, 1992.

Neuman, W. Russell: *The Future of the Mass Audience.* New York: Cambridge University Press, 1991.

INDEX

A Current Affair, 347
A&E (*see* Arts and Entertainment)
A&M (record company), 182
ABC (*see* Capital Cities-ABC)
ABC Radio Networks, 307
Abel, John, 70
AC-DC, 86
Accelerated amortization, 236
Accelerated Cost Recovery System (ACRS), 235
Access time (television), 114
Accuratings, 384
Acquisition:
 existing facility, 195–203
 new broadcast facility, 203–220
 (*See also* Analysis, market; Capitalization and acquisition financing)
Action News, 55, 112
Adelphia, 178
Advanced television research laboratory, 18
Advance Publications, 187
Advertiser-supported cable networks, 140
Affiliate relations, 326–327
Affiliates, network, 123
Affiliation agreement, 325
Affirmative action, 265–267
AFL-CIO, 287
AFM (*see* American Federation of Musicians)
Afternoon drive (time), 85
AFTRA (*see* American Federation of Television and Radio Artists)

Agencies, talent (*see* Talent representatives, managers, and agencies)
Air Supply, 88
Aircheck (*see* Bandsweep)
Alfred Hitchcock Presents, 345
Alger, Horatio, 41
All channel receiver bill, 106
All in the Family, 303
All Things Considered, 82
AM Chicago, 338
American Airlines, 60
American Broadcasting Company (*see* Capital Cities-ABC)
American Demographics, 222
American Federation of Musicians (AFM), 286
American Federation of Television and Radio Artists (AFTRA), 286, 287
American Gladiators, 117
American Guild of Variety Artists (AGVA), 287
American Movie Classics (AMC), 55, 328, 345
American Radio, 209, 222, 244
American Society of Composers, Authors and Publishers (ASCAP), 94
American Urban Network, 307
America's Most Wanted, 56
AM-FM combos, 96
AM radio (*see* Radio, AM)
Amortization, 234–237
Amos and Andy, 79
Analysis:
 investment, 244

Analysis (*Cont.*):
 market, 220–223
 regression, 210–212
 trend, 244
Analyst, financial, 206
Anheuser-Busch, 58
Annual reports, 283
Antenna-2 (France), 186
Apple Computers, Inc., 34, 189, 409
Aptitude tests, 269
Arbitration (*see* Negotiation)
Arbitron Ratings Company, 223, 381
 local market report, 384
ARD (Germany), 186
Arista, 182
Arsenio, 115
Arts & Entertainment (cable network), 119, 147, 331, 302, 357, 372
ASCAP (*see* American Society of Composers, Authors, and Publishers)
Ascertainment studies, 380
Assets, 229
 fixed, 230–232
 intangible, 232–233
 quick, 230
 tangible, 230
Associated Actors and Artists of America (AAAA), 287
Assumption of existing indebtedness, 217
AT&T, 18, 32, 54, 57, 95, 155, 189
AT&T Commercial Finance, 217

419

420 INDEX

Atari, 68
ATC (*see* Time Warner)
Atlanta Braves, 140
Audience:
  flow, 347
  fragmentation, 40
  research, 40, 378–402
Audience Research and Development (ARD), 388, 394
Audition tapes, 268, 393
Auditorium tests, 393
Avails, 93, 353, 364
Average quarter hour (AQH), 390
Axel Springer Verlag, 186

Baby boom, 411
Bakker, Jim, 335
Ball, Lucille, 43
Bandsweep, 312–313
Bank of Boston, 215
Banner (rep firm), 93
Barnouw, Eric, 32, 104
Barriers to entry, industrial, 56, 57
  absolute cost, 56
  product differentiation, 57
  scale economy, 57
Barter, 120, 121
Barter syndication, 93, 334
Baruch, Ralph, 134
Basic accounting equation, 229
Bastard, The, 108
BBC (*see* British Broadcasting Company)
Beach Boys, the, 86
Beatles, the, 86, 88, 303, 411
Beavis and Butt-head, 348, 407
Bell Atlantic, 54, 159
Bell South, 54, 159
Benefits, employee, 278–280
Bennett, Tony, 88
Benny, Jack, 43, 79
Berle, Milton, 13, 105
Berlusconi, Silvio, 185, 186
Bertelsmann (Germany), 182, 185, 186, 406
Best predictors (of value), 210
Best time available (BTA), 361
Betamax, 57
Beverly Hillbillies, The, 106
Beverly Hills 90210, 55, 56, 108, 113

Bewitched, 345
BFMA (*see* Broadcast Financial Management Association)
Billboard, 309
Binding arbitration, 288
Blackburn and Company, 204
Black Entertainment Television, 44, 140, 328, 345
Blair (rep firm), 353, 356
BMI (*see* Broadcast Music Incorporated)
Bob and Ray, 303
Bobby Poe's Music Survey, 309
Bolton, Michael, 88
Bonneville International Corporation, 67, 178, 292, 294
Box Tops, the, 86
Boy King, The, 338
Bozo Show, The, 335
Brand identification, 55
Bread, 88
Breaks (break-outs), 379
Breneman Review, 309
Brinkley, David, 106
British Broadcasting Company (BBC), 186
Broadcast Cable Financial Management Association (BCFM), 30, 91, 229, 279
Broadcast Investment Analysts, 222, 244
Broadcast Music Incorporated (BMI), 94
Broadcast Promotion and Mailing Executives (BPME), 375–376
Broadcasting and Cable, 30, 196, 209, 264
Broadcasting and Cable Marketplace, 43, 177, 197, 221, 388
Brokaw, Tom, 394
Brokers, media, 206
Brooks, Mel, 33
Bruskin Associates, 388
Buchanan, Pat, 308
Buena Vista Cable Television, 44
Burger King, 58, 353
Burkhart/Douglas, 318
Burn-out, 206, 305
Burns and Allen Show, The, 81
Burt Sherwood and Associates, 206
Buyer concentration, 57, 58

C-SPAN, 55, 328
Caballero Network, 307
Cable:
  build-up (phase), 133
  customer service representatives (CSRs), 150
  (*See also* Television, cable)
Cable Advertising Directory, 223
Cable Communications Act of 1984, 61, 156, 173, 266
Cable Communications Act of 1992, 38, 61, 139, 142, 150, 198–199, 266
Cable consultants, 199
Cable Networks, Inc., 355
Cable News Network (CNN), 55, 61, 70, 109, 137, 140, 144, 145, 149, 270, 291, 303, 328, 340, 346, 357, 358, 383, 406
Cable Television (*see* Television, cable)
Cable Television Administration and Marketing Society (CTAM), 376
Cable Television Advertising Bureau (CAB), 223, 387
Cable TV Financial Databook, 209, 222
Cablevision, 30
Cablevision Industries (MSO), 180
Cablevision Systems, Inc., 25, 143, 180, 324
Cable World, 262
Cactus Broadcasting, 177
Caesar, Sid, 33
Cagney and Lacey, 113, 331
Campaign (advertising), 353
Canal Plus, 187
Capital Cities-ABC, 4, 6, 10, 43, 53, 54, 55, 61, 63, 66, 67, 79, 105–106, 113, 114, 116, 119, 120, 170, 173, 178, 187, 244, 355
Capitalization and acquisition financing, 214–219
Capitol, 182
Carey, Mariah, 88
Carnegie Commission on Educational Television, 109
Carnegie II, 111
Carolco films, 189

INDEX   421

Carpenters, the, 88
Carr, Vicki, 88
Cash Box, 309
Cash flow, 213–214
CATV (*see* Community-Antenna Television)
Caves, Richard, 51
CBS (*see* Columbia Broadcasting Systems, Inc.)
CBS Evening News, 112
CBS Radio Networks, 307
CBS Records, 66, 182
CD-ROM, 406, 408
Cellular telephones, 63, 69
Channel capacity, 14
Chapman Associates, 206
Character (requirement of ownership), 168
Chase Manhattan, 215
Chayefsky, Paddy, 105
Cheers, 330, 343, 344, 346
Chermi Communications, 177
China Beach, 331
Chris-Craft Industries, 178
Chrysler Capital, 217
Chung, Connie, 394
Churn (subscriber disconnects), 137
Cinemax, 15, 16, 55, 57, 58, 140, 328, 345, 369
Citibank, 215
Citizenship (requirement for ownership), 167
Civic involvement, 168
Civil War, The, 111
Clark, Dick, 79
Clifford, Jack, 134
Clinton, Bill, 157, 233
Cluster Plus, 396
CNBC (*see* Consumer News and Business Channel)
CNN (*see* Cable News Network)
Coca, Imogene, 33
Cole, Natalie, 88
Collective bargaining, 288–290 (*See also* Negotiation)
Columbia Broadcasting Systems, Inc., 6, 10, 41, 43, 54, 55, 61, 63, 66, 79, 104–106, 113, 114, 116, 119, 120, 173, 178, 355, 357
Columbia Pictures, 6
Columbia Records, 6

Combined Communications, 214
Combos (radio station combinations), 80
Comcast, 54, 172, 180, 244
Comedy Central, 185
Communication, internal, 281–284
Communication Workers of America (CWA), 153, 287
Communications Act of 1934, 219
Communications audit, 282
Community-access cable television, 141, 304, 339
Community-antenna television (CATV), 14, 133
Compact disc (CD), 38, 85, 94
Comparative method (of valuation), 209–210
Compensation, employee, 276–278
Competitive renewal, 201, 202
Computer applications, 244–248
  accounting, 247
  distributed (batch-processing) systems, 247
  microcomputers, 247–248
  on-line (time-sharing) systems, 247
COMSAT Corporation, 4, 16
Concealment, 400
Concentration, 188, 189
Concentration ratios, 52–54
Consecutive week discount, 361
Construction permit, 197, 198
Consumer demand, 4
Consumer News and Business Channel (CNBC), 55, 61
Continental Cablevision, 54, 172, 180, 188, 214, 340
Contractual negotiations (*see* Negotiations)
Convergence, 7, 409
Converters, 14
Cook Inlet, 191, 213
Cops, 56
Corporate culture, 9
Corporation for Public Broadcasting (CPB), 82, 110, 111, 302
Cosby Show, The, 55, 113, 123, 343, 344
Cosell, Howard, 394
Cost per ratings point, 391

Cost per thousand (CPM), 390, 391
Costs, fixed, 58, 59
Courtroom Television Network (Court TV), 185, 329
Coventure, 7, 189
Cox Enterprises (group owner), 54, 66, 157, 170, 180, 190, 244, 283
CPB (*see* Corporation for Public Broadcasting)
Creative Artists Agency (CAA), 285
Cronkite, Walter, 106
Crook and Chase, 344
Crosby, Bing, 79
Cross-ownership, 66
Crossfire, 344
CTAM (*see* Cable Television Administration and Marketing Society)
Cume (cumulative index), 390
CWA (*see* Communication Workers of America)

DAB (*see* digital audio broadcasting)
Dallas, 56
DAT, 10
Dave's World, 347
Dayne, Taylor, 90
Dayparts, 84, 315, 342
Days and Nights of Molly Dodd, The, 331
Daytime, 85, 115
DBS (*see* Direct-broadcast satellite)
Deception (in research), 401
Decisional roles (of managers), 31
Deep Space Nine, 327
Deferred compensation, 329
DeForest, Lee, 5
Delegation, organizational, 26
Delta Air Lines, 60
Demand growth, 59, 60
Demographics, 84, 220, 309, 379
Demographics USA, 222
Depeche Mode, 87
Depreciation, 234, 235
Depression, Great, 39, 78, 82

Designated market area (DMA), 389
Designing Women, 113, 331
DGA (*see* Directors Guild of America)
Diamond, Neil, 88
Dick Van Dyke, 347
Digital Audio Broadcasting (DAB), 13, 38, 39, 57, 59, 63, 69, 81
Digital Media, 189
Diller, Barry, 24
DIR Communications, 308
Direct mail, 375
Direct-broadcast satellite (DBS), 4, 10, 16, 17, 62, 63, 69, 166
Directors Guild of America (DGA), 286
DirecTV, 16
Disciplinary action, 274, 276
Discovery Networks, 117, 145, 328, 331, 357, 372
Disney (corporation), 12, 43, 178, 185
Disney (studios), 106
Disney Channel, The, 15, 16, 140, 141, 369
Distant-signal stations (cable TV), 140
Diversification, industrial, 65, 66
Diversification of management background (ownership), 169
Divorce Court, 343
Dolan, Charles, 25
Donahue, 113, 121, 335, 337, 343, 348
Donahue, Phil, 347
Donnelley Information Services
Doors, The, 87, 319
Downconverters, 16
Downsizing (of media workforce), 252
Dragnet, 345
Drexel Burnham, 217
Drops (cable TV), 14
DuMont (television network), 105
Duopoly, 97, 171
Dylan, Bob, 87
Dynasty, 55

E-mail, 248, 283
Early fringe (television), 115
Econometrics (regression analysis), 210, 212
Economic Recovery Act of 1981, 235
Ed Sullivan Show, The, 13, 113
Edison, Thomas A., 5, 41
Educational television (*see* Television, noncommercial)
EFM Media, 308
Eisner, Michael, 43
Electromagnetic spectrum, 196
Electronic Media, 30, 196, 264
Electronic Media Ratings Council (EMRC), 398
Electronic newsgathering (ENG), 107
Electronics Industries Association of America (EIAA), 98
Emmis Broadcasting, 212, 292–293
Employee assistance programs, 280
Employees:
  above-the-line, 286
  below-the-line, 286
  (*See also* Management, telecommunications; Personnel management)
Employment:
  cable, 155–157
    salaries, 156
  radio, 98, 99
  television, 126–129
    salaries, 128
Encore, 55
Entertainment software firms, 10
Entertainment and Sports Network (ESPN), 55, 66, 137, 140, 144, 145, 149, 291, 328, 357, 358, 383
Entertainment Tonight, 113, 114, 121, 333, 335, 344, 355
Equal Employment Opportunity (EEO), 64, 263, 265
Equal Employment Opportunity Commission (EROC), 252, 265
Equity, employment, 64
Erving, Julius, 191
Esparaza, Montezuma, 44

ESPN (*see* Entertainment and Sports Network)
Evaluation methods, employee, 272–273
  (*See also* Performance measurement and appraisal)
Evening (daypart), 85
Evening Shade, 347
Evergreen, 178
Ewing, Patrick, 191
Exclusivity, 312
Expenses:
  cable, 149–153
    capital expenditures, 149
    general and administrative, 150
    marketing, 151
    program, 149–150
    technical, 150–151
  radio
    salaries, 99–101
    trends, 94, 95
  television, 123–125
    advertising expenditures, 121
    general and administrative, 124
    program and production, 123, 124
    sales, 124, 125
    technical costs, 124
External capital, 214, 217

Face The Nation, 336
Facilitator (focus group), 392
Family Channel, The, 55
Family Feud, 113, 332, 344
Family Matters, 342
Farber, Barry, 308
Farnsworth, Philo T., 5
Fayol, Henri, 23
FCC (*see* Federal Communications Commission)
Federal Communications Bar Association (FCBA), 197
Federal Communications Commission, (FCC), 11, 18, 61, 62, 64, 66, 67, 78, 82, 91, 96, 98, 104, 109, 197, 252
  All Channel Receiver Bill, 106
  Common Carrier Division, 167

Federal Communications Commission, (FCC) (*Cont.*):
  Equal Employment Opportunity Trend Report, 98
  freeze on TV licenses, 13
  hearing, 197
  minorities and women, 64, 265
  model EEO program, 265–267
  Second Report and Order, 373
  spectrum management by, 196
  vertical integration regulations, 66
Federal Mediation and Conciliation Service (FMCS), 288
Federal Trade Commission, 66, 188
Feedback, 28
Fessenden, Reginald, 5
Fetzer, John E., 41
Fibber McGee and Molly, 79
Fiber optics, 149, 155
Field and Stream, 66
Financial Accounting Standards Board (FASB), 237, 283
Financial analyst, 206
Financial interest and syndication rules (Fin/Syn), 66, 328
Financial planning and projections, 243, 244
Financial reporting methods, 237, 243
  balance sheets and P & L statements, 238, 240
  liquidity and leverage, 241–242
  margins and ratios, 240–241
  profit and return, 242–243
Financing (*see* Capitalization and acquisition financing)
Fininvest (Italy), 185, 186
First Media Corporation, 191
First National of Chicago, 215
First-run syndication, 331
Flight (advertising), 352
  (*See also* Campaign)
FM radio (*see* Radio, FM)
FMCS (*see* Federal Mediation and Conciliation Services)
FMR Associations, 388
Focus groups, 319, 392–393

Ford Foundation, 111
Ford Theater, 107
Format, 303
Format wheel, 313
Formats, radio, 40, 85–89
  adult contemporary, 86, 89
  album-oriented rock (AOR), 86, 87, 89
    classic rock, 86, 87
    progressive, alternative rock, 87
  AM trends, 87, 88
  black/urban contemporary, 88
    rap, 88
    salsa, 88
  classic rock, 86, 87
  contemporary hit radio, 88
  country music, 87, 89
  easy listening, 88
  FM, 88
  middle of the road (MOR), 88, 89
  nostalgia, 88
  news-only, 88, 89
  news-talk, 87, 88
Fox Broadcasting Company, 6, 54, 55, 58, 66, 67, 108, 113, 116, 120, 123, 126, 172, 178, 187, 213, 214, 244, 355, 406
Franchise agreements (cable), 199, 230
Franchise negotiation, 201
Freed, Alan, 79
Frequency discount, 361
Frequency search, 196
Friendly, Fred W., 107
Fringe time periods, 343
Frontline, 112, 373
Full House, 331, 347

Galavision, 355
Gannett (group owner), 67, 126, 178, 187, 283
Gavin Report, The, 309
Gaylord Broadcasting Corporation, 126, 170
General Electric, 32, 54, 61, 178, 183, 187
General Electric Capital, 217

General Foods (network advertiser), 58
General Instruments, 18
General ledger, 238
General Magic, 189
General partnership (*see* Partnerships, general and limited)
Genesis, 319
Genre, 303
Geodemographics, 396
Georgia Radio News Service, 308
Geraldo, 113, 332, 337, 343
Get Smart, 345
Gillet, George, 216
Gilligan's Island, 343
Global giants, 183
Global village, 406
Globalization, 406–408
Godfrey, Arthur, 43
Golda, 108
Golden Girls, 331, 343
Goldenson, Leonard, 24, 43
Goldmark, Peter, 43
Goldwyn, Samuel, 24
Good Morning America, 115, 337
Good Morning Seattle, 338
Goodwill, 232–233
Gore, Al, 407, 410
Grandfathered combinations, 171
Granite Broadcasting, 191
Grateful Dead, The, 411
Graveyard (radio time period), 85
Green Acres, 106, 345
Greene, Harold, 155
Griffith, Andy, 411
Gross impressions, 391
Gross profit margin, 239
Gross ratings points (GRP), 391
Group W (*see* Westinghouse)
Gruner + Jahr (publishing company), 185
GTE, 54
Guilds and unions, 35, 286–292
Guns 'n' Roses, 86

Hachette (France), 186
Hammocking, 387
Handbooks, employee, 281–282
Happy Days, 106

424 INDEX

Hard Copy, 331, 346
Hardesty-Puckett, 216
Hardware (equipment), 8
Harrington, Righter and Parsons (HRP) (rep firm), 356
Harvey, Paul, 306
Havas (France), 186
Hawkeye Network, 308
HBO (*see* Home Box Office)
HDTV (*see* High-definition television)
Head end, 14
Headline News, 55, 140, 150
Heller Financial, 217
Herman's Head, 347
Heston, Charleton, 105
High-definition television (HDTV), 10, 18, 39, 56, 57, 59, 67, 69, 201, 329
Highly leveraged transactions (HLTs), 215–216
Hit radio era, 79–80
Hitachi, 54
Hits, 309
Home Box Office (HBO), 15, 16, 25, 55, 57, 58, 67, 107, 140, 153, 185, 304, 328, 345, 357
Home Improvement, 331, 355
Home video, 18, 19
Homes passed (cable), 134, 136
Honeymooners, The, 346
Hook (radio), 393
Hope, Bob, 79
Hot clock, 313
Hourly percentage (Equivalent hour), 119
Hubbard, Stanley, 16
Hundt, Reed, 407
Huntley, Chet, 106
Hyping (ratings), 399

I Love Lucy, 55
IATSE (*see* International Alliance of Theatrical and Stage Employees)
IBEW (*see* International Brotherhood of Electrical Workers)
IBM, 34
Ideological invasion, 6
Image studies, 381

Impulse PPV, 136
Imus, Don, 306
In Living Color, 55, 108, 113
In The Heat of the Night, 411
Independent stations (television), 123
Infinity Broadcasting Company, 54, 178, 213, 308
Information superhighway, 405
Informed consent, 390–401
Insertion, 144
Inside Edition, 333
Integration:
  horizontal, 65, 67
  vertical, 66, 67
Integration of ownership and management, 168
Intelligence tests, 269
Interactive capability, 14
Interconnection (cable advertising), 144, 358
Interep, 93
Internal capital, 214, 217, 218
Internal Revenue Service (IRS), 90, 233, 235
International Alliance of Theatrical and Stage Employees (IATSE), 31, 287, 289
International Brotherhood of Electrical Workers (IBEW), 154, 286, 287, 289
International Creative Management (ICM), 285
International Directory of Telecommunications, 222
International Radio and Television Society, 44
Internship programs, 264
Interpersonal roles (of managers), 29, 30
  figurehead, 29
  leader, 29
  liaison, 29
INTV (*see* Association of Independent Television Stations)
Inventory (commercial), 352
  control, 365
Investing in Radio, 360
Investing in Television, 209
Investment analysis, 244
INXS, 319
IQ (testing), 269

Island (record company), 182
Itochu (Japan), 189

Jack Benny Show, The, 79
Jacor, 178
Jazz Alive, 374
Jefferson Pilot, 66
Jennings, Peter, 394
Jeopardy, 114, 333, 345
Job descriptions (in telecommunications), 259–263
Jobs, Steven, 34
Joel, Billy, 88
Johnson, Lyndon B., 109
Johnson, Robert W., 44
Joint venture, 7, 189, 190
Jones Intercable Inc., 174, 180
Jones, Quincy, 191
Journey, 319
Joyner, Tom, 306
Junk bonds, 215

KABC-TV (Los Angeles), 178
Kallen, Lucille, 33
Kagan, Paul, 137, 147, 205, 222, 244
Karmazin, Mel, 178
Katz Radio Group, 93, 356
KBTL (Houston), 320
KDKA (Pittsburgh), 11
Kelley, Grace, 105
KGO-TV (San Francisco), 178
KILT-AM/FM (Houston), 172
King, Larry, 308
Kluge, John, 217
KMDY (Los Angeles), 89
KMOX (St. Louis), 11
Kohlberg-Kravis-Roberts, 216, 217
Koplovitz, Kay, 44, 134
Koppel, Ted, 285, 346
KPAL (Little Rock), 320
KRTH-FM (Los Angeles), 212
KTLA (Los Angeles), 140, 178, 338
KTRK-TV (Houston), 178
KTTV (Los Angeles), 67, 126
KTVT (Fort Worth), 140
KTWV (Los Angeles), 320

KWGN (Denver), 126, 178, 336
KXEW (South Tucson, AZ), 177

Labor relations (see Negotiation)
Labor unions (see Guilds and unions)
Lansbury, Angela, 411
Large-screen televisions, 38
Larry King Live, 344, 348, 384
Larry Sanders Show, The, 303
Late fringe (television), 114
Late Night, 55
Later, 55
Laverne and Shirley, 106
Lawson, Jennifer, 110, 111
Lazard Freres, 216
LBS Entertainment, 6
Lead-in (programming), 347
Lease management agreement (LMA), 172
Leased access, 304, 340
Led Zepplin, 87, 319
Lehman Brothers, 216
Lemmon, Jack, 105
Letterman, David, 116, 292, 346
Leverage, 241–242
Levin, Gerald, 30, 134
Liabilities, 229, 233–234
License application and renewal, broadcast, 198
Liddy, G. Gordon, 308
Lifetime, 66, 140, 145, 372
Limbaugh, Rush, 306, 308
Limited partnership (see Partnerships, general and limited)
Listenership, radio, 84–86
Little House on the Prairie, 55
Little Rascals, The, 105, 236
Live From The Improv, 339
Live With Regis and Kathy Lee, 338, 348
Local area networking (LAN), 283
Local monopoly, 166, 167
Local ordinances, 200
Local origination, 304, 324
Local ownership, 168
Local spot sales, 90
Local TV sales, 122
Loggins, Kenny, 411
Love and War, 347

Love Connection, The, 346
Low-power television (LPTV), 13, 14, 57, 62, 196
Lucy Show, The, 347

M*A*S*H, 113, 303
MacGyver, 113
MacNeil/Lehrer NewsHour, 344
Made-for-television movies, 338
Madonna, 88, 285
Magid, Frank, N., Associates, 394
Major Market Radio (rep firm), 93
Major radio groups, 177–178
Major television groups, 178
Malone, John, 30, 180
Management, 22
  approaches:
    functional school, 22
    human performance, 22
    management science, 22
    personnel (see Personnel management)
    program (see Program management)
  roles, 23
    decisional, 29, 31
    informational, 29, 30, 31
    interpersonal, 29, 30
  skills, 23
    conceptual, 24
    human, 24
    technical, 24
    (See also Personnel management)
  telecommunications:
    controlling, 28, 29
    functions, 23
    innovating, 29
    organizing, 26
    planning, 25, 26
    staffing, 27
Managers, talent (see Talent representatives, managers, and agencies)
Mankiewicz, Frank, 82
Mannheim Steamroller, 320
Mansfield, Harry, 272
Mantovani, 411
Marconi, Guglielmo, 5, 77
Margins, financial, 240, 241

Mark and Brian, 306
Market analysis, 220–223
Market conduct, 52, 60
Market Evaluations, Inc., 394
Market performance, 52, 62, 63
  efficiency, 63
  equity, 64
  planning and growth, 65–67
  stability, 63, 64
Market size (in valuation), 211
Market Statistics Inc., 222
Market structure, 51
  product differentiation, 54, 55, 56
  seller concentration, 52
    monopoly, 52, 53
    oligopoly, 52, 53
Married with Children, 55, 108, 331, 347
Marshall, Garry, 302
Martin, Mary, 105
Marx, Groucho, 43
Mary Tyler Moore, 347
Masterpiece Theater, 303, 373
Matlock, 411
Matsushita (Japan), 6, 54, 182, 185, 189, 406
Matthau, Walter, 105
Maxwell, Robert, 24
Mayer, Louis B., 24
MCA/Universal, 6, 182
McCaw Cellular, 216
McDonald's, 58
McGaven-Guild, 356
McGregor, Douglas, 32
MCI, 57
McKinsey and Company, 414
McLuhan, Marshall, 406
Media brokers, 206
Media buyer, 354
Media Mark (research firm), 388
Media Report to Women, 190
Media, internal, 283
Media valuation, 208–214
Mediation, 288
Mega-reps (radio), 356
Mellencamp, John, 86
Mellon Bank, 215
Melrose Place, 56
Metro survey area (MSA), 389
Metromedia, 170, 217, 244
Milken, Michael, 216

Minorities and women (in telecommunications), 40, 190
  ownership, 190, 191
Minority Small Business Enterprise Committee, 191
Moderator (focus group), 392
Monday Night Football, 55
Moneyline, 344
Montovani, 88
Moonlighting, 331
Morgan Guarantee Trust, 215
Mork and Mindy, 106, 345
Morning drive (time), 85
Morning Edition, 82, 374
Morrissey, 87
Mortal Kombat, 407
Motion Picture Association of America (MPAA), 30
Mottola, Tony, 88
Movie Channel, The, 15, 16, 55, 57, 140, 149, 185, 345
Mr. Rogers' Neighborhood, 302
Multichannel, multipoint distribution service (MMDS), 10, 15, 16, 62
  FCC, 62
  scale-economy barriers, 57
Multiple dwelling unit (MDU), 15
Multiple system operator (MSO), 61, 65, 134, 177–181, 387
Multiples, 212
Murder She Wrote, 115, 411
Murdoch, Rupert, 6, 24, 108, 173, 178, 185, 187, 217
Murphy Brown, 113, 120, 123, 342, 347
Murray the K., 79
Murrow, Edward R., 55, 105
Museum of Television & Radio, 41
Music rotation, 313
Music sweeps, 313
Music Television (MTV), 16, 55, 140, 145, 217, 318, 358, 369, 383
Mutual Broadcasting System, 6, 79, 307
My Mother the Car, 106
Myhren, Trygve, 134

Nashville Network, 140
Nashville Now, 344

National Association of Broadcast Engineers and Technicians (NABET), 31, 287, 289
National Association of Broadcasters (NAB), 44, 70, 91, 96–98, 126, 197, 198, 279, 388, 414
National Association of Television Program Executives (NATPE), 30, 326, 328, 330
National Black Media Coalition (NBMC), 27
National Broadcasting Company (NBC), 6, 10, 25, 54, 55, 61, 63, 78, 79, 104–106, 113, 114, 116, 119, 120, 173, 183, 185, 277, 355, 357
National Cable TV Association, 223
National Information Superhighway, 157
National League of Cities, 170
National Public Radio (NPR), 82–83, 302
National spot sales, 90
Nations Bank, 215
NATPE (*see* National Association of Television Program Executives)
NBC (*see* National Broadcasting Company)
NBC Radio, 307
Negotiated renewal, 201
Negotiated split (in programming), 327
Negotiation, 282–292
Net profit, 242
Network advertising, 90, 123
Network compensation, 79, 93, 118, 119–121
Network erosion, 108
Network origination, 325
Network radio, 78, 79
Network share, 139
Network station rate (NSR), 119
Network-affiliate relationship, 106, 108
  (*See also* Affiliate relations)
Networking (radio), 92
New Kids on the Block, The, 88
New Line, 38
New world order, the, 406
New York Post, 173

Newhart, 347
Newhouse Broadcasting, 178
News Corporation, 185, 187, 406
  (*See also* Fox Broadcasting)
Newton (computer), 409
Nick at Night, 55, 331, 345, 412
Nickelodeon, 16, 55, 185, 217, 303, 412
Nielsen, A.C. Company, 59, 139, 144, 223, 381
  local metered markets, 384
  peoplemeter, 144, 383
  "Viewers-In-Profile" (VIP), 383
Nightline, 346
Nintendo, 68
Noncommercial television (*see* Television, noncommercial)
Northern Exposure, 113, 120, 123, 342, 347
NOVA, 373
NPR (*see* National Public Radio)
Number of television homes (in valuation), 211
NY1, 141
Nynex, 54, 156

O'Connor, Carroll, 411
O'Neal, Shaquille, 285
O+Os (*see* Owned-and-operated stations)
Off-net syndication, 331
Off-network (programs), 123
Office of Depreciation Analysis, 235
Oligopolies, 60, 61, 69
Olympics, 61
101 Strings, 88, 411
Operating income margin, 241
Operating income statement, 238
Operating plans, 167, 170
Operator-originated channels, 340
Oprah!, 113, 121, 333, 335, 343, 348, 381
Ouchi, William, 32
Overbuild, 203
Overhead expenses, 119
Overnight (programming), 346
Oversaturation (of programs), 338
Ovitz, Michael, 285

Owned-and-operated stations (O+Os), 178
Owner's equity, 229
Ownership, telecommunications, 165–191
  conglomerates, 181
  corporate, 176
  criteria, 167–170
  cross-ownership rules, 173–175
  general and limited partnerships, 175, 176
  mom-and-pop stations, 174, 175
  multinational corporations, 182, 186
  multiple station ownership, 170, 172
  multiple system operators, 177–180
  one-to-a-market, 171
  stand-alone, 176, 177

Paid programming, 335
Paine-Webber, 216
Paley, William S., 24, 41, 43
Panasonic, 182 (*see* also Matsushita)
Paragon (research firm), 318, 388
Paramount, 185, 212, 214 (*see* also Viacom)
Paramount Pictures, 16
Paramount Studios, 114
Parents, 185
Passive peoplemeter, 396
Paul Bunyon Broadcasting Co., 177
Paul Harvey News and Comment, 92, 384
Pay cable, 15, 117, 137, 138, 140, 141
  premium or tiered channels, 135–137
  (*See also* Pay-per-view; Television, cable)
Pay-per-view (PPV), 15, 135, 138, 139
Paycheck, Johnny, 98
PBS (*see* Public Broadcasting System)
Pearl Jam, 87, 319, 411
Pension and welfare (P & W), 287

People's Court, 113, 343
Performance tests, 196, 270
Personal Communications Services (PCS), 159, 200
Personal digital assistants (PDA), 405, 406
Personality test, 269
Personnel administration (in telecommunications), 253–255, 303–304
Personnel management, 251–294
  compensation, benefits and assistance, 276–281
  core departments, 258–259
  disciplinary action and termination, 274–276
  evaluation methods, 272–273
  functions, 253–254
  internal communications, 281–284
  interviewing, audition tapes, and screening procedures, 267–271
  job descriptions, 259–263
  promotion, 273, 274
  recruiting, 264, 265
  table of organization, 254, 258
  theory X, 32, 33, 258
  theory Y, 32, 33, 34
  theory Z, 32, 34, 35, 36
  training, 271, 272
  (*See also* Management, telecommunications)
Peter Pan, 105
Petition to deny, 198
Petry Television, 353, 356
Petty, Tom, 86
Philco Playhouse, 105
Philips (Netherlands), 54, 182, 189
Pioneer (Japan), 189
Planning (*see* Financial planning and projections)
Planning, organizing, staffing, directing, coordinating, reporting, and budgeting (POSDCORB), 23
Playlist, 309
Pledge drives, 373
Pollock, Clark, 22
Polygram, 182
Portable peoplemeter, 397
Power programming, 348
Pozner, Vladimir, 347

PPV (*see* Pay-per-view)
Preemptions, 326
Premium channels, 15
Premium household, 135
Premium service (cable), 135, 136, 137
Presley, Elvis, 79
Price fixing, 60
Price is Right, The, 113
Price-setting policies, 60
Prime time (television), 114
Prime time access rule (PTAR), 343
Prior experience, 169
Proctor & Gamble (network advertiser), 58
Product differentiation, 54–56
Product image, 55
Product life-cycle theory, 67, 68, 69, 70
  decline, 70
  growth, 69
  introduction, 68, 69
  maturity, 69
Product protection, 365
Product separation, 365
Profit-and-loss statements (P+Ls), 238
Profits and profit margins:
  cable, 151, 152
  radio, 95, 96, 97
  television, 125, 126
Program management, 322–350
  personnel administration, 303–304
  program acquisition, 304
  program evaluation, 305
  program origination, 304
  program scheduling, 304–305
Program testing, 380, 395
Programming:
  cable, 139–141
    local origination, 141
    must-carry, 139
  television, 113
    local origination, 113, 324, 325
    network, 113, 325
    syndication, 113, 330
Programming formats (*see* Formats, radio)
Projections (*see* Financial planning and projections)
Promotion:

Promotion (*Cont.*):
　activities
　　cable, 371
　　public broadcast, 372–375
　　radio, 370
　　television, 371
　audience, 269
　department, 366
　director, 368
　public relations, 369–370
　salaries, 367
　sales, 369
　scope of, 366
Proprietary regression equations, 210
Psychographics, 220, 380, 396
Public Broadcasting Act of 1967, 109
Public Broadcasting System (PBS), 110–111, 117
Public interest, convenience or necessity, 166
Pulse of Radio, The, 309

Quality circle, 35
QUBE system, 14
Queen City Broadcasting, 191
Quick ratio, 241

R.E.M., 87
Radio:
　AM, 4, 11, 12, 79, 80, 81, 82, 86, 87, 89, 96, 97
　AM stereo, 12
　classifications, 11–12
　dayparts, 84
　FM, 4, 11, 80, 81, 82, 88, 92, 96, 97
　format, 80
　hit radio, 78, 80
　localism, 81
　network affiliation, 78, 79
　network era, 78, 79
　noncommercial, 82, 83, 89
　salaries, 99–101
　(*See also* Broadcasting, radio; Profits and profit margins, radio; Revenues, radio)
Radio and Records, 309

Radio Advertising Bureau (RAB), 223, 261, 375, 387
Radio Corporation of America (RCA), 25, 33, 41
　(*See also* General Electric)
Radio expenses (*see* Expenses, radio)
Radio financial report, 91
Radio formats (*see* Formats, radio)
Radio listening (*see* Listenership, radio)
Radio Only, 309
Radio's All-Dimensional Audience Research (RADAR), 384
Raitt, Bonnie, 88
Rand Corporation, 188
Rate cards (radio), 361
Rate guides (television), 361–362, 364
Rate setting, 359
Rather, Dan, 394
Rating (definition), 389
Ratings (audience), 125
Ratings distortion, 397
Ratios, financial, 240, 241
Raytheon, 33
RBOCs (*see* Regional Bell Operating Companies)
RCA (*see* Radio Corporation of America)
RCA Records, 182
Reach, 390
Reagan revolution, 205
Reagan, Ronald, 291, 372
Real Personal, 347
Reality programming, 333
Red Hot Chili Peppers, 89, 319
Redstone, Sumner, 217
Regional Bell operating companies (RBOCs), 3, 6, 19, 54, 132, 155, 156–158, 252, 406
Regression analysis (econometrics), 210–212
Reiner, Carl, 33
Remote control devices, 38, 39
Remotes (broadcast), 304
Ren and Stimpy, 113
Reporting, financial (*see* Financial reporting methods)

Republic (rep firm), 93, 331
Request for proposal (RFP), 200, 201
Rescue 911, 56
Research:
　accuracy, 397
　auditorium, 393
　call-out, 400
　cost effectiveness, 398–399
　debriefing, 400
　ethics, 399
　focus group, 380
　program testing, 380, 395
　psychobiological, 396
　ratings, 125, 379
　(*See also* Audience, Analysis, Ratings)
Research and development (R&D), 5
Research Group, The, 318
Retention marketing (cable), 134
Return on investment, 243
Revenue per subscriber, 212
Revenues:
　cable television, 141–147
　　advertising, 141–147, 356–358
　　pay-per-view, 145
　　subscriber fees, 142
　radio, 92, 93
　　barter (trade), 93
　　networking, 92
　television
　　advertising, 121–123
　　network compensation, 118–121
Reverse leveraged buyout, 218
Reymer and Associates, 394
Right of first refusal (programming), 326
Roc, 347
Rockwell, Norman, 42
Rogers, Kenny, 88
Rolling Stones, The, 87
Ronstadt, Linda, 88
Roots (miniseries), 61, 106, 116, 303
Rose, Pete, 308
Rose, Reginald, 105
Roseanne, 55, 302, 331, 335, 347, 357
Ross, Steven, 181

Rough cut, 395
Rule, Elton, 34, 279
Rule of 7s, 170
Rule of 12s, 170
Rule of 18s, 170
Run-of-schedule (ROS), 361
Rush Limbaugh (program), 92, 384

Sajak, Pat, 394
Sales and Marketing Management, 222
Sally Jessy Raphael, 337
Sanders, Bonnie, 44
Sarnoff, David, 3, 19, 24, 25, 32, 41
Satellite master-antenna television (SMATV), 10, 15
Satellite Music Network, 6, 66, 308
Saturday Night Live, 339
Scanning lines, 18
Scarcity issue, 166
Scatter (advertising), 355
Scholastic Aptitude Test (SAT), 210
Scientific Atlanta, 189
Screen Actors Guild (SAG), 286
Sears (advertiser), 353
Securities and Exchange Commission (SEC), 66, 283
Security Pacific, 215
See It Now, 105
Sega (Japan), 68, 189
Seinfeld, 331
Seller concentration, 52, 54
Seltel (rep firm), 353, 356
Serling, Rod, 105
Sesame Street, 302
Shamrock Broadcasting, 12, 178
Share (defined), 389
Sheen, Bishop Fulton J., 105
Showtime (pay televison channel), 15, 16, 55, 57, 67, 139, 140, 153, 304, 345
Silk Stalkings, 346
Silver King, 178
Simmons Market Research Bureau (SMRB), 223
Simon, Neil, 33, 105
Simon, Paul, 88

Simpsons, The, 55, 108, 113, 123, 302, 342
Sinatra, Frank, 88
60 Minutes, 55, 112, 120, 344, 379
Skelton, Red, 105
Skew (audience), 379
Small format videotape, 107
Smashing Pumpkins, 411
SMATV (*see* Satellite master-antenna television)
Smith, Kate, 79
Smulyan, Jeff, 292
Social Darwinism, 24
Sony Corporation (Japan), 6, 54, 66, 182, 185, 189, 406
 product differentiation by, 57
Sound hour, 313
Southern Spot Sales, 355
Southwestern Bell, 54, 158, 173
Spectaculars, 105
Spectrum fees, 234
Sports Channel, 15
Sports Channel America, 55
SportsCenter, 344, 346, 348
Spot market, 355
Spot TV sales, 122
Spots (comercial), 352
Springsteeen, Bruce, 86
Sprint, 57
SRDS (*see* Standard Rate and Data Service)
Stability (management theory), 37
Stanford Research Institute (SRI), 396
 (*See also* VALS)
Standard and Poor's Industry Surveys, 244
Standard Metropolitan Statistical Area (SMSA), 389
Standard Rate and Data Service (SRDS), 93, 223, 361, 362, 364
Star Search, 113
Star Trek, 346, 405
Star Trek: The Next Generation, 327
Station Representatives (reps), 355
Statistical Abstract of the United States, 220
Statistical regression, 210

Stereotype, 42
Stern, Howard, 178, 308, 326, 394
Stieglitz, Alfred, 42
Strategic alliance (*see* Joint venture)
Strategic Research, 318
Streisand, Barbra, 88
Strip syndication (stripping), 348
Structuralization, organizational, 26
Studio One, 108
Study of Media and Markets, 223
Stunting, 348
Subscribers, basic (cable), 134–137
Sullivan, Ed, 105
Sum-of-the-runs (amortization), 237
Super Bowl telecast, 58
Superstations (*see* Distant-signal stations)
Supremes, The, 86
Survey of Buying Power, 222
Susquehanna Broadcasting Company, 178
Swaggart, Jimmy, 335
Sweeps period (television), 116, 383
Swift, Tom, 41
Syndication, 113, 330

Table of assignments, 196
Taft, 170
Talent representatives, managers, and agencies, 285–286
Talentmasters, 317
Talk-America, 185
Tax Reform Act of 1986, 235
Taxi, 346
Technophobia, 248
Teenage Mutant Ninja Turtles, 38, 117
Tele-Communications Inc. (TCI), 30, 54, 58, 156, 157, 172, 173, 174, 180, 181, 187, 189, 214, 216, 244, 283, 324
Telecommunications, definition of, 3

Telecommunications management (see Management, telecommunications)
Telecommunications ownership (see Ownership, telecommunications)
Telemarketing, 375
Telephone Industry Directory, 222
Telerep (rep firm), 353, 356
Televangelists, 335
Television (TV):
  cable, 14, 69, 117
    advertising, 356–358
    community-access, 141
    growth, 132–139
    subscriber fees and base, 134, 135, 136, 137
    subscribers, basic, 134, 135, 136, 137
    (See also Employment, cable; Expenses, cable; Pay cable; Pay-per-view; Profits and profit margins, cable; Programming, cable; Revenues, cable television)
  commercial, 104–109
    era of maturity, 108, 109
    golden age, 104, 105
    independent stations, 107
    network heyday, 105–107
    UHF, 10, 13
    VHF, 10, 13
    (see also Employment, cable; Expenses, television; Profits and profit margins, television; Programming, television; Revenues, television)
  noncommercial, 109, 111
Television Bureau of Advertising (TVB), 30, 223, 261
Television Factbook, 197, 221
Television receive-only earth station (TVRO), 19
Termination, employee, 274–276
Texaco Star Theater, The, 13
Theory X, 32, 33, 258
Theory Y, 32, 33, 34
Theory Z, 32, 34, 35, 36
This Week with David Brinkley, 336
Thorn EMI, 182
Three Stooges, The, 105, 236
Three's Company, 106
Tiered channels, 15
Time buyer, 354
Time spent listening (TSL), 311, 390
Time Warner, 16, 25, 30, 43, 54, 58, 140, 141, 152, 156, 157, 172, 180, 181, 182, 185, 187, 189, 214, 283, 324, 327, 328, 340, 406
Times Mirror, 187
Tipsheets, 309
Tisch, Lawrence, 66
Titanic disaster, 41
Toast of the Town, 13
Today Show, The, 55, 105, 115, 334, 337
Tolkin, Mel, 33
Tonight Show, The, 55, 105, 115, 334, 346
Toshiba (Japan), 189
Total survey area (TSA), 389
Townsend, Ronald, 248
Tracey Ullman Show, The, 108
Training, employee, 271, 272
Transfer of control, 219, 220
Trend analysis, 244
Tribune, 126, 178, 244
Turner, Ted, 24, 25, 134, 185
Turner Broadcasting Systems, 58, 150, 185, 188, 216, 328, 331, 357, 388
  (See also WTBS)
Turner Network Television (TNT), 145, 357
Turnover (T/O), 331
TVB (see Television Bureau of Advertising)
TV Dimensions, 220
TV-Q, 392
TVRO (see Television receive-only earth station)
20th Century Fox, 67
20/20, 112
21 Jump Street, 108

U.S. Supreme Court, 30
U2, 317
UHF Stations, 10
Underwriting, 373
Unistar, 6
United (air carrier), 60
United States Satellite Broadcasting (USSB), 16
United Stations, 308
Univision, 178
Unsolved Mysteries, 56
Up-front (advertising), 355
US West, 157, 173
USA Network, 44, 55, 140, 149, 357, 383
USA Today, 187
User fees, 234

Valenti, Jack, 30
Valuation, media, 208–214
Values and Life-Styles (VALS) program, 396
VCR (see Videocassette recorder)
Venture capital, 214
Veronis, Shuler and Associates, 244
Vertical integration, 406–408
  (See also Concentration)
VH-1 (program service), 16, 55, 140
VHF stations, 10
Viacom, 6, 16, 54, 140, 169, 172, 185, 188, 216, 217, 331, 412
  (See also Paramount)
Video, home (see Home video)
Video dial tone (VDT), 157
Video disc players, 19
Video game, 38
Video on demand, 149
Videocassette recorder (VCR), 18, 38, 52, 59–60, 68, 69, 117, 125, 405, 408
Virtual Reality, 407, 415

WABC-TV (New York), 178
Waller Capital, 216
War and Remembrance, 61
Warner Brothers, 106, 114
Warner Communications (see Time-Warner)
Warner-Amex Cable Communications, 14, 152
Watermark, 308

WBZ (Boston), 11
WCBS-AM (New York), 178
WCBS-FM (New York), 167
WCPT-AM (Washington, DC), 172
WCXR-FM (Washington, DC), 172
Weather Channel, 336
Weaver, Sylvester L. "Pat", 105
Weekend Edition, 374
Westinghouse, 32, 93, 171, 172, 178, 355
Westinghouse Credit Corporation, 217
Westwood One, 6, 307
WFAN (New York), 212, 320
WFOX-FM (Atlanta), 12
WFRV-TV (Green Bay, Wisc.), 178
WGA (*see* Writers Guild of America)
WGBS-TV (Philadelphia), 214
WGN (Chicago), 6, 126, 140, 178, 331, 335, 336
WGNX (Atlanta), 178, 336
WGRL (Rochester), 340

Wheel of Fortune, 113, 114, 333, 344, 355, 381
Who, The, 411
Wide World of Sports, 55
WIFR (Rockford, IL), 362
William Morris (agency), 285
Wilson Phillips, 88
Window dressing, 168
Winds of War, The, 61
Winfrey, Oprah, 191, 394
Wireless cable, 16
Wisconsin Network, 308
Witt, Katarina, 66
WJOK (Gaithersburg, Md.), 89
WKBD (Detroit), 212
WKGW-TV (Buffalo, NY), 171
WLS-TV (Chicago), 178
WMZQ-AM/FM (Washington, DC), 172
WNYW (New York), 67, 126, 173
Woman's Day, 66, 185
Women in Communications, 27
Wonder Years, The, 335
World Administrative Radio Conference (WARC), 12

World Guide to Television and Programming, 222
WPHL (Philadelphia), 178
WPIX (New York), 140, 178
WPVI-TV (Philadelphia), 178
WQRF (Rockford, IL), 362
Writer's Guild of America (WGA), 286
WSB (Atlanta), 190, 338
WTBS (Atlanta), 6, 26, 140, 145, 149, 291, 357, 388
WTNH (New Haven, Conn), 191
WTTG (Washington), 67
WTXF (Philadelphia), 214
WVBK-AM (Herndon, VA), 98
WWOR (New York), 6, 140

Yount, Paul, 98
Your Show of Shows, 33

Zap, 19
Zenith Corporation, 6, 18
Zip, 19
Zworykin, Vladimir, 5, 43